Lecture Notes in Mathematics 1507

Editors:
A. Dold, Heidelberg
B. Eckmann, Zürich
F. Takens, Groningen

K. Hulek T. Peternell M. Schneider
F.-O. Schreyer (Eds.)

Complex Algebraic Varieties

Proceedings of a Conference held in
Bayreuth, Germany, April 2–6, 1990

Springer-Verlag
Berlin Heidelberg New York
London Paris Tokyo
Hong Kong Barcelona
Budapest

Editors

Klaus Hulek
Fachbereich Mathematik
Universität Hannover
Welfengarten 1
W-3000 Hannover, Germany

Thomas Peternell
Michael Schneider
Frank-Olaf Schreyer
Mathematisches Institut
Universität Bayreuth
Postfach 101251
W-8580 Bayreuth, Germany

Mathematics Subject Classification (1991): 14DXX, 14EXX, 14FXX, 14JXX, 14KXX, 32LXX, 53CXX

ISBN 3-540-55235-9 Springer-Verlag Berlin Heidelberg New York
ISBN 0-387-55235-9 Springer-Verlag New York Berlin Heidelberg

Printed in Germany

Typesetting: Camera ready by author
Printing and binding: Druckhaus Beltz, Hemsbach/Bergstr.
46/3140-543210 - Printed on acid-free paper

Preface

This volume contains the proceedings of the conference "Complex Algebraic Varieties" held in Bayreuth from April 2-6, 1990.

The main topic of this meeting was the classification of algebraic varieties including related topics, such as vector bundles. Most of the contributions are closely related to talks given by one of the authors at the conference.
All the papers are original research articles.

The conference was organized as part of the activities of the DFG research programme "Komplexe Mannigfaltigkeiten". We should like to thank the DFG for the financial support which made this conference possible. We are grateful to the University of Bayreuth for their hospitality.

This conference profited greatly from the political changes of the last two years. It was one of the first conferences in Algebraic Geometry at which mathematicians from Eastern countries could participate without restrictions. This contributed greatly to a fruitful exchange of ideas during the meeting.

January 1992

K. Hulek

Th. Peternell

M. Schneider

F.-O. Schreyer

Table of Contents

ANNULATION DU H^1 POUR LES FIBRÉS EN DROITES PLATS

Arnaud Beauville
Mathématiques - Bât. 425
Université Paris-Sud
F-91 405 Orsay Cedex

Introduction

Soit X une variété kählérienne compacte, et soit $\mathrm{Pic}^\tau(X)$ la variété des fibrés en droites holomorphes sur X dont la première classe de Chern est nulle dans $H^2(X,\mathbf{C})$. Lorsque L est un élément général de $\mathrm{Pic}^\tau(X)$, des hypothèses assez faibles sur X entraînent l'annulation des espaces de cohomologie $H^i(X,L)$ pour $i < \dim(X)$ [G-L 1]. On est ainsi amené à s'intéresser à la sous-variété $S^i(X)$ de $\mathrm{Pic}^\tau(X)$ formée des fibrés en droites L tels que $H^i(X,L) \neq 0$. Green et Lazarsfeld ont prouvé récemment [G-L 2] que *les composantes irréductibles de* $S^i(X)$ *sont des translatés de sous-tores complexes de* $\mathrm{Pic}^\tau(X)$. En fait, des discussions avec F. Catanese nous ont conduit à proposer une conjecture plus précise :

Conjecture.- *Les composantes de* $S^i(X)$ *sont des translatés de tores complexes par des points d'ordre fini.*

La conjecture est vraie pour les composantes de dimension ≥ 1 de $S^1(X)$: cela résulte de la description précise de ces composantes donnée au §2 (et qui corrige l'énoncé analogue dans [B], qui contient une erreur). Nous proposons dans cet exposé une stratégie pour prouver que *les points isolés de* $S^1(X)$ *sont de torsion.* L'idée de base est de considérer, plutôt que le groupe $\mathrm{Pic}^\tau(X)$, le groupe $H^1(X,\mathbf{C}^*)$ des faisceaux localement constants de rang 1 sur X. On montre au §3 que le sous-ensemble $\Sigma^1(X)$ de $H^1(X,\mathbf{C}^*)$ formé des faisceaux $\mathcal{L}$ tels que $H^1(X,\mathcal{L}) \neq 0$ est étroitement relié à $S^1(X)$. On peut considérer $H^1(X,\mathbf{C}^*)$ comme le groupe $\mathrm{Hom}(\pi_1(X),\mathbf{C}^*)$ des caractères de $\pi_1(X)$; le point clé est que les points isolés de $\Sigma^1(X)$ correspondent à des caractères *unitaires.* On voit de plus, en utilisant l'action de $\mathrm{Aut}(\mathbf{C})$ sur $H^1(X,\mathbf{C}^*)$, que les valeurs de ces caractères sont des nombres algébriques. Si en outre ce sont des *entiers* algébriques, un lemme célèbre de Kronecker permet de conclure. Malheureusement je ne sais pour l'instant prouver cette intégralité que sous des hypothèses restrictives, par exemple quand *le groupe dérivé* $D(\pi_1(X))$ *est de type fini.* Plus que ce résultat, dont l'intérêt est limité par l'aspect artificiel de l'hypothèse, c'est la méthode qui me semble intéressante; je continue à espérer qu'elle permettra de prouver la conjecture pour $S^1(X)$.

§1. Fibrations sur une courbe

Dans tout cet article, la lettre X désigne une variété kählérienne compacte (connexe). Nous appellerons *fibration* de X sur une courbe B tout morphisme surjectif $p : X \longrightarrow B$ de X sur une courbe lisse B, à fibres connexes. Dans ce paragraphe, nous considérons une fibration $p : X \longrightarrow B$; nous allons étudier la relation entre les groupes de Picard de X, de B et des fibres de p.

Soit b un point de B. Le diviseur p^*b s'écrit $\sum n_\alpha D_\alpha$, où les diviseurs D_α sont irréductibles et réduits; le p.g.c.d. m des entiers n_α est appelé *multiplicité* de la fibre p^*b. Si $m > 1$, on peut écrire $p^*b = mF$, où F est un diviseur effectif; on dit alors que la fibre p^*b est multiple.

Soient $m_1 F_1, \dots, m_s F_s$ les fibres multiples de p.

Lemme 1.1.- *Soit* V *un diviseur de* X *dont les composantes soient contenues dans les fibres de* p. *Pour que la classe de* V *dans* $H^2(X, \mathbf{C})$ *soit nulle, il faut et il suffit que l'on puisse écrire* $V = p^*\delta + \sum k_i F_i$, *où* δ *est un diviseur de degré* 0 *sur* B *et les* k_i *des entiers vérifiant* $\sum k_i/m_i = 0$.

Notons ω la classe dans $H^2(X, \mathbf{C})$ d'une forme de Kähler sur X. Si n désigne la dimension de X, notons $\int_X$ l'isomorphisme canonique de $H^{2n}(X, \mathbf{C})$ sur $\mathbf{C}$. Soit NS(X) le sous-groupe de $H^2(X, \mathbf{C})$ formé des classes de diviseurs; on le munira de la forme bilinéaire symétrique définie par $(\alpha.\beta) = \int_X \omega^{n-2} \wedge \alpha \wedge \beta$. En vertu du théorème de l'indice de Hodge, elle est de signature $(+1, -1, \dots, -1)$.

Soit F une fibre de p. La classe de F dans NS(X) est non nulle, et de carré nul; par suite, la forme bilinéaire induite sur l'orthogonal de F dans NS(X) est négative. Notons $(D_\alpha)_{\alpha \in I}$ la famille des composantes irréductibles de F, et n_α la multiplicité de D_α, de sorte que le diviseur F est égal à $\sum n_\alpha D_\alpha$. Considérons la matrice $(D_\alpha . D_\beta)_{(\alpha,\beta) \in I \times I}$. D'après ce qui précède, elle est négative, et son noyau contient l'élément $n = (n_\alpha)_{\alpha \in I}$. Pour $\alpha \neq \beta$, le produit $(D_\alpha . D_\beta)$ est positif, et même strictement positif si D_α rencontre D_β : en effet ce nombre s'obtient en intégrant la forme positive ω^{n-2} sur l'espace analytique $D_\alpha \cap D_\beta$. Comme la fibre F est connexe, il en résulte qu'il n'existe pas de partition $I = J \cup K$ de I telle qu'on ait $(D_\alpha . D_\beta) = 0$ pour $\alpha \in J$, $\beta \in K$. Un lemme classique d'algèbre linéaire (cf. par exemple [Bo], Ch. V, §3, n°5, lemme 4) affirme alors que le noyau de la matrice $(D_\alpha . D_\beta)$ est $\mathbf{R}n$. Par suite, tout diviseur D de la forme $\sum k_\alpha D_\alpha$ (avec $k_\alpha \in \mathbf{Z}$) vérifie $(D.D) \leq 0$, et $(D.D) = 0$ si et seulement si D est multiple rationnel de F.

Il en résulte que tout diviseur vertical V de carré nul peut s'écrire $V = p^*\delta + \sum k_i F_i$ (rappelons que $m_1 F_1, \dots, m_s F_s$ sont les fibres multiples de p). Quitte à modifier les k_i, on peut supposer $\deg(\delta) = 0$. La classe de V dans $H^2(X, \mathbf{C})$ est alors k fois celle de F, avec $k = \sum k_i/m_i$, d'où le lemme. ■

Proposition 1.2.- *Soit* L *un élément de* $\mathrm{Pic}^\tau(X)$. *Les conditions suivantes sont équivalentes:*

a) *Le faisceau* p_*L *n'est pas nul.*

b) *La restriction de* L *à toute fibre lisse de* p *est triviale.*

c) *La restriction de* L *à une fibre lisse de* p *est triviale.*

d) *Il existe un élément* L_0 *de* $\mathrm{Pic}^0(B)$ *et des entiers* $k_1, \ldots, k_s$, *satisfaisant à* $\sum k_i/m_i = 0$, *tels que* L *soit isomorphe à* $p^*L_0 \otimes \mathcal{O}_X(\sum k_i F_i)$.

a) $\Leftrightarrow$ b) : Notons U le plus grand ouvert de B au-dessus duquel p est lisse. Sous l'hypothèse b), le théorème de changement de base implique que le faisceau p_*L est localement libre de rang 1 au-dessus de U, d'où a).

Inversement, si la restriction de L à une fibre lisse F de p n'est pas triviale, l'espace $H^0(F, L_{|F})$ est nul (puisque F est kählérienne). Les théorèmes de semi-continuité et de changement de base entraînent alors que le faisceau p_*L est concentré sur un nombre fini de points; comme il est sans torsion, il est nécessairement nul.

b) $\Leftrightarrow$ c) : Pour $b \in U$, notons F_b la fibre (lisse) $p^{-1}(b)$. Les groupes $\mathrm{Pic}^\tau(X)$ et $\mathrm{Pic}^\tau(F_b)$ s'identifient à $H^1(X, \mathbf{S}_1)$ et $H^1(F_b, \mathbf{S}_1)$ respectivement, et l'application de restriction $\mathrm{Pic}^\tau(X) \longrightarrow \mathrm{Pic}^\tau(F_b)$ s'identifie à l'application naturelle $H^1(X, \mathbf{S}_1) \longrightarrow H^1(F_b, \mathbf{S}_1)$. Il en résulte que le noyau de cette application est indépendant du point b, d'où l'implication c) $\Rightarrow$ b). L'implication opposée est triviale.

b) $\Leftrightarrow$ d) : L'implication d) $\Rightarrow$ b) est claire. Sous l'hypothèse b), l'homomorphisme canonique $p^*p_*L \longrightarrow L$ est un isomorphisme au-dessus de $p^{-1}(U)$. Il en résulte que l'on peut écrire L sous la forme $\mathcal{O}_X(\sum k_\alpha D_\alpha)$, où chaque diviseur D_α est une composante d'une fibre de p. On conclut alors à l'aide du lemme 1.1. ∎

Proposition 1.3.- *Soit* L *un élément de* $\mathrm{Pic}^\tau(X)$, *tel que* p_*L *ne soit pas nul. Ecrivons* $L = p^*L_0 \otimes \mathcal{O}_X(\sum k_i F_i)$, *où* L_0 *est un fibré en droites sur* B *et* $k_1, \ldots, k_s$ *des entiers satisfaisant à* $0 \leq k_i < m_i$ *et* $\sum k_i/m_i \in \mathbf{N}$. *Le faisceau* p_*L *est alors isomorphe à* L_0.

Posons $N = \mathcal{O}_X(\sum k_i F_i)$. Il s'agit de prouver que p_*N est isomorphe à $\mathcal{O}_B$; en utilisant la suite exacte

$$0 \rightarrow N(-F_i) \longrightarrow N \longrightarrow N_{|F_i} \rightarrow 0 ,$$

et en raisonnant par récurrence sur l'entier $\sum k_i$, il suffit de prouver que le faisceau $N_{|F_i}$ n'a pas de section globale non nulle. Or ce faisceau est égal à $\mathcal{O}_{F_i}(k_i F_i)$, et le faisceau $\mathcal{O}_{F_i}(F_i)$ est d'ordre m_i dans $\mathrm{Pic}(F_i)$ (cf. par exemple [BPV], lemme III.8.3, où l'hypothèse que X est une surface n'est pas utilisée dans la démonstration). On conclut à l'aide du lemme 1.4 ci-dessous. ∎

Lemme 1.4.- *Pour tout élément non nul* L *de* $\mathrm{Pic}^\tau(F_i)$, *on a* $H^0(F_i, L) = 0$.

Supposons que $H^0(F_i, L)$ contienne une section non nulle s. Notons D le plus grand diviseur $\leq F_i$ tel que $s_{|D} = 0$, et posons $E = F_i - D$; on a $E \neq 0$. La section s définit par restriction une section du faisceau inversible $L(-D)_{|E}$, qui ne s'annule sur aucune composante de E. On en déduit une suite exacte

$$0 \longrightarrow \mathcal{O}_E \longrightarrow L(-D)|_E \longrightarrow \mathcal{T} \longrightarrow 0 ,$$

où le support Z de $\mathcal{T}$ est de codimension 2 dans X. En considérant les secondes classes de Chern on obtient $c_1(D).c_1(E) = -\mathrm{cl}(Z)$. On en déduit, avec les notations de la démonstration du lemme 1.1, $(D.E) \leq 0$ dans $NS(X)$, et il n'y a égalité que si $\mathcal{T}$ est nul. En écrivant $E = F_i - D$ on obtient $(D.D) \geq 0$, ce qui entraîne $D = 0$ (*loc. cit.*). La section s définit alors un isomorphisme de $\mathcal{O}_{F_i}$ sur L, ce qui contredit l'hypothèse sur L. ■

Nous noterons $\mathrm{Pic}^{\tau}(X,p)$ le noyau de l'homomorphisme $\mathrm{Pic}^{\tau}(X) \to \mathrm{Pic}^{\tau}(F)$, où F désigne une fibre lisse quelconque de p (d'après la prop. 1.2, ce noyau est indépendant du choix de F). C'est un sous-groupe de Lie fermé du groupe de Lie (complexe) $\mathrm{Pic}^{\tau}(X)$.

Notons $\Gamma^{\tau}(p)$ le sous-groupe de $\oplus \mathbf{Z}/(m_i)$ formé des éléments $(\dot{k}_1, \dots, \dot{k}_s)$ tels que $\sum k_i/m_i$ soit entier, et $\varphi : \mathrm{Pic}^{\tau}(X,p) \longrightarrow \Gamma^{\tau}(p)$ l'homomorphisme qui associe à la classe d'un diviseur $p^*D + \sum k_i F_i$ l'élément $(\dot{k}_1, \dots, \dot{k}_s)$.

Proposition 1.5.- *La suite*

$$0 \longrightarrow \mathrm{Pic}^0(B) \xrightarrow{p^*} \mathrm{Pic}^{\tau}(X,p) \xrightarrow{\varphi} \Gamma^{\tau}(p) \longrightarrow 0$$

est exacte.

En d'autres termes, le groupe de Lie complexe $\mathrm{Pic}^{\tau}(X,p)$ a pour composante neutre la sous-variété abélienne $p^*\mathrm{Pic}^0(B)$, et son groupe des composantes connexes est isomorphe à $\Gamma^{\tau}(p)$. Il est donc isomorphe (non canoniquement) au produit $\mathrm{Pic}^0(B) \times \Gamma^{\tau}(p)$.

Considérons l'homomorphisme $\psi : \Gamma^{\tau}(p) \longrightarrow \mathrm{Pic}^{\tau}(X,p)/p^*\mathrm{Pic}^0(B)$ qui associe à $(\dot{k}_1, \dots, \dot{k}_s)$ la classe de $\sum k_i F_i$. La prop. 1.2 exprime qu'il est surjectif; il s'agit de prouver qu'il est injectif. Soient donc $k_1, \dots, k_s$ des entiers tels que le faisceau $L = \mathcal{O}_X(\sum k_i F_i)$ appartienne à $p^*\mathrm{Pic}^0(B)$. Le faisceau $L|_{F_i}$ est alors trivial pour tout i. Mais d'autre part ce faisceau est égal à $\mathcal{O}_{F_i}(k_i F_i)$, et l'on a déjà vu que le faisceau $\mathcal{O}_{F_i}(F_i)$ est d'ordre m_i dans $\mathrm{Pic}(F_i)$. On en déduit que k_i est multiple de m_i, ce qui prouve notre assertion.

(1.6) Désignons par $\mathrm{Pic}^0(X,p)$ l'intersection de $\mathrm{Pic}^{\tau}(X,p)$ avec $\mathrm{Pic}^0(X)$, et par $\Gamma^0(p)$ l'image de $\mathrm{Pic}^0(X,p)$ dans $\Gamma^{\tau}(p)$. Comme le quotient $\mathrm{Pic}^{\tau}(X)/\mathrm{Pic}^0(X)$ s'identifie au sous-groupe de torsion de $H^2(X, \mathbf{Z})$, le groupe $\Gamma^0(p)$ est défini par la suite exacte

$$0 \longrightarrow \Gamma^0(p) \longrightarrow \Gamma^{\tau}(p) \longrightarrow \mathrm{Tors}\, H^2(X, \mathbf{Z}) .$$

Pour toute fibre lisse F de p, on a une suite exacte (non canonique)

$$0 \longrightarrow p^*\mathrm{Pic}^0(B) \times \Gamma^0(p) \longrightarrow \mathrm{Pic}^0(X) \longrightarrow \mathrm{Pic}^0(F) ,$$

et $\Gamma^0(p)$ s'identifie au groupe des composantes connexes de $\mathrm{Pic}^0(X,p)$. Par dualité de Cartier, le dual de Pontrjagin de $\Gamma^0(p)$ s'identifie au groupe des composantes connexes du noyau de l'homomorphisme canonique $\mathrm{Alb}(F) \longrightarrow \mathrm{Alb}(X)$.

Remarque 1.7.- Soit F une fibre lisse de p dans X . La prop. 1.5 peut s'exprimer de manière purement topologique, par la suite exacte

$$0 \longrightarrow H^1(B,S_1) \times \Gamma^\tau(p) \longrightarrow H^1(X,S_1) \longrightarrow H^1(F,S_1) \ .$$

Par dualité de Pontrjagin, on en déduit une suite exacte

$$H_1(F,\mathbf{Z}) \longrightarrow H_1(X,\mathbf{Z}) \longrightarrow H_1(B,\mathbf{Z}) \times \hat{\Gamma}^\tau(p) \longrightarrow 0 \ ,$$

où $\hat{\Gamma}^\tau(p)$ désigne le dual de Pontrjagin de $\Gamma^\tau(p)$.

Plus canoniquement, cela revient à dire que l'homologie du complexe

$$H_1(F,\mathbf{Z}) \longrightarrow H_1(X,\mathbf{Z}) \longrightarrow H_1(B,\mathbf{Z})$$

est isomorphe à $\hat{\Gamma}^\tau(p)$ (on peut bien sûr obtenir ces résultats directement par une étude topologique simple, cf. [S]).

On obtient de même une suite exacte

$$H_1(F,\mathbf{Z})_\ell \longrightarrow H_1(X,\mathbf{Z})_\ell \longrightarrow H_1(B,\mathbf{Z}) \times \hat{\Gamma}^0(p) \longrightarrow 0$$

(où pour tout Z-module M , on note M_ℓ le quotient de M par son sous-module de torsion).

Exemple 1.8.- Commençons par quelques remarques générales. Soit Y une variété kählérienne compacte sur laquelle opère un groupe fini G ; supposons pour simplifier que le quotient Y/G soit lisse. Notons $\mathrm{Pic}^G(Y)$ le plus grand sous-tore complexe de $\mathrm{Pic}^0(Y)$ sur lequel G opère trivialement. Comme $H^1(Y/G, \mathcal{O}_{Y/G})$ s'identifie au sous-espace des éléments G-invariants de $H^1(Y,\mathcal{O}_Y)$, l'homomorphisme $\pi^* : \mathrm{Pic}^0(Y/G) \longrightarrow \mathrm{Pic}^G(Y)$ est *surjectif ;* son noyau est le sous-groupe de $\mathrm{Hom}(G,\mathbf{C}^*)$ formé des caractères qui sont triviaux sur les stabilisateurs des points de Y .

Soit maintenant F une variété kählérienne compacte, sur laquelle le groupe G opère librement ; soient B une courbe et $\beta : \tilde{B} \longrightarrow B$ un revêtement (ramifié) galoisien de groupe G . Le groupe G opère librement sur le produit $\tilde{B} \times F$; notons X la variété quotient et $\pi : \tilde{B} \times F \longrightarrow X$ l'application canonique. La première projection définit un morphisme $p : X \longrightarrow B$; les fibres de p au-dessus des points de ramification de β sont multiples, les autres fibres sont isomorphes à F .

D'après ce qui précède, l'homomorphisme

$$\pi^* : \mathrm{Pic}^0(X) \longrightarrow \mathrm{Pic}^G(\tilde{B}) \times \mathrm{Pic}^G(F)$$

est surjectif; son noyau $\hat{G}$ s'identifie au groupe $\mathrm{Hom}(G,\mathbf{C}^*)$. On en déduit que le groupe $\mathrm{Pic}^0(X,p)$ est égal à $\hat{G} + p^*\mathrm{Pic}^0(B)$. Supposons en outre que le sous-groupe distingué de G engendré par les stabilisateurs des points de $\tilde{B}$ soit égal à G (par exemple que l'une des fibres de β soit réduite à un point), de sorte que l'homomorphisme $\beta^* : \mathrm{Pic}^0(B) \longrightarrow \mathrm{Pic}^G(\tilde{B})$ est injectif; alors l'intersection $\hat{G} \cap p^*\mathrm{Pic}^0(B)$ est réduite à 0 , et *le groupe* $\Gamma^0(p)$ *est isomorphe à* $\hat{G}$. Observons qu'il est facile de choisir β de façon qu'on ait $\Gamma^0(p) = \Gamma^\tau(p)$.

Cet exemple est en fait un cas très particulier d'une situation générale :

Proposition 1.9.- *Il existe un diagramme commutatif*

$$(1.9)\qquad \begin{array}{ccc} \tilde{X} & \xrightarrow{\ \pi\ } & X \\ {\scriptstyle \tilde{p}}\downarrow & & \downarrow{\scriptstyle p} \\ \tilde{B} & \xrightarrow{\ \beta\ } & B \end{array} ,$$

où π *est un revêtement étale,* β *un morphisme fini et* $\tilde{p}$ *une fibration, tel qu'on ait* $\pi^* \operatorname{Pic}^\tau(X,p) \subset \tilde{p}^* \operatorname{Pic}^0(\tilde{B})$.

Soit G le dual de Pontryagin de $\Gamma^\tau(p)$; il s'identifie au quotient de $\oplus\, \mathbf{Z}/(m_i)$ par le sous-groupe des éléments $(\dot{n}, \dots, \dot{n})$ $(n \in \mathbf{Z})$. On désignera par e_i la classe dans G de l'élément dont la i-ième composante est $\dot{1}$ et dont les autres composantes sont nulles. Soit m le p.p.c.m. des m_i ; notons d_i l'entier m/m_i et δ_i le p.g.c.d. des entiers d_j pour $j \neq i$. On vérifie facilement que l'ordre de e_i dans G est m_i/δ_i .

Notons B' le complémentaire des b_i dans B ; soit o un point de B' . Le groupe $\pi_1(B', o)$ est engendré par des éléments $a_1, \dots, a_{2g}, t_1, \dots, t_s$, soumis à la relation $(a_1,a_{g+1})\dots(a_g,a_{2g})\, t_1\dots t_s = 1$; le générateur t_i est obtenu en reliant o à un petit cercle autour de b_i . Définissons un homomorphisme $\varphi : \pi_1(B',o) \longrightarrow G$ en choisissant arbitrairement $\varphi(a_i)$, et en posant $\varphi(t_i) = e_i$. On en déduit un revêtement $\beta : \tilde{B} \longrightarrow B$ galoisien de groupe G ; chaque point de $\tilde{B}$ au-dessus de b_i a un indice de ramification égal à l'ordre de e_i dans G , soit m_i/δ_i .

Notons $\tilde{X}$ la normalisation du produit fibré $X \times_B \tilde{B}$ et $\pi : \tilde{X} \longrightarrow X$, $\tilde{p} : \tilde{X} \longrightarrow \tilde{B}$ les morphismes déduits des deux projections. Le morphisme π est étale : il suffit en effet de le vérifier en codimension 1 , autrement dit au-dessus d'un point assez général de F_i , auquel cas cela résulte d'un calcul simple et bien connu. La fibre de $\tilde{p}$ au-dessus d'un point général $\tilde{b}$ de $\tilde{B}$ est isomorphe à $p^{-1}(\beta(\tilde{b}))$, donc connexe.

Pour tout i , on a

$$\tilde{p}^*\beta^* b_i \;=\; \frac{m_i}{\delta_i} \sum_{\tilde{b} \in \beta^{-1}(b_i)} \tilde{p}^* \tilde{b} \;=\; m_i\, \pi^* F_i \quad ;$$

on en déduit que chaque fibre $\tilde{p}^* \tilde{b}$ a pour multiplicité δ_i , et que le faisceau $\pi^* \mathcal{O}_X(\delta_i F_i)$ appartient à $\tilde{p}^* \operatorname{Pic}(\tilde{B})$. Soit alors L un élément de $\operatorname{Pic}^\tau(X,p)$; écrivons $L = \pi^* M(\sum k_i F_i)$, avec $\sum k_i/m_i \in \mathbf{Z}$. Cette dernière relation signifie que m divise $\sum k_i d_i$; comme δ_i est premier avec d_i , on en déduit que δ_i divise k_i . Par suite le faisceau $\pi^* L$ provient d'un faisceau inversible sur $\tilde{B}$, nécessairement de degré 0 . ∎

Proposition 1.10.- *Soit* S *une composante de* $\operatorname{Pic}^\tau(X,p)$; *soient* $k_1, \dots, k_s$ *des entiers tels que* $0 \le k_i < m_i$, *et que les éléments de* S *soient de la forme* $p^* L_0 \otimes \mathcal{O}_X(\sum k_i F_i)$, *avec* $L_0 \in \operatorname{Pic}(B)$.

a) *Pour tout* $L \in S$, *on a* $\dim H^1(X,L) \ge g(B) - 1 + \sum k_i/m_i$.

b) *Supposons de plus que* X *soit une* surface. *On a alors*

$$\dim H^1(X,L) \;=\; g(B) - 1 + \sum k_i/m_i$$

pour tous les éléments L *de* S *sauf un nombre fini.*

Posons $L = p^*L_0 \otimes \mathcal{O}_X(\sum k_i F_i)$. Compte tenu de la prop. 1.3, la suite spectrale de Leray fournit une suite exacte

$$0 \longrightarrow H^1(B,L_0) \longrightarrow H^1(X,L) \longrightarrow H^0(B,R^1p_*L) \longrightarrow 0 \ .$$

On a $\dim H^1(B,L_0) \geq g(B)-1+\sum k_i/m_i$, et il y a égalité si $L \neq \mathcal{O}_X$. Cela entraîne a) ; pour démontrer b) , il suffit de prouver qu'on a $H^0(B,R^1p_*L) = 0$ sauf pour un nombre fini d'éléments L de S . Considérons le diagramme (1.9). Le faisceau L est facteur direct de $\pi_*\pi^*L$, donc R^1p_*L est facteur direct de $\beta_*R^1\tilde{p}_*(\pi^*L)$, et l'espace $H^0(B,R^1p_*L)$ s'identifie à un sous-espace de $H^0(\tilde{B}, R^1\tilde{p}_*(\pi^*L))$. Ecrivons $\pi^*L = \tilde{p}^*M$, avec $M \in \mathrm{Pic}^0(\tilde{B})$ (prop. 1.9); le faisceau $R^1\tilde{p}_*(\pi^*L)$ est isomorphe à $M \otimes R^1\tilde{p}_*\mathcal{O}_{\tilde{X}}$.

D'après [F] , le fibré $R^1\tilde{p}_*\mathcal{O}_{\tilde{X}}$ admet une décomposition en somme directe $R^1\tilde{p}_*\mathcal{O}_{\tilde{X}} = \bigoplus_{i=0}^{r} F_i$, où le dual de F_0 est ample, et où F_i est stable de degré 0 pour $i \geq 1$. On en déduit que $M \otimes R^1\tilde{p}_*\mathcal{O}_{\tilde{X}}$ admet une section non nulle si et seulement si M est le dual de l'un des F_i , d'où la proposition. ■

Le résultat b) permet de corriger les assertions de [B] sur le système paracanonique des surfaces (voir ci-dessous). Supposons que X soit une surface et que S soit une composante de $\mathrm{Pic}^0(X,p)$; en suivant la démonstration du th. 2 de *loc. cit.*, on obtient que la variété des diviseurs effectifs D tels que $\mathcal{O}_X(D) \otimes K_X^{-1} \in S$ est de dimension $\chi(\mathcal{O}_X) + 2g(B) - 2 + \sum_{k_i \geq 1} \frac{m_i - k_i}{m_i}$. C'est une composante du systeme paracanonique $\{K_X\}$ si et seulement si cette dimension est $\geq p_g$.

§2. Le sous-ensemble $S^1(X)$ de $\mathrm{Pic}^\tau(X)$

L'énoncé suivant est une généralisation du lemme de Castelnuovo-De Franchis (qui correspond au cas $L = \mathcal{O}_X$); il est énoncé incorrectement dans [B] . Je reprends ici, en la corrigeant, la démonstration de [B] .

Proposition 2.1.- *Soient* L *un élément de* $\mathrm{Pic}^\tau(X)$, *et* ω *une 1-forme holomorphe non nulle sur* X . *Les conditions suivantes sont équivalentes :*

(i) *La suite*

$$H^0(X,L) \xrightarrow{\wedge\omega} H^0(X,\Omega^1_X \otimes L) \xrightarrow{\wedge\omega} H^0(X,\Omega^2_X \otimes L)$$

n'est pas exacte;

(ii) *Il existe un morphisme* p *de* X *sur une courbe* B *de genre* ≥ 1 *tel que la forme* ω *provienne par image réciproque de* $H^0(B,\Omega^1_B)$, *et que l'on ait :*

- $L \in \mathrm{Pic}^\tau(X,p)$ *si* $g(B) \geq 2$;
- $L \in \mathrm{Pic}^\tau(X,p) - p^*\mathrm{Pic}^0(B)$ *si* $g(B) = 1$.

Supposons la condition (i) satisfaite. Il existe alors une forme non nulle $\alpha \in H^0(X, \Omega^1_X \otimes L)$ telle que $\alpha \wedge \omega = 0$; si en outre $L = \mathcal{O}_X$, on peut prendre α non proportionnelle à ω. La relation $\alpha \wedge \omega = 0$ signifie qu'il existe une section méromorphe φ de L telle qu'on ait $\alpha = \omega \otimes \varphi$. Soit ∇ la connexion holomorphe unitaire sur L. La théorie de Hodge entraîne $\nabla\alpha = 0$ et $d\omega = 0$, d'où $\omega \wedge \nabla\varphi = 0$ dans $\Omega^2_X \otimes L$. On en déduit qu'il existe une fonction méromorphe f sur X telle que $\nabla\varphi = f\omega \otimes \varphi$. Appliquant de nouveau la connexion intégrable ∇, on obtient $df \wedge \omega = 0$.

Comme le diviseur de φ n'est pas nul par hypothèse, la forme $\varphi^{-1}\nabla\varphi = f\omega$ a des pôles, de sorte que la fonction f n'est pas constante; considérons-la comme une application méromorphe de X dans $\mathbf{P}^1$. D'après le théorème de Hironaka, il existe un morphisme $\varepsilon : \hat{X} \longrightarrow X$, composé d'un nombre fini d'éclatements, et un morphisme $\hat{f} : \hat{X} \longrightarrow \mathbf{P}^1$ qui prolonge $f \circ \varepsilon$. Soit $\hat{f} : \hat{X} \longrightarrow B \xrightarrow{p} \mathbf{P}^1$ la factorisation de Stein de $\hat{f}$. La relation $df \wedge \omega = 0$ implique que la restriction de $\varepsilon^*\omega$ à une fibre générale de p est nulle, donc que $\varepsilon^*\omega$ provient par image réciproque d'une forme méromorphe ω_0 sur B ; comme $p^*\omega_0$ est holomorphe, la forme ω_0 est nécessairement holomorphe. On a donc $g(B) \geq 1$, ce qui entraîne *a posteriori* que f était partout définie : on peut prendre $\hat{X} = X$ et $\varepsilon = \mathrm{Id}_X$.

Comme $\alpha = \omega \otimes \varphi$, il existe un diviseur effectif E sur B tel que le diviseur des pôles de φ soit contenu dans p^*E ; l'espace $H^0(X, L(p^*E))$, isomorphe à $H^0(B, p_*L \otimes \mathcal{O}_B(E))$, est donc non nul. Il en résulte que le faisceau p_*L n'est pas nul, ce qui implique $L \in \mathrm{Pic}^\tau(X,p)$ (prop. 1.2). Pour achever de prouver (ii), il reste à éliminer le cas $g(B) = 1$ et $L = p^*L_0$, avec $L_0 \in \mathrm{Pic}^0(B)$. Dans ce cas la forme $\alpha = \omega \otimes \varphi$ provient comme ci-dessus d'une forme holomorphe $\alpha_0 \in H^0(B, \Omega^1_B \otimes L_0)$; comme $g(B) = 1$, cela implique $L_0 = \mathcal{O}_B$ et α proportionnelle à ω, contrairement à l'hypothèse.

Supposons maintenant la condition (ii) satisfaite. Ecrivons L sous la forme $p^*L_0(\sum k_i F_i)$, où $m_1F_1, \dots, m_sF_s$ sont les fibres multiples de p, et $0 \leq k_i < m_i$ pour tout i. Soit D_0 le diviseur sur B somme des points correspondant aux fibres F_i telles que $k_i \geq 1$. On a $\deg(L_0) = -\sum k_i/m_i$, d'où

$$\deg(L_0(D_0)) = \sum_{k_i \geq 1} \frac{m_i - k_i}{m_i} \geq 0,$$

et de plus $\deg(L_0(D_0)) > 0$ si $g(B) = 1$. Par suite l'espace $H^0(B, \Omega^1_B \otimes L_0(D_0))$ contient un élément non nul α_0 ; si de plus L_0 est trivial et que tous les k_i sont nuls (ce qui entraîne $g(B) \geq 2$), on peut prendre α_0 non proportionnelle à ω_0. Alors $\alpha = p^*\alpha_0$ est une section holomorphe de $\Omega^1_X \otimes L$, non proportionnelle à ω si $L = \mathcal{O}_X$, et l'on a $\alpha \wedge \omega = 0$. Cela achève la démonstration de la proposition. ∎

Théorème 2.2.- *Notons* $(p_i : X \longrightarrow B_i)_{i \in I}$ *la famille des fibrations de* X *sur une courbe de genre* ≥ 1. *La sous-variété* $S^1(X)$ *de* $\mathrm{Pic}^\tau(X)$ *formée des faisceaux inversibles* L *tels que* $H^1(X, L) \neq 0$ *est réunion des sous-groupes* $\mathrm{Pic}^\tau(X, p_i)$ *pour*

$g(B_i) \geq 2$, *des sous-variétés* $\mathrm{Pic}^\tau(X,p_i) - p_i^* \mathrm{Pic}^0(B_i)$ *pour* $g(B_i) = 1$, *et d'un nombre fini de points isolés.*

Soit L un élément non isolé de $S^1(X)$. En vertu du corollaire 1.9 et du lemme 2.6 de [G-L 1] , il existe une 1-forme holomorphe non nulle ω sur X telle que la suite

$$H^0(X, L^{-1}) \xrightarrow{\wedge\omega} H^0(X, \Omega^1_X \otimes L^{-1}) \xrightarrow{\wedge\omega} H^0(X, \Omega^2_X \otimes L^{-1})$$

ne soit pas exacte. La prop. 2.1 implique alors que L^{-1} , et donc aussi L , sont des éléments de $\mathrm{Pic}^\tau(X,p)$; de plus L n'appartient pas à $p^* \mathrm{Pic}^0(B)$ si B est de genre 1 .

Inversement, soit $p : X \longrightarrow B$ une fibration, et soit S une composante de $\mathrm{Pic}^\tau(X,p)$; la prop. 1.10, a) entraîne $S \subset S^1(X)$, sauf lorsque $g(B) = 1$ et $S = p^* \mathrm{Pic}^0(B)$. ∎

Corollaire 2.3.- *L'intersection* $S^1(X) \cap \mathrm{Pic}^0(X)$ *est réunion des ensembles* $\mathrm{Pic}^0(X,p_i)$ *pour* $g(B_i) \geq 2$, $\mathrm{Pic}^0(X, p_i) - p_i^* \mathrm{Pic}^0(B_i)$ *pour* $g(B_i) = 1$, *et d'un nombre fini de points isolés.* ∎

Remarque 2.4.- Soit $p : X \longrightarrow B$ une fibration. Pour que le groupe $\mathrm{Pic}^\tau(X,p)$ soit distinct de $p^* \mathrm{Pic}^0(B)$, il faut et il suffit que le groupe $\Gamma^\tau(p)$ ne soit pas trivial, autrement dit que *deux des multiplicités* m_i *ne soient pas premières entre elles.* Si $g(B) = 1$, c'est la condition nécessaire et suffisante pour que la fibration p fournisse des composantes de $S^1(X)$ non réduites à un point.

§3. Le sous-ensemble $\Sigma^1(X)$ de $H^1(X, \mathbb{C}^*)$

(3.1) Nous supposerons choisi un point o de X , et poserons $\pi_1(X) = \pi_1(X,o)$. Nous allons nous intéresser au groupe $H^1(X, \mathbf{C}^*)$. Ce groupe paramètre les classes d'isomorphisme de chacun des objets suivants :

a) *Les caractères de* $\pi_1(X)$ (c'est-à-dire les homomorphismes de $\pi_1(X)$ dans $\mathbf{C}^*$).

b) *Les faisceaux localement constants de* **C**-*espaces vectoriels de dimension* 1 *sur* X (nous dirons simplement *faisceaux localement constants de rang* 1).

c) *Les couples* (L,∇), *où* L *est un faisceau inversible sur* X *et* $\nabla : L \longrightarrow \Omega^1_X \otimes L$ *une connexion holomorphe sur* X (une telle connexion est toujours de courbure nulle).

(3.2) Notons $\mathbf{S}_1$ le cercle unité dans $\mathbf{C}$. Le sous-groupe $H^1(X, \mathbf{S}_1)$ de $H^1(X, \mathbf{C}^*)$ correspond aux objets suivants :

a) Les caractères χ qui sont *unitaires* (c'est-à-dire à valeurs dans $\mathbf{S}_1$);

b) Les faisceaux localement constants $\mathscr{L}$ de rang 1 qui sont *unitaires,* c'est-à-dire qui admettent une forme hermitienne positive non nulle (à valeurs dans le faisceau constant $\mathbf{C}_X$).

c) Les couples (L,∇_u), où ∇_u désigne la connexion (holomorphe) *unitaire* sur L, associée à la métrique hermitienne de courbure nulle sur L (qui est unique à une constante près).

Nous munirons le groupe $H^1(X,\mathbf{C}^*)$ de sa topologie naturelle. L'homomorphisme $(L,\nabla)\longmapsto L$ fait apparaître ce groupe comme extension du groupe de Lie complexe $\mathrm{Pic}^\tau(X)$ par l'espace vectoriel $H^0(X,\Omega_X^1)$. En tant qu'extension de groupes de Lie *réels,* cette extension est scindée : on en obtient une section canonique en associant à un fibré L le couple (L,∇_u).

Soit $\mathcal{L}$ un faisceau localement constant *unitaire* sur X. Une partie de la théorie de Hodge classique s'étend sans changement à la cohomologie de X à coefficients dans $\mathcal{L}$. Nous utiliserons en particulier les faits suivants :

(3.3) Soit $L=\mathcal{L}\otimes_{\mathbf{C}}\mathcal{O}_X$, et soit ∇_u la connexion unitaire sur L. La suite spectrale

$$E_1^{pq} = H^q(X,\Omega_X^p\otimes L) \quad\Rightarrow\quad H^{p+q}(X,\mathcal{L})$$

associée au complexe de de Rham de (L,∇_u) dégénère en E_1. En particulier, les homomorphismes $H^q(\nabla_u): H^q(X,\Omega_X^p\otimes L)\longrightarrow H^q(X,\Omega_X^{p+1}\otimes L)$ sont nuls.

(3.4) Choisissons une métrique hermitienne plate sur L, d'où un isomorphisme antilinéaire $L\longrightarrow L^{-1}$. On déduit de cet isomorphisme et de la conjugaison des formes harmoniques un isomorphisme antilinéaire

$$H^q(X,\Omega_X^p\otimes L)\longrightarrow H^p(X,\Omega_X^q\otimes L^{-1}).$$

Soit $\omega\in H^0(X,\Omega_X^1)$, et $\bar\omega$ la classe dans $H^1(X,\mathcal{O}_X)$ de la forme conjuguée; le diagramme

$$\begin{array}{ccc} H^q(X,\Omega_X^p\otimes L) & \xrightarrow{\ \cup\bar\omega\ } & H^{q+1}(X,\Omega_X^p\otimes L) \\ \big\downarrow & & \big\downarrow \\ H^p(X,\Omega_X^q\otimes L^{-1}) & \xrightarrow{\ \wedge\omega\ } & H^p(X,\Omega_X^{q+1}\otimes L^{-1}) \end{array}$$

est commutatif.

Proposition 3.5.- *Soit* $\mathcal{L}$ *un faisceau localement constant de rang* 1 *sur* X, *et soit* (L,∇) *le fibré à connexion correspondant.*

a) *Supposons* $\mathcal{L}$ *unitaire* (c'est-à-dire $\nabla=\nabla_u$). *La condition* $H^1(X,\mathcal{L})\neq 0$ *équivaut alors à* $L\in S^1(X)\cup -S^1(X)$.

b) *Supposons que* $\mathcal{L}$ *ne soit pas unitaire. La condition* $H^1(X,\mathcal{L})\neq 0$ *équivaut à :*

(∗) *Il existe une fibration* $p:X\longrightarrow B$ *et une forme non nulle* $\omega_0\in H^0(B,\Omega_B^1)$ *telles qu'on ait*

- $L\in\mathrm{Pic}^\tau(X,p)$ *si* $g(B)\geq 2$, $L\in\mathrm{Pic}^\tau(X,p)-p^*\mathrm{Pic}^0(B)$ *si* $g(B)=1$;
- $\nabla=\nabla_u+p^*\omega_0$.

Considérons la suite spectrale (3.3)

$$E_1^{pq} = H^q(X, \Omega_X^p \otimes L) \quad \Rightarrow \quad H^{p+q}(X, \mathcal{L}) .$$

Si $\mathcal{L}$ est unitaire, elle dégénère en E_1 ; on en déduit une suite exacte

$$0 \rightarrow H^0(X, \Omega_X^1 \otimes L) \longrightarrow H^1(X, \mathcal{L}) \longrightarrow H^1(X, L) \rightarrow 0 .$$

Comme les espaces $H^0(X, \Omega_X^1 \otimes L)$ et $H^1(X, L^{-1})$ sont conjugués (3.4), l'assertion a) en résulte.

Supposons que $\mathcal{L}$ ne soit pas unitaire; posons $\nabla = \nabla_u + \omega$, où ω est une 1-forme holomorphe non nulle sur X . Comme on a $H^q(\nabla_u^p) = 0$, la différentielle d_1 de la suite spectrale n'est autre que le cup-produit avec ω . La suite spectrale fournit alors une suiteexacte

$$0 \rightarrow E_2^{10} \longrightarrow H^1(X, \mathcal{L}) \longrightarrow E_2^{01} ,$$

où E_2^{10} est l'homologie du diagramme

$$H^0(X, L) \xrightarrow{\wedge\omega} H^0(X, \Omega_X^1 \otimes L) \xrightarrow{\wedge\omega} H^0(X, \Omega_X^2 \otimes L)$$

et E_2^{01} le noyau de la flèche $H^1(\omega) : H^1(X, L) \longrightarrow H^1(X, \Omega_X^1 \otimes L)$.

Si le couple (L, ∇) satisfait à (*), l'espace E_2^{10} n'est pas nul (prop. 2.1), et il en est de même de $H^1(X, \mathcal{L})$. Supposons inversement $H^1(X, \mathcal{L}) \neq 0$. Cela entraîne que E_2^{10} ou E_2^{01} n'est pas nul. Dans le premier cas, (L, ∇) satisfait à (*) d'après la prop. 2.1.

Il reste à prouver que (L, ∇) satisfait à (*) lorsque E_2^{01} n'est pas nul, c'est-à-dire lorsque l'homomorphisme $H^1(\omega)$ n'est pas injectif. Par conjugaison (3.4), cela signifie qu'il existe un élément non nul α de $H^0(X, \Omega_X^1 \otimes L^{-1})$ tel que la classe de $\alpha \wedge \bar{\omega}$ dans $H^1(X, \Omega_X^1 \otimes L^{-1})$ soit nulle. Posons $\beta = \alpha \wedge \omega$; on a $\beta \in H^0(X, \Omega_X^2 \otimes L^{-1})$, et la classe de $\beta \wedge \bar{\beta}$ dans $H^2(X, \Omega_X{}^2)$ est nulle. Si κ est une forme de Kähler sur X , on a donc $\int \beta \wedge \bar{\beta} \wedge \kappa^{n-2} = 0$. Montrons que cela entraîne $\beta = 0$. Soit x un point de X ; il existe un système de coordonnées locales $(z_1, \ldots, z_n)$ sur X tel que κ coïncide en x avec la forme $\sum dz_i \wedge d\bar{z}_i$. La forme β s'écrit dans ces coordonnées $\sum b_{ij}\, dz_i \wedge dz_j$, où (b_{ij}) est une matrice antisymétrique formée de sections de L^{-1} . Notons $\| \ \|$ la norme associée à la métrique hermitienne plate sur L^{-1} ; un calcul facile donne (au point x)

$$\beta \wedge \bar{\beta} \wedge \kappa^{n-2} = c \Big(\sum_{i,j} \| b_{ij} \|^2 \Big) \kappa^n , \qquad \text{avec } c = \frac{-1}{n(n-1)} ,$$

de sorte que la relation $\int \beta \wedge \bar{\beta} \wedge \kappa^{n-2} = 0$ entraîne $\beta = 0$.

Ainsi on a $\alpha \wedge \omega = 0$, ce qui conduit à deux possibilités :

- ou bien la suite $H^0(X, L^{-1}) \xrightarrow{\wedge\omega} H^0(X, \Omega_X{}^1 \otimes L^{-1}) \xrightarrow{\wedge\omega} H^0(X, \Omega_X{}^2 \otimes L^{-1})$ n'est pas exacte, et le couple (L^{-1}, ω) satisfait à (*) ; il en est alors de même de (L, ω) .
- ou bien $L = \mathcal{O}_X$, et α est proportionnelle à ω . Dans ce cas la classe de $\omega \wedge \bar{\omega}$ dans $H^1(X, \Omega_X^1)$ est nulle; si κ est une forme de Kähler sur X , cela entraîne $\int \omega \wedge \bar{\omega} \wedge \kappa^{n-1} = 0$.

On voit alors comme ci-dessus que cela implique $\omega = 0$, ce qui est impossible. ■

Corollaire 3.6.- *Le sous-ensemble* $\Sigma^1(X)$ *de* $H^1(X, \mathbf{C}^*)$ *formé des faisceaux localement constants de rang un* $\mathcal{L}$ *sur* X *tels que* $H^1(X,\mathcal{L}) \neq 0$ *est réunion :*

- d'une famille finie de sous-ensembles de la forme $\alpha \cdot p^* H^1(B, \mathbf{C}^*)$ *, où* α *est un élément d'ordre fini de* $H^1(X,\mathbf{C}^*)$ *et* $p : X \longrightarrow B$ *une fibration sur une courbe de genre* ≥ 1 ;

- d'un nombre fini de faisceaux unitaires. ■

§4. Le sous-ensemble $\Sigma^1(G)$ de $\mathrm{Hom}(G,\mathbb{C}^*)$

Soit G un groupe de type fini. Nous désignerons par $\hat{G}$ le groupe $\mathrm{Hom}(G,\mathbf{C}^*)$ des caractères de G , muni de la topologie usuelle. Pour tout caractère χ de G , nous noterons $\mathbf{C}_\chi$ le G-module $\mathbf{C}$ muni de l'action de G définie par χ . Nous allons nous intéresser au sous-ensemble $\Sigma^1(G)$ de $\hat{G}$ formé des caractères χ tels que $H^1(G,\mathbf{C}_\chi) \neq 0$.

Le lemme suivant est certainement bien connu :

Lemme 4.1.- *Soit G un groupe commutatif de type fini, et* χ *un caractère non trivial de* G . *On a* $H^i(G,\mathbf{C}_\chi) = 0$ *pour tout* $i \geq 0$.

Notons T le sous-groupe de torsion de G , et L le groupe quotient G/T . On dispose d'une suite spectrale de Hochschild-Serre

$$E_2^{pq} = H^p(L, H^q(T, \mathbf{C}_\chi)) \quad \Rightarrow \quad H^{p+q}(G,\mathbf{C}_\chi) .$$

L'espace $H^q(T,\mathbf{C}_\chi)$ est nul sauf si $q = 0$ et $\chi_{|T} = 1$; il suffit donc de prouver le lemme lorsque G est un $\mathbf{Z}$-module libre (de type fini).

Posons $A = \mathbf{C}[G]$; l'espace $H^p(G,\mathbf{C}_\chi)$ s'identifie à $\mathrm{Ext}_A^p(\mathbf{C}_1,\mathbf{C}_\chi)$. Choisissons une base $(e_1, \dots , e_n)$ de G ; l'anneau A s'identifie à l'anneau des polynômes de Laurent $\mathbf{C}[e_1, e_1^{-1}, \dots , e_n, e_n^{-1}]$. Une résolution libre du A-module $\mathbf{C}_1$ est fournie par le complexe de Koszul

$$0 \to \wedge^n(A^n) \longrightarrow \dots\dots \longrightarrow \wedge^2(A^n) \longrightarrow A^n \longrightarrow A \longrightarrow \mathbf{C}$$

associé à la suite régulière $(e_1-1, \dots , e_n-1)$. Les espaces $\mathrm{Ext}_A^p(\mathbf{C}_1,\mathbf{C}_\chi)$ sont donc les espaces de cohomologie du complexe

$$0 \to \mathbf{C} \xrightarrow{\varepsilon} \mathbf{C}^n \xrightarrow{\wedge\varepsilon} \wedge^2(\mathbf{C}^n) \longrightarrow \dots\dots \longrightarrow \wedge^n(\mathbf{C}^n) \to 0 ,$$

avec $\varepsilon = (\chi(e_1)-1, \dots , \chi(e_n)-1)$. Comme χ n'est pas trivial, le vecteur ε n'est pas nul, et le complexe ci-dessus est exact, ce qui démontre le lemme. ■

Revenons au cas d'un groupe G de type fini quelconque. Notons $D(G)$ son groupe dérivé, et G_{ab} le quotient $G/D(G)$; le groupe G opère sur $D(G)$ par conjugaison.

Proposition 4.2.- *Pour tout caractère non trivial* χ *de* G, *l'espace* $H^1(G,\mathbf{C}_\chi)$ *s'identifie à l'espace des homomorphismes* $u : D(G) \longrightarrow \mathbf{C}$ *qui satisfont à*

$$u(g.x) = \chi(g)\,u(x) \qquad \textit{pour } g \in G\,,\ x \in D(G)\,.$$

Autrement dit, $H^1(G,\mathbf{C}_\chi)$ est le composant isotypique de type χ de la représentation de G sur l'espace vectoriel complexe $\mathrm{Hom}\,(D(G),\mathbf{C})$. Ainsi $\Sigma^1(G)$ est l'ensemble des caractères qui apparaissent dans cette représentation.

La suite spectrale de Hochschild-Serre

$$E_2^{pq} = H^p(G_{ab}, H^q(D(G),\mathbf{C}_\chi)) \implies H^{p+q}(G,\mathbf{C}_\chi)$$

donne lieu en bas degré à une suite exacte

$$0 \longrightarrow H^1(G_{ab},\mathbf{C}_\chi) \longrightarrow H^1(G,\mathbf{C}_\chi) \longrightarrow \mathrm{Hom}\,(D(G),\mathbf{C}_\chi)^G \longrightarrow H^2(G_{ab},\mathbf{C}_\chi)\,;$$

l'action de G sur $\mathrm{Hom}\,(D(G),\mathbf{C}_\chi)$ est donnée par la formule

$$(g.u)(x) = \chi(g)\,u(g^{-1}x) \qquad \text{pour } g \in G\,,\ x \in D(G)\,,\ u \in \mathrm{Hom}\,(D(G),\mathbf{C}_\chi)\,.$$

Si χ n'est pas trivial, on déduit du lemme 4.1 un isomorphisme $H^1(G,\mathbf{C}_\chi) \xrightarrow{\sim} \mathrm{Hom}\,(D(G),\mathbf{C}_\chi)^G$. La proposition en résulte. ∎

Lorsque G est le groupe fondamental d'une variété kählérienne compacte X, le groupe $\hat{G}$ s'identifie naturellement à $H^1(X,\mathbf{C}^*)$: à un caractère χ de G correspond un faisceau localement constant de rang un $\mathcal{L}_\chi$ sur X. On a un isomorphisme canonique

$$H^1(G,\mathbf{C}_\chi) \xrightarrow{\sim} H^1(X,\mathcal{L}_\chi)\,;$$

par suite, l'ensemble $\Sigma^1(G)$ coïncide avec l'ensemble $\Sigma^1(X)$. En vertu du cor. 3.6, il est réunion d'une partie continue $\Sigma^1_c(G)$, formée de translatés de sous-groupes de la forme $\mathrm{Hom}\,(G/H,\mathbf{C}^*)$ où H est un sous-groupe distingué de G, et d'une partie finie $\Sigma^1_i(G)$ formée de caractères *unitaires*.

Proposition 4.3.- *Soit* $\chi \in \Sigma^1_i(G)$, *et soit* $g \in G$. *Le nombre complexe* $\chi(g)$ *est un nombre algébrique, dont tous les conjugués sont de module* 1. *S'il est entier sur* $\mathbf{Z}$, *c'est une racine de l'unité.*

Soit σ un automorphisme de $\mathbf{C}$. L'application $z \mapsto \sigma(z)$ définit un isomorphisme $\mathbf{Z}[G]$-linéaire de $\mathbf{C}_\chi$ sur $\mathbf{C}_{\sigma\circ\chi}$; par suite le sous-ensemble $\Sigma^1(G)$ de $\hat{G}$ est stable par l'action de $\mathrm{Aut}\,(\mathbf{C})$. Il résulte aussitôt de la définition de $\Sigma^1_c(G)$ que cet ensemble est stable par $\mathrm{Aut}\,(\mathbf{C})$; il en est donc de même de $\Sigma^1_i(G)$. On en déduit que l'ensemble des conjugués de $\chi(g)$ est fini, donc que $\chi(g)$ est un nombre algébrique, et que tous ces conjugués sont de module 1. La dernière assertion résulte alors d'un lemme bien connu de Kronecker. ∎

Corollaire 4.4.- *Soit* M *un* $\mathbf{Z}[G]$*-module, de type fini sur* $\mathbf{Z}$. *Soit* χ *un élément de* $\Sigma^1_i(G)$; *on suppose qu'il existe un vecteur non nul* v *de* $M\otimes\mathbf{C}$ *satisfaisant à* $g.v = \chi(g)\,v$ *pour tout* $g \in G$. *Alors* χ *est d'ordre fini.*

Tout élément g de G induit un endomorphisme $\mathbb{Z}$-linéaire de M, dont les valeurs propres sont des entiers algébriques en vertu du théorème de Cayley-Hamilton. Il résulte alors de la proposition que $\chi(g)$ est une racine de l'unité pour tout g dans G, donc que χ est d'ordre fini puisque G est de type fini. ∎

Proposition 4.5.- *On suppose que le $\mathbb{Z}$-module* $D(G)/D^2(G)$ *est de type fini. Alors les ensembles* $S^1(X)$ *et* $\Sigma^1(X)$ *sont finis, et leurs éléments sont d'ordre fini.*

La prop. 4.2 entraîne que l'ensemble $\Sigma^1(G)$ est fini, et que le $\mathbb{Z}[G]$-module $\mathrm{Hom}(D(G)/D^2(G), \mathbb{Z})$ vérifie les hypothèses du cor. 4.3 ; il en résulte que $\Sigma^1(X)$ est formé d'éléments d'ordre fini. Compte tenu de la prop. 3.5, il en est de même de $S^1(X)$. ∎

Exemple 4.6.- Supposons qu'on ait $\dim H^1(X, \mathcal{O}_X) = 1$. L'application d'Albanese définit alors une fibration α de X sur une courbe elliptique E ; *supposons que cette fibration n'ait pas de fibre multiple* (ou plus généralement, que les multiplicités de ses fibres soient premières entre elles deux à deux). Soient F une fibre lisse de α et o un point de F. Comme la suite

$$\pi_1(F, o) \longrightarrow \pi_1(X, o) \longrightarrow \pi_1(E, \alpha(o)) \rightarrow 0$$

est exacte, le noyau de $\pi_1(\alpha)$ est de type fini. On déduit alors de la suite exacte

$$1 \rightarrow D(G) \longrightarrow \mathrm{Ker}\, \pi_1(\alpha) \longrightarrow \mathrm{Tors}\, H_1(X, \mathbb{Z}) \rightarrow 0$$

que D(G) est de type fini.

Ce résultat s'applique par exemple aux surfaces avec $q = p_g = 1$ étudiées dans [C–C] : on vérifie en effet en utilisant la prop. 1.9 que la fibration d'Albanese a au plus une fibre multiple, sauf dans un cas avec $K^2 = 8$ où la fibration est isotriviale et où il est facile de décrire explicitement l'ensemble $S^1(X)$.

Remarque 4.7.- La prop. 4.5 admet la généralisation suivante, qui se démontre par la même méthode : si H est un sous-groupe distingué de G tel que le $\mathbb{Z}$-module $D(H)/D^2(H)$ soit de type fini, les caractères de $\Sigma^1(G)$ dont la restriction à H n'est pas triviale sont d'ordre fini. Ce résultat s'applique par exemple si X admet une fibration $p : X \rightarrow B$ sans fibres multiples, dont la fibre générale F est un tore complexe (ou plus généralement, une variété telle que l'image de $\pi_1(F)$ dans $\pi_1(X)$ soit abélienne).

Il me paraît toutefois plus intéressant de considérer d'abord le cas où X n'a pas de telles fibrations, et de comprendre dans quelle mesure l'hypothèse sur D(G) est restrictive. Je ne connais pour l'instant aucun exemple où elle n'est pas vérifiée.

BIBLIOGRAPHIE

[B] A. BEAUVILLE : *Annulation du* H^1 *et systèmes paracanoniques sur les surfaces.* J. reine angew. Math. **388** (1988), 149-157.

[Bo] N. BOURBAKI : *Groupes et algèbres de Lie, ch. 4 à 6* (2ème éd.). Masson, Paris (1981).

[BPV] W. BARTH, C. PETERS, A. VAN DE VEN : *Compact complex surfaces.* Ergebnisse der Math. **4** (3. Folge), Springer-Verlag, Berlin Heidelberg New York Tokyo (1984).

[C-C] F. CATANESE, C. CILIBERTO : *Surfaces with* $p_g = q = 1$. Dans "Problems in the theory of surfaces and their classification", Symposia math. **32**, Academic Press (1991).

[F] T. FUJITA : *The sheaf of relative canonical forms of a Kähler fibre space over a curve.* Proc. Japan Acad. **54** (1978), 183-184.

[G-L 1] M. GREEN, R. LAZARSFELD : *Deformation theory, generic vanishing theorems, and some conjectures of Enriques, Catanese and Beauville.* Invent. math. **90** (1987), 389-407.

[G-L 2] M. GREEN, R. LAZARSFELD : *Higher obstructions to deforming cohomology groups of line bundles.* J. A.M.S. **4** (1991), 87-103.

[S] F. SERRANO : *Multiple fibres of a morphism.* Comment. Math. Helvetici **65** (1990), 287-298.

RESULTS ON VARIETIES WITH MANY LINES AND THEIR APPLICATIONS TO ADJUNCTION THEORY

by

Mauro C. Beltrametti, Andrew J. Sommese and Jarosław A. Wiśniewski

Contents

Introduction. Let L be an ample line bundle on a smooth connected n-dimensional projective manifold X. It follows from a theorem of Kawamata that if K_X is not nef there is a positive rational number τ, the *nef value* of (X, L), such that $K_X+\tau L$ is nef but not ample. Kawamata also showed that there is a morphism $\Phi : X \to Y$ with connected fibers onto a normal projective variety such that $N(K_X+\tau L) \approx \Phi^*H$ where H is ample and N is a positive number such that $N\tau$ is an integer. We refer to Φ as the morphism associated to $K_X+\tau L$, that is Φ is the morphism with connected fibers and normal image defined by $m(K_X+\tau L)$ for $m >> 0$.

In this article we show that if there are sufficiently many "lines" on X relative to L then Φ must contract these lines and further Φ is a Mori contraction. We also show a converse to this. These results have many applications to projective manifolds and adjunction theory which we develop in this paper.

Before we give a detailed description of the paper we would like to tell the history leading to the paper.

The first two authors and M.L.Fania recently completed a paper [BFS1] which among other results gives a complete structure theory for n-folds (X, L) with L very ample, $n \geq 6$, and the Kodaira dimension of $K_X \otimes L^{n-3}$ not equal to n. In applying the results to the study of the dual variety, the first two authors of this paper were led to the following result.

Theorem. *Assume that* L *is very ample on* X *and there is a rational curve* ℓ *on* X *with normal bundle* $\mathcal{N}_\ell^X$ *spanned*, $L \cdot \ell = 1$ *and* $-K_X \cdot \ell := t \geq n/2+1$. *Then the nef value of* (X, L) *is equal*

to t, i.e. K_X+tL *is nef but not ample, and the morphism associated to* K_X+tL *contracts all the lines.*

In particular if L is very ample on X an the defect def(X, L) = k is positive, i.e. the codimension of the dual variety of X in $\mathbf{P}(\Gamma(L))^\vee$ is $k+1 \geq 2$, then $K_X+((n+k)/2+1)L$ is nef and the morphism associated to $K_X+((n+k)/2+1)L$ contracts all the linear $\mathbf{P}^k$'s that arise in the "standard way" from def(X, L) = k > 0.

The proof of the results above was based on two ideas. By Kawamata's rationality theorem [KMM] there is a fibre F of Φ with $\dim F \geq t-1$. If the family of lines through a point $x \in F$ is sufficiently large then by counting dimension in $\mathbf{P}(\mathcal{T}^*_{X|x})$ we conclude that there is a line in the fibre. This argument is based on the fact that two lines through x with the same tangent direction are the same.

At the Bayreuth conference they were discussing this result with the third author and wondering if it could still be true with L ample and spanned. Wiśniewski quickly pointed out that we were using a primitive version of the "non-breaking" Lemma (1.4.3) and by using it directly and Mori theory much more should be true. In particular he had shown one of Mukai's conjectures (i.e. if $K_X^{-1} \approx tL$ for L ample and $t \geq n/2+1$ then either $\mathrm{Pic}(X) \cong \mathbf{Z}$ or $X \cong \mathbf{P}^{n/2} \times \mathbf{P}^{n/2}$) using a related set of ideas.

In the next few days at the conference this paper developed. The results considerably extend the above theorem (and imply Φ is a Mori contraction). The results make clear that varieties with lines are ubiquitous as the adjunction theoretic building blocks of pairs (X, L) in the "stable range".

Let us describe the theorems in detail.

(*) Let τ be the nef value of (X, L) where L is ample on X and let $\Phi : X \to Y$ be the morphism associated to $K_X+\tau L$. Assume that for each point $x \in X$ there exists a curve ℓ on X with $x \in \ell$, $L\cdot\ell = 1$, the normalization of ℓ is $\mathbf{P}^1$ and $\nu := -K_X\cdot\ell - 2$.

Theorem (2.1). *With the notation as above, assume (*). Then either* $\nu = \tau-2$, $\Phi(\ell)$ *is a point and* $\dim Y \leq n-\nu-1$ *or* $\nu < \tau-2$ *and* $\tau+\nu \leq n$ *and in particular* $\nu < (n-2)/2$.

Theorem (2.2). *Assume (*). If* $\nu < \tau-2$ *and* $\tau+\nu = n$ *then X is the product of projective spaces* $\mathbf{P}^{\tau-1} \times \mathbf{P}^{\nu+1}$.

Theorem (2.3). *Assume (*). If* $\nu \geq (n-2)/2$ *then* $\nu = \tau-2$ *and, unless* $(X, L) \cong (\mathbf{P}^{n/2} \times \mathbf{P}^{n/2}, \mathcal{O}(1))$, Φ *is a Mori contraction, in particular* $\mathrm{Pic}(X) \cong \Phi^*\mathrm{Pic}(Y) \oplus \mathbf{Z}[L]$.

Note that in case Y is a point the theorem above follows from [W3].

This has the application to the discriminant locus or dual variety as an immediate corollary (2.4). See [BFS2] for a detailed study of the consequences of this result.

Theorem (2.5). *Assume (*). If* $\nu \geq (n-3)/2$, *then either* $\nu < \tau-2$ *and* $X \cong \mathbf{P}^{\tau-1} \times \mathbf{P}^{\nu+1}$, *or* $\nu = \tau-2$ *and one of the following is true:*

(2.5.1) $\dim Y \leq n-\nu-1$, ℓ *is an extremal rational curve and* $\Phi = \mathrm{cont}_R$, *the fiber type contraction of the extremal ray* $R = \mathbf{R}_+[\ell]$, *or*

(2.5.2) $\Phi : X \to Y$ *factors as* π *composed with* φ, $\Phi = \pi\circ\varphi$, *where* $\varphi : X \to W$ *is a* $\mathbf{P}^{\tau-1}$ *bundle over a smooth variety W of dimension* $\tau = (n+1)/2$. *Furthermore Y is a smooth curve. Denoting* $\mathcal{E} = \varphi_*L$, *there is a rank* τ *vector bundle* $\mathcal{F}$ *over Y such that* $W \cong \mathbf{P}(\mathcal{F})$ *and* $\mathcal{E}_\Delta \cong \oplus^\tau \mathcal{O}_{\mathbf{P}^{\tau-1}}(1)$ *for any fiber* Δ $(\cong \mathbf{P}^{\tau-1})$ *of* $W \to Y$,

(2.5.3) *Y is a point and X is a Fano manifold of index* $\tau = (n+1)/2$, *and Picard number* $\rho(X) \geq 2$. *Such X are classified in* [W4].

The condition (*) is not needed in a stable range of dimensions. Indeed we have a relative form of the Mukai conjecture proved in [W3] when Y is a point.

Theorem (3.1.1). *Let L be an ample line bundle on X, a smooth connected n-dimensional projective manifold. Let* τ *be the nef value of* (X, L) *and* $\Phi : X \to Y$ *the morphism associated to* $K_X+\tau L$. *Assume* Φ *is not birational. Then if* $\tau > (\dim X - \dim Y + 1)/2$ *there exists a family of lines* ℓ *on X that cover X and satisfy* $K_X \cdot \ell = -\tau$. *If* $\tau \geq n/2+1$ *then* Φ *is a Mori contraction, unless* $(X,L) \cong (\mathbb{P}^{n/2} \times \mathbb{P}^{n/2}, \mathcal{O}(1))$.

A second application is to the theory of scrolls. In adjunction theory (X, L) as above is defined to be a *scroll* if $\tau = \dim X - \dim Y + 1$. The general fiber of a scroll is $(\mathbb{P}^{\tau-1}, \mathcal{O}_{\mathbb{P}^{\tau-1}}(1))$.

Theorem (3.2.1). *Let L be an ample line bundle on X and* (X, L) *a scroll* $p : X \to Y$ *over Y with* τ *the nef value of* (X, L). *If* $\tau \geq \dim Y$ *then p is a Mori contraction and in particular there are no divisorial fibers.*

The first two authors conjectured that a scroll is a bundle if $\tau \geq \dim Y$. If $\tau \geq \dim Y+2$ and L is very ample this follows from a theorem of Ein ([E2], (1.7)). If $\dim Y = 1$ or 2 and L is spanned this was shown by Sommese ([S2], (3.3)). See also Fujita ([Fu1], (2.12)).

As a consequence of (3.2.1) we prove the conjecture for L very ample and $\dim Y = 3$.

There is also a result for quadric fibrations in (3.3.1). Combined with the main result of [BS1] we are led to the following general conjecture subsuming the above conjecture and result.

Conjecture. Let L be an ample line bundle on X. Let τ be the nef value of (X, L) and $\Phi : X \to Y$ the morphism associated to $K_X+\tau L$. Assume $\tau > (\dim X - \dim Y + 1)/2$ and $\tau \geq \dim Y$. Then Φ is a flat morphism, a Mori contraction and Y is smooth. ■

Finally an appendix by the first two authors is added with some implications of these ideas worked out for adjunction theory. These results are technical but powerful. Let us give one illustrative statement which follows from an application of these ideas to the main result of [BFS1].

Theorem (A.4.2). *Let L be a very ample line bundle on a connected n-dimensional projective manifold X. Assume* $n \geq 8$. *Assume that the Kodaira dimension of* $K_X+(n-3)L$ *is positive and not equal to n. Then* $K_X+(n-3)L$ *is nef and the morphism* $\Phi : X \to Y$ *associated to* $K_X+(n-3)L$ *is a Mori contraction.*

The analogous results for $\tau \geq n-3$ hold and suggest for reaching "stable" results in adjunction theory that would subsume many known results. We have made a few conjecture that the above results give strong evidence for.

All three authors would like to thank the University of Bayreuth for making our collaboration possible. The first two authors would also like to thank the University of Genova and the University of Notre Dame for their support. The second author would also like to thank the National Science Foundation (DMS 89-21702) for their financial support.

Finally, we thank Miss Cinzia Matrì for her fine typing.

§ 0. Background material.

We work over the complex field C. By *variety* (*n-fold*) we mean an irreducible reduced projective scheme V of dimension n. We denote its structure sheaf by $\mathcal{O}_V$.

If V is normal, the dualizing sheaf, K_V, is defined to be $j_*K_{Reg(V)}$ where $j : Reg(V) \to V$ is the inclusion of the smooth points of V and $K_{Reg(V)}$ is the canonical sheaf of holomorphic n-forms. Note that K_V is a line bundle if V is Gorenstein.

Let $\mathcal{L}$ be a line bundle on a normal variety V. $\mathcal{L}$ is said to be *numerically effective* (*nef*, for short) if $\mathcal{L}\cdot C \geq 0$ for all effective curves C on V and in this case $\mathcal{L}$ is said to be *big* if $c_1(\mathcal{L})^n > 0$ where $c_1(\mathcal{L})$ is the fist Chern class of $\mathcal{L}$.

Let Div(V) be the group of *Cartier divisors* on V and Pic(V) the group of line bundles. We usually don't distinguish between a Cartier divisor D and its associated line bundle $\mathcal{O}_V(D)$. Let $Z_{n-1}(V)$ denote the group of Weil divisors, i.e. the free abelian group generated by prime divisors on V. An element of $Z_{n-1}(V) \otimes Q$ (respectively $Div(V) \otimes Q$) is called a Q-*divisor* (respectively a Q-*Cartier divisor*). We also say that a divisor $D \in Z_{n-1}(V)$ is *e* -Cartier if *e* is the smallest positive integer such that $eD \in Div(V)$. Two elements $D, D' \in Z_{n-1}(V) \otimes Q$ are said to be Q-*linearly equivalent*, denoted by $D \approx D'$, if there exists a positive integer m such that $mD, mD' \in Z_{n-1}(V)$ and that mD and mD' are linearly equivalent in the ordinary sense. Two elements $D, D' \in Z_{n-1}(V) \otimes Q$ are said to be Q-*numerical equivalent*, denoted by $D \sim D'$, if there exists a positive integer m such that $mD, mD' \in Z_{n-1}(V)$ and that mD and mD' are numerical equivalent in the ordinary sense. We say that a divisor $D \in Z_{n-1}(V)$ is *ample* (respectively *nef* or *big*) if mD is an ample (respectively nef or big) Cartier divisor for some positive integer m.

If γ is a subcycle of V and $D \in Z_{n-1}(V) \otimes Q$ with $mD \in Div(V)$ for some integer m, then the intersection symbol $D\cdot\gamma$ stands for $(mD\cdot\gamma)/m$.

For any divisor $D \in Z_{n-1}(V)$ we shall denote by $\mathcal{O}_V(D)$ the associated reflexive sheaf of rank 1. Note that the correspondence

$$Z_{n-1}(V)/\approx \ \to \{\text{reflexive sheaves of rank } 1\}/\cong$$

given by $D \to \mathcal{O}_V(D)$ is a bijection. Recall that, for any $D, D' \in Z_{n-1}(V)$,

$$\mathcal{O}_V(D_1+D_2) \cong (\mathcal{O}_V(D_1) \otimes \mathcal{O}_V(D_2))^{**}, \text{ the double dual.}$$

Abuses and further notation. Linear equivalence classes of Weil divisors on V and isomorphism classes of reflexive sheaves of rank 1 are used with little (or no) distinction. Hence we shall freely switch from the multiplicative to the additive notation and vice versa. E.g. if L is a rank 1 reflexive sheaf on V we use in the Appendix the notation K_V+L with the meaning $K_V+L = (\mathcal{O}_V(K) \otimes \mathcal{O}_V(D))^{**}$, where $L \cong \mathcal{O}_V(D)$ with $D \in Z_{n-1}(V)$ and $K \in Z_{n-1}(V)$ is the *canonical divisor* of V defined by $\mathcal{O}_{Reg(V)}(K) = \Lambda^n\Omega^1_{Reg(V)}$, where Reg(V) is the nonsingular locus of V. Also, for a positive integer m, mK_V stands for $mK_V = (\mathcal{O}_V(K)^{\otimes m})^{**}$.

We fix some more notation (here $\mathcal{L} = \mathcal{O}_V(D)$ denotes a rank 1 reflexive sheaf).

$h^i(\mathcal{L})$, the complex dimension of $H^i(V, \mathcal{L})$,

$\chi(\mathcal{L}) = \sum (-1)^i h^i(\mathcal{L})$, the Euler characteristic of $\mathcal{L}$,

$\Gamma(\mathcal{L})$ = the space of the global sections of $\mathcal{L}$. We say that $\mathcal{L}$ is *spanned* if it is spanned by $\Gamma(\mathcal{L})$,

$|\mathcal{L}|$, the complete linear system associated to $\mathcal{L}$,

$\mathcal{T}_V$, the tangent bundle of V, for V smooth,

(0.1) Assumption. Throughout this paper it will be assumed, unless otherwise stated, that X is a smooth connected variety of dimension $n \geq 2$ and L is an ample line bundle on X.

(0.2) Let X be as in (0.1). Define

$Z_1(X)$ = the free abelian group generated by reduced irreducible curves,

$N_1(X) = \{Z_1(X)/\sim\} \otimes \mathbf{R}$,

$\rho(X) = \dim_{\mathbf{R}} N_1(X)$, the *Picard number* of X,

NE(X) = the convex cone in $N_1(X)$ generated by the effective 1-cycles; $\overline{NE}(X)$ = the closure of NE(X) in $\mathbf{R}^{\rho(X)}$ in the usual Euclidean topology.

A part of Mori's theory of extremal rays is to be used throughout the paper. We will use freely the notion of extremal rays, extremal rational curves as well as the basic theorems such as Cone Theorem and Contraction Theorem. We refer the reader to [M2], [KMM] and [W1].

In particular we will denote by $\phi = \mathrm{cont}_R : X \to Y$, the morphism given by the contraction of an extremal ray R. We also simply refer to ϕ as *Mori contraction.* We say that ϕ is of fiber type, or R is numerically effective, if $n > \dim Y$. If γ is a 1-dimensional cycle in X we will denote by $\mathbf{R}_+[\gamma]$, where $\mathbf{R}_+ = \{x \in \mathbf{R}, x \geq 0\}$, or $[\gamma]$ its class in $\overline{NE}(X)$. Let us recall a few facts we use.

(0.2.1) Let ℓ be an extremal rational curve on X. Then the normalization is $\mathbf{P}^1$, $\mathbf{R}_+[\ell]$ is an extremal ray and $1 \leq -K_X\cdot\ell \leq n+1$.

(0.2.2) The *length* $\ell(R)$ of an extremal ray R is defined as

$$\ell(R) = \min\{-K_X\cdot C,\ C \text{ rational curve and } [C] \in R\}.$$

Let E be the *locus* of R, that is the locus of curves whose numerical classes are in R. If the contraction $\phi = \mathrm{cont}_R$ of R has a nontrivial fiber of positive dimension d we have the following result of Wiśniewski [W2], (1.1).

(0.2.2.1) $$\dim E \geq n + \ell(R) - d - 1.$$

Finally if $\phi = \mathrm{cont}_R$, $R = \mathbf{R}_+[\ell]$, one has an exact sequence

(0.2.3) $$0 \to \mathrm{Pic}(Y) \xrightarrow{\Phi^*} \mathrm{Pic}(X) \xrightarrow{\cdot\ell} \mathbf{Z}.$$

(0.3) Some special varieties. Let (X, L) be as in (0.1). We say that (X, L) is a *scroll* (respectively a *quadric fibration*) over a normal variety Y of dimension m if there exists a surjective morphism with connected fibers $p : X \to Y$ and an *ample* line bundle $\mathcal{L}$ on Y such that $K_X + (n-m+1)L \approx p^*\mathcal{L}$ (respectively $K_X + (n-m)L \approx p^*\mathcal{L}$). We say that X is a *Fano variety* of *index* i = index(X) if $-K_X$ is ample and i is the largest integer such that $K_X \approx -iH$ for some ample line bundle H on X.

We say that (X, L) is a $\mathbf{P}^d$ *bundle* over a smooth variety Y if there exists a surjective morphism $p : X \to Y$ such that all fibers F of p are $\mathbf{P}^d$ and $L_F \cong \mathcal{O}_{\mathbf{P}^d}(1)$. We refer to [BS2] for comparing this classical definition and the previous adjunction theoretic definition of scroll.

(0.4) The nef value. Let (X, L) be as in (0.1). Assume that K_X is not nef. We say that

$$\tau = \min\{t \in \mathbf{R},\ K_X + tL \text{ is nef}\}$$

is the *nef value* of (X, L) or simply of L. Note that $0 < \tau < +\infty$ since K_X is not nef and L is ample. From the Kawamata rationality theorem [KMM], (4.1) we know that τ is a rational number and the Kawamata-Shokurov basepoint free theorem applies to say that $m(K_X + \tau L)$ defines a morphism for an integer $m >> 0$, say Φ. If m is large enough we can assume that Φ has connected fibers and normal image. ■

Let us recall the following general fact we need.

(0.5) Lemma. *Let* X *be a projective variety and let* $\psi : X \to Y$ *be a morphism with connected fibers onto a normal variety* Y. *Then* ψ *is an isomorphism if* $\rho(X) = \rho(Y)$.

Proof. Assume otherwise and let F be a positive dimensional fiber of ψ. The pullback ψ^* gives an injection of Pic(Y) into Pic(X). Note that all pullbacks are trivial on F and therefore not ample. Thus there is an ample line bundle, $\mathcal{L}$, on X which is non trivial on the fiber F though all line bundles pulled back are trivial on F. Now note that if $\rho(X) = \rho(Y)$, then, up to torsion, there is a line bundle on Y that pulls back to $\mathcal{L}'$ which is numerically equivalent to some rational multiple of the ample line bundle $\mathcal{L}$. Thus the restriction of $\mathcal{L}'$ to F is a torsion line bundle. This contradicts the Nakai ampleness criterion. Q.E.D.

We also need the following result of Fujita.

(0.6) Theorem (Fujita, [Fu2]). *Let* $\mathcal{E}$ *be an ample vector bundle on a smooth projective variety* Y *of dimension* m *such that rank* $(\mathcal{E}) \geq m$. *Then* $K_Y \otimes \det\mathcal{E}$ *is ample except in the following cases*:

(1) $(Y,\mathcal{E}) \cong (\mathbf{P}^m, \oplus^{m+1}\mathcal{O}_{\mathbf{P}^m}(1))$;

(2) $(Y,\mathcal{E}) \cong (\mathbf{P}^m, \oplus^{m}\mathcal{O}_{\mathbf{P}^m}(1))$;

(3) $(Y,\mathcal{E}) \cong (\mathbf{P}^m, \mathcal{O}_{\mathbf{P}^m}(2) \oplus (\oplus^{m-1}\mathcal{O}_{\mathbf{P}^m}(1)))$;

(4) $(Y,\mathcal{E}) \cong (\mathbf{P}^m, \mathcal{T}_{\mathbf{P}^m})$;

(5) $(Y,\mathcal{E}) \cong (Q, \oplus^{m}\mathcal{O}_Q(1))$, Q *smooth hyperquadric in* $\mathbf{P}^{m+1}$;

(6) *there is a rank* m *vector bundle* $\mathcal{F}$ *over a smooth curve* C *such that* $Y \cong \mathbf{P}(\mathcal{F})$ *and* $\mathcal{E}_\Delta \cong \oplus^m\mathcal{O}_{\mathbf{P}^{m-1}}(1)$ *for any fiber* $\Delta (\cong \mathbf{P}^{m-1})$ *of* $Y \to C$. ■

For any further background material we refer to [M2], [W1] and [BFS1].

§ 1. Some results from Mori theory.

In this section we recall some results from Mori's theory we need in the sequel.

(1.1) Let X be a smooth connected projective variety over **C** of dimension n and let L be an *ample* line bundle on X.

(1.2) Lemma. *Let* K_X *be not nef and let* C *be an extremal rational curve on* X, $R = \mathbf{R}_+[C]$. Let $\Phi : X \to Y$ *be a morphism with connected fibers onto a normal variety* Y. *If* $\dim\Phi(C) = 0$, *then* Φ *factors through* $\phi = \mathrm{cont}_R : X \to Z$. *In particular if* $\rho(X) = \rho(Y)+1$, *then* $\Phi = \mathrm{cont}_R$, *that is* $Y \cong Z$.

Proof. Let $W = \mathrm{Im}(\Phi, \phi) \subset Y \times Z$. Then we have a commutative diagram

$$\begin{array}{ccc} X & \xrightarrow{(\Phi,\phi)} & W \\ \phi \downarrow & \swarrow \alpha & \\ Z & & \end{array}$$

where α is the restriction to W of the projection on the 2nd factor. Note that α has connected fibers since ϕ does. We claim that α has no positive dimensional fibers. Otherwise $\rho(W) \geq \rho(Z) + 1$ by Lemma (0.5) and therefore, since $\rho(X) = \rho(Z) + 1$, we would have $\rho(W) \geq \rho(X)$. Thus $\rho(X) = \rho(W)$ and so $(\Phi, \phi) : X \to W$ has no positive dimensional fibers. This contradicts the fact that C is contracted by (Φ, ϕ). Then we conclude that α is an isomorphism. Let π be the composition of α^{-1} with the restriction to W of the projection of $Y \times Z$ onto Y. Therefore $\Phi = \pi \circ \phi$, so we are done. Q.E.D.

Let t be a positive integer such that K_X+tL is nef. Then $m(K_X+tL)$ is spanned for $m >> 0$, by the Kawamata-Shokurov basepoint free theorem.

(1.3) Lemma (Key-Lemma). *Let* K_X *be not nef and let* C *be an extremal rational curve on* X, $R = \mathbf{R}_+[C]$. *Let* t *be a positive rational number such that* K_X+tL *is nef and let* Φ *be the morphism with connected fibers and normal image associated to* $|m(K_X+tL)|$ *for* $m >> 0$. *If* $\dim\Phi(C) = 0$ *then either*

(1.3.1) $\Phi = \mathrm{cont}_R$, *or*

(1.3.2) *there exist two distinct extremal rays* R_1, R_2 *such that* Φ *contracts all the curves* $C \in R_1 \cup R_2$.

Proof. Let $\phi = \mathrm{cont}_R$. By Lemma (1.2) either we are done or Φ factors as $\Phi = \pi\circ\phi$ with π not isomorphism. Then there exists a curve, B, contracted by Φ, which is not contracted by ϕ. Clearly $[B] \notin R$, otherwise $\dim\phi(B) = 0$. By using the Mori cone theorem, for an arbitrary positive ε we can write $B = \sum \lambda_i \ell_i + \beta$ in $\overline{NE}(X)$, where $\lambda_i \in \mathbf{R}_+$, ℓ_i are extremal rational curves, $i = 1,\ldots, s$, and β is a 1- cycle in $\overline{NE}(X)$, which satisfies the condition $K_X\cdot\beta \geq -\varepsilon L\cdot\beta$. Since K_X+tL is nef with $t > 0$ and $(K_X+tL)\cdot B = 0$ we find with $\varepsilon < t$ that $(K_X+tL)\cdot\ell_i = 0$ for all i and $\beta = 0$ in $\overline{NE}(X)$. If there are (at least) two distinct not numerically equivalent ℓ_i's, say ℓ_1, ℓ_2, we are done by taking $R_1 = \mathbf{R}_+[\ell_1]$, $R_2 = \mathbf{R}_+[\ell_2]$. Otherwise we can assume $B \sim \lambda\ell$, for $\lambda \in R$, and ℓ extremal rational curve. So we are done by taking $R_1 = R$, $R_2 = \mathbf{R}_+[\ell]$. Q.E.D.

(1.3.3) Remark. With the assumptions as in (1.3), let $\tau > 0$ be a rational number such that $K_X + \tau L$ is nef but not ample. Then the morphism associated to $|m(K_X + \tau L)|$, m>>0, is the contraction of the extremal face (in the sense of [KMM])

$$(K_X + \tau L)^\perp \cap \overline{NE}(X) - \{0\}$$

where "$\perp$" means the orthogonal complement. This is a consequence of [KMM].

(1.4) Families of rational curves. By a *non-breaking family* T *of rational curves* on X we understand the following. T is an irreducible compact variety (parameter space) and there exists a variety $V \subset T \times X$ with projections $p : V \to X$, $q : V \to T$ which are proper. Moreover the map q is assumed to be equidimensional with all fibers reduced and irreducible of dimension 1 and for any (closed) point $t \in T$, the curve $C_t = p(q^{-1}(t))$ is a rational curve on X, i.e. with normalization isomorphic to $\mathbf{P}^1$. We assume moreover that for $t_1 \neq t_2$ the curves C_{t_1} and C_{t_2} are distinct.

We will say that X is *dominated* by a non-breaking family T of rational curves if the map p is onto. Note that in this case, for any $x \in X$ there exists a point $t \in T$ such that $C_t \ni x$. For any point $x \in X$ we denote

$$T_x = q(p^{-1}(x)).$$

Then by the dimension of T at a point $x \in X$ we understand the dimension (of an irreducible component of maximal dimension) of T_x.

The following results follow from Mori's "breaking up" technique (see [M1]).

(1.4.1) Lemma (Mori [M1]). *Let* T *be a non-breaking family of rational curves on* X. *Then, under the above notation, the restriction map* $p : q^{-1}(T_x) \to X$ *is finite to one away from* $p^{-1}(x)$ *for any* $x \in X$.

Proof. See [M1], p. 599.

(1.4.2) Example-Remark. Let C_0 be a rational curve and L an ample line bundle on X such that $L\cdot C_0 = 1$. Assume also that $K_X\cdot C_0 < 0$. Then as in [M1] or [I], or as in the Appendix of [W1] we construct a non-breaking family T of rational curves which are deformations of C_0 (our present notation is consistent with this of [W1] and [W3]). Note that the non-breakingness of such a family is provided by the minimality of the intersection $L\cdot C_0$ (see Appendix of [W1]). If the family T dominates X, i.e. if the map $p : V \to X$ is onto, then the dimension of the family at any point of X is at least $-K_X\cdot C_0 - 2$ (see [M1], Prop. 3). Then by (1.4.1) the dimension of the locus of curves from the family T which pass through a given point is $\geq -K_X\cdot C - 1$.

(1.4.3) Lemma ("Non-breaking Lemma") ([W3]). *Let* T *be a non-breaking family of rational curves on* X *and let* $\varphi : X \to Y$ *be a morphism with connected fibers onto a projective normal variety* Y. *Let* F *be an irreducible component of a positive dimensional fiber. Under the notation above, assume that there exists a curve* C_t *from the family* T *such that* $C_t \cap F \neq \emptyset$ *and* C_t *not contained in* F. *Then for any smooth point* $x \in C_t \setminus C_t \cap F$ *we have*

$$\dim(p(q^{-1}(T_x)) \cap F) = 0.$$

In particular if T *dominates* X, *then the above is true for any irreducible component* F *of any fiber of* φ.

Proof. We can choose a smooth point x on C_t which is not contained in F. If the dimension of the intersection above is positive, then we take a compact curve B' contained in T_x such that $\dim(p(q^{-1}(B')) \cap F) \geq 1$. We normalize B' to get a smooth curve B and by base change we obtain a family $S' = V\times_T B$ of rational curves parameterized by B whose normalization is a ruled surface $\pi : S \to B$ (see 1.14 in [W1]). The ruled surface S admits a map $\psi : S \to X$ (the composition of p with the base change and the normalization) which contracts a section, say B_0, to x. By the assumptions on C_t every image under ψ of fibers of π meets F, but it is not contained in it. Therefore we can choose an irreducible curve F_0 on S which is mapped by ψ into F and then contracted to a point $y = \varphi(F)$, $y \neq \varphi(x)$, by the map φ. On the ruled surface S we have the following intersections (see [M1], p. 599 or [I], p. 460).

$$F_0^2 \leq 0 \ , \ B_0^2 < 0 \ , \ F_0\cdot B_0 = 0 \quad \text{and} \quad (F_0 - aB_0)^2 = 0$$

for some $a > 0$, which lead to a contradiction.

If T dominates X, then for any irreducible component of any fibre of φ we can choose a curve C_t from the family T satisfying the condition $C_t \cap F \neq \emptyset$. So we are done. Q.E.D.

(1.4.4) Lemma. *Let* X *be dominated by a non-breaking family* T *of rational curves whose dimension at every point is at least* d. *Assume that there exists a morphism* $\varphi : X \to Y$ *with connected fibers onto a projective normal variety* Y. *If* φ *contracts any curve* C_t *from the family then* $\dim Y \leq n-d-1$.

Proof. If there is a curve C_t from the family T which is contracted by φ, then since the family is non-breaking it is easy to see that all curves in the family are contracted. In this case since X is dominated by T we have that for any fiber F of φ, $\dim F \geq \dim D_x \geq d+1$, where D_x is the locus of the deformations of C_t containing a smooth point $x \in C_t$. Thus $\dim Y \leq n-d-1$. Q.E.D.

(1.4.5) Lemma. *Let* X, T, d, $\varphi: X \to Y$ *be as in* (1.4.4). *Assume that* φ *does not contract curves from the family* T. *Then any fiber of* φ *is of dimension* $\leq n - d - 1$ *and in particular* $\dim Y \geq d + 1$. *Moreover, if there exists a fiber* F *of* φ *such that* $\dim F = n - d - 1$ *then*

$$NE(X) = NE(F) + \mathbf{R}_+[C_t]$$

where NE(F) *denotes a subcone of* NE(X) *spanned by curves contained in* F, *and* C_t *is any curve from the family* T.

Proof. Let F be any fibre of φ and let T(F) denote the subvariety of T parameterizing curves meeting F. Note that $T(F) = q(p^{-1}(F))$. Since no curve C_t with $t \in T$ belongs to F we see that the restriction map, $q_{p^{-1}(F)}$ is finite to one. Thus since q has one dimensional fibers

$$\dim(q^{-1}(T(F))) = \dim(q^{-1}(T(F)) \setminus p^{-1}(F))$$

and

$$\dim T(F) + 1 = \dim(q^{-1}(T(F))) = \dim(p^{-1}(F)) + 1.$$

Since all fibers of p are of dimension $\geq d$ we see that $\dim(p^{-1}(F)) \geq d + \dim F$. By the above $\dim(q^{-1}(T(F)) \setminus p^{-1}(F)) \geq \dim F + d + 1$. From (1.4.3) it follows that the restriction of p to $q^{-1}(T(F)) \setminus p^{-1}(F)$ is finite to one. From this we conclude that $\dim X \geq \dim F + d + 1$, with equality implying that T(F) dominates X, i.e. $p(q^{-1}(T(F)) = X$.

To prove the second part of the lemma let us take an irreducible curve C in X. We may assume that C is not from the family T. We are to show that C is numerically equivalent to $aC_t + bf$ for some effective 1-cycle f contained in F and non-negative real numbers a, b. Since the family T(F) dominates X we can find an irreducible curve inside T(F) which parameterizes curves meeting both F and C. As in the proof of (1.4.3) we produce a ruled surface $S \to B$ with a map $\psi: S \to X$ which maps any line ℓ of the ruling birationally onto a curve C_t which meets both F and C. Therefore S contains two irreducible curves: F_0, such that $\psi(F_0)$ is contained in F, and C_0, such that $\psi(C_0) = C$, and moreover none of these curves is equivalent to ℓ. Clearly, on the surface S the curve C_0 is equivalent to $a\ell + bF_0$ and we will be done if we show that a and b are non-negative. This is apparent if we intersect C_0 with the line ℓ and with a line bundle H which is a pull-back of an ample line bundle on Y. Indeed, in the former case $\ell \cdot \ell = 0$, $\ell \cdot F_0 > 0$, $\ell \cdot C_0 > 0$ so that $b > 0$ whilst in the latter case $H \cdot F_0 = 0$, $H \cdot C_0 \geq 0$ and $H \cdot \ell > 0$ so that $a \geq 0$. Q.E.D.

§ 2. Results on the relation between nef values and dimensions of families of lines.

Let X be a smooth connected projective variety of dimension n. Let L be an *ample* line bundle on X. The notation are as in (1.4).

(2.0) Assumptions. We shall assume that the pair (X, L) satisfies the following conditions.

(2.0.1) There exists a line ℓ relative to L, i.e. ℓ is a rational curve with normalization $\mathbf{P}^1$ such that $L \cdot \ell = 1$ and $K_X \cdot \ell < 0$.

(2.0.2) Let T be a non-breaking family of rational curves which are deformations of ℓ (compare with (1.4.1)). We shall assume that the family T dominates X, i.e. there exists a line from the family through each point of X.

(2.0.3) For any point $x \in X$, let $T_x := q(p^{-1}(x))$. Then

$$\dim T_x \geq \nu := -K_X \cdot \ell - 2.$$

(2.0.4) Remark. In the special case when ℓ is smooth, we can further show the following facts, but we don't need these results in this paper.

(2.0.4.1) Let t be a general point of T and ℓ_t the curve from the family corresponding to t. Then the normal bundle $\mathcal{N}_{\ell_t}$ of ℓ_t in X is spanned.

(2.0.4.2) By the adjunction formula

$$\nu = -K_X\cdot\ell - 2 = \deg\mathcal{N}_\ell \geq 0$$

and $\deg\mathcal{N}_\ell = \Sigma a_i$ where $\mathcal{N}_\ell \cong \oplus^i\mathcal{O}_{P^1}(a_i)$, $a_i \geq 0$. Note that $\nu = h^0(\mathcal{N}_\ell(-1))$, the dimension of the space of deformations of ℓ that contain a point $x \in \ell$. ■

Let τ be the nef value of L. From now on, we will denote by $\Phi : X \to Y$ the morphism with connected fibers and normal image associated to $|m(K_X+\tau L)|$ for some m >> 0. Then $K_X+\tau L \approx \Phi^*\mathcal{L}$ for some ample line bundle $\mathcal{L}$ on Y. We will simply refer to Φ as the morphism associated to $K_X+\tau L$.

The following is the main result of this section.

(2.1) Theorem. *Let* X *be a smooth connected projective variety of dimension* n, L *an ample line bundle on* X *and* ℓ *a line on* X *relative to* L. *Let* T *be a non-breaking family of rational curves which are deformations of* ℓ *and suppose that* (X, L), ℓ, T *satisfy the assumptions* (2.0). *Let* τ *be the nef value of* L *and* $\Phi : X \to Y$ *the morphism associated to* $K_X+\tau L$. *Let* $\nu = -K_X\cdot\ell-2$. *Then either*

(2.1.1) $\nu = \tau - 2$, $\dim\Phi(\ell) = 0$ *and* $\dim Y \leq n-\nu-1$, *or*

(2.1.2) $\nu < \tau - 2$, $\tau + \nu \leq n$ *and in particular* $\nu < (n-2)/2$.

Proof. Since $K_X+\tau L$ is nef, we have $(K_X+\tau L)\cdot\ell = -\nu-2+\tau \geq 0$, i.e. $\nu \leq \tau-2$. If $\nu = \tau-2$, then, $(K_X+\tau L)\cdot\ell = -\nu-2+\tau = 0$, so $\dim\Phi(\ell) = 0$. Note that $\dim Y \leq n-\nu-1$ by Lemma (1.4.4). So we are in (2.1.1).

Let $\nu < \tau-2$ and assume $\tau + \nu > n$. Then $\tau - 2 > (n-2)/2$ and hence

(2.1.2.1) $$\tau > n/2 + 1$$

Let μ be any extremal rational curve contracted by Φ (see (1.3) and (1.3.3)). We claim that

$$L\cdot\mu = 1.$$

Indeed, if not, we would have $2\tau \leq \tau L\cdot\mu = -K_X\cdot\mu \leq n+1$ which contradicts (2.1.2.1). From $(K_X+\tau L)\cdot\mu = 0$ and $L\cdot\mu = 1$ we see that

$$-K_X\cdot\mu = \tau = \text{length}(R).$$

where $R = R_+[\mu]$. Let F be a positive dimensional fibre of the contraction of R. Let E be the locus of R, i.e. the locus of curves whose numerical classes in $\overline{NE}(X)$ belong to R. From (0.2.2.1) we know that

$$n \geq \dim E \geq n + \tau - \dim F - 1$$

and hence

(2.1.2.2) $$\dim F \geq \tau - 1.$$

Note that ℓ is not contained in F. Otherwise $(K_X+\tau L)\cdot\ell = 0$ which contradicts the assumption $\nu < \tau-2$. It thus follows that there are no curves from the family T which are contracted by Φ. Furthermore by the assumption (2.0.2) we can assume that $\ell \cap F \neq \emptyset$ and we can choose a smooth point x of ℓ, $x \notin F$. Moreover, by the non-breaking Lemma (1.4.3) we get

$$\dim(D_x \cap F) = 0$$

where $D_x = p(q^{-1}(T_x))$ is the locus of the deformations of ℓ containing x and hence $\dim D_x \geq -K_X\cdot\ell - 1$ by (1.4.2). Then (2.1.2.2) and the present assumption $\tau+\nu > n$ lead to the contradiction

$$\dim(D_x \cap F) \geq \dim D_x + \dim F - \dim X \geq -K_X\cdot\ell - 2 + \tau - n = \nu + \tau - n > 0.$$

Thus we conclude that $\tau + \nu \leq n$ and hence in particular $\nu < (n-2)/2$. Q.E.D.

In the situation of (2.1.2), if $\tau + \nu = n$, we can prove more.

(2.2) Theorem. *Let* X *be a smooth connected projective variety of dimension* n, L *an ample line bundle on* X *and* ℓ *a line on* X *relative to* L. *Let* T *be a non-breaking family of rational curves which are deformations of* ℓ *and suppose that* (X, L), ℓ, T *satisfy the assumptions* (2.0). *Let* τ *be the nef value of* L *and* $\Phi : X \to Y$ *the morphism associated to* $K_X+\tau L$. *Let* $\nu = -K_X \cdot \ell - 2$. *If* $\nu < \tau - 2$ *and* $\tau + \nu = n$ *then* X *is the product of projective spaces* $\mathbf{P}^{\tau-1} \times \mathbf{P}^{\nu+1}$.
Proof. The proof is similar to this of (2.1). Let R and F be as in the proof of (2.1), i.e. $R = \mathbf{R}_+[\mu]$ is an extremal ray with μ an extremal rational curve satisfying $(K_X + \tau L)\cdot\mu = 0$, and F is a positive dimensional fiber of the contraction, φ, of R. Since $\tau = n - \nu > n/2 + 1$ we conclude that $L\cdot\mu = 1$ and $\text{length}(R) = -K_X \cdot \mu = \tau$. By (1.4.5), $\dim F \le n - d - 1$ where $d = \dim T_x = \dim q(p^{-1}(x))$, and where T_x is the subvariety of T corresponding to lines passing through $x \in X$. By (1.4.2), $d \ge -K_X \cdot \ell - 2 = \nu$. Thus $\dim F \le n - \nu - 1 = \tau - 1$.

Let E be the locus of R. By (0.2.2.1), $\dim E \ge n - \text{length}(R) - \dim F - 1 \ge n$. Thus φ is of fiber type. From (0.2.2.1) we conclude also that $\dim F \ge \tau - 1$,and hence that $\dim F = \tau - 1$ for every fibre of φ. Since $K_X + \tau L$ is trivial on F, it is clear from a well known result of Kobayashi-Ochiai that $(F, L_F) \cong (\mathbf{P}^{\tau-1}, \mathcal{O}_{\mathbf{P}^{\tau-1}}(1))$ for a general fibre, F, of φ. Thus by a result of Fujita ([Fu1], (2.12)), $\varphi : X \to W$ is a $\mathbf{P}^{\tau-1}$ bundle. Note that $\dim W = n - \tau + 1 = \nu + 1$. Recall also

(2.2.1) $$\tau - 1 > n/2 > \nu + 1.$$

By the canonical bundle formula for a projective bundle $K_X \otimes L^\tau \cong \varphi^*(K_W \otimes \det(\mathcal{E}))$ where $\mathcal{E} = \varphi_* L$. Note by (2.2.1) that $\text{rank}\,\mathcal{E} = \tau > \nu + 2 = \dim W + 1$. Thus $K_W \otimes \det(\mathcal{E})$ is ample by [Fu2]. In particular since Φ factors as $\pi \circ \varphi$ and $K_X \otimes L^\tau$ is the pullback of an ample line bundle under Φ we conclude that π is an isomorphism, i.e. Φ is a $\mathbf{P}^{\tau-1}$ bundle and a scroll in the sense of (0.3). Let ℓ be a line in the family T. From (1.4.5), $NE(X) = R + \mathbf{R}_+[\ell]$. Hence we see that ℓ is an extremal rational curve and that the contraction $\alpha : X \to V$ associated to $R' = \mathbf{R}_+[\ell]$ is of fiber type since T dominates X. Moreover V is not a point since $\rho(X) = 2$ by (1.4.5) and $\rho(X) - \rho(V) = 1$ by the sequence (0.2.3) associated to α. Let T' be the family of deformations of μ. Note that $d' = \dim T'_x \ge -K_X \cdot \mu - 2 = \tau - 2$. Then by (1.4.5), $\dim F' \le n - \tau + 1 = \nu + 1$ for any positive dimensional fiber F' of α. On the other hand it follows from $L\cdot\ell = 1$ that $\text{length}(R') = -K_X \cdot \ell = \nu + 2$. From this and (0.2.2.1) we conclude that

$$n + \dim F' \ge n + \nu + 1.$$

So $\dim F' = \nu + 1$. We know that $K_X + (\nu + 2)L$ is trivial on F' and we conclude that a general fiber of α is $(\mathbf{P}^{\nu+1}, \mathcal{O}_{\mathbf{P}^{\nu+1}}(1))$. From the Fujita result quoted above we conclude that α is a $\mathbf{P}^{\nu+1}$ bundle.

We claim that X is biholomorphic to $\mathbf{P}^{\tau-1} \times \mathbf{P}^{\nu+1}$. To see this consider the product map $g = (\alpha, \varphi) : X \to V \times W$. Note that the map is finite to one. Indeed if it isn't then there would be an irreducible curve, C, that went to a point under g. Such a curve C would be contained in a fibre $\varphi^{-1}(w)$ for some $w \in W$. Since $\text{Pic}(\mathbf{P}^{\tau-1}) \cong \mathbf{Z}$, it would follow that $\alpha(\varphi^{-1}(w)) = \alpha(C) = v$ which would imply that $\varphi^{-1}(w)$ is contained in $\alpha^{-1}(v)$ for some $v \in V$. This is absurd for dimension reasons.

We have just shown that $\alpha : \varphi^{-1}(w) \to V$ for $w \in W$ and $\varphi : \alpha^{-1}(v) \to W$ for v in V are finite to one surjections. Since W, V are smooth, and since $\varphi^{-1}(w) \cong \mathbf{P}^{\tau-1}$ and $\alpha^{-1}(v) \cong \mathbf{P}^{\nu+1}$ we conclude from a theorem of Lazarsfeld ([L], Theorem (4.1)) that $W \cong \mathbf{P}^{\nu+1}$ and $V \cong \mathbf{P}^{\tau-1}$. Using this we see that it suffices by Zariski's "Main Theorem" to show that the product map $(\alpha, \varphi) : X \to \mathbf{P}^{\tau-1} \times \mathbf{P}^{\nu+1}$ is one to one. To see this assume otherwise. Then we would have a fiber F ($\cong \mathbf{P}^{\tau-1}$) of φ which is mapped k to one to V ($\cong \mathbf{P}^{\tau-1}$) under α where $k > 1$. If we can show that $\alpha^*\mathcal{O}_V(1) \approx L_F \approx \mathcal{O}_{P^{\tau-1}}(1)$ we will have a contradiction to $k > 1$. To show this note that by the canonical bundle formula for the bundle α we have

$$K_X \otimes L^{\nu+2} \approx \alpha^*(K_V \otimes \det(\mathcal{F}))$$

where $\mathcal{F} = \alpha_*L$. Since L is ample, $\mathcal{F}$ is an ample vector bundle of rank $\nu + 2$. By restriction to a line in V we see that $\det(\mathcal{F}) \approx \mathcal{O}_V(b)$ where $b \geq \nu + 2$. Thus

$$\mathcal{O}_F(\nu + 2 - \tau) \approx K_F \otimes L_F^{\nu+2} \approx \alpha_F^*\mathcal{O}_V(b - \tau).$$

So $\nu + 2 - \tau = \lambda(b - \tau)$ where $\alpha_F^*\mathcal{O}_V(1) \approx \mathcal{O}_F(\lambda)$. Since $b - \tau \geq \nu + 2 - \tau$ we conclude that $\lambda = 1$ unless $\nu + 2 - \tau = 0$. This contradicts the assumption $\nu < \tau - 2$. Q.E.D.

(2.2.2) Remark. Notation and assumptions as in (2.1). If $\nu + \tau < n$ then $\nu + \tau \leq n - 1$ and hence in particular $\nu < (n - 3)/2$. Indeed if $n - 1 < \nu + \tau < n$ then τ is not integral. Thus (2.1) gives $\nu < \tau - 2$, but then $n + 1 < 2\tau$ so as in the proof of (2.1) we have for an extremal rational curve μ, that $L\cdot\mu = 1$. Hence we have that $\tau = \text{length}(R_+[\mu])$, which is a contradiction. ■

Theorem (2.1) above has two main consequences.

(2.3) Theorem. *Let X be a smooth connected projective n-fold, L an ample line bundle on X and ℓ a line relative to L. Let T be a non-breaking family of rational curves which are deformations of ℓ and suppose that (X, L), ℓ, T satisfy the assumptions (2.0). Let τ be the nef value of L and $\Phi : X \to Y$ be the morphism associated to $K_X + \tau L$. Let $\nu = -K_X\cdot\ell - 2$. If $\nu \geq (n-2)/2$ then $\nu = \tau - 2$, ℓ is an extremal rational curve, $\Phi = \text{cont}_R$ the fiber type contraction of the extremal ray $R = R_+[\ell]$ and $\text{Pic}(X) \cong \Phi^*\text{Pic}(Y) \oplus \mathbb{Z}[L]$ unless (X, L) $\cong (\mathbb{P}^{n/2} \times \mathbb{P}^{n/2}, \mathcal{O}(1))$.*

Proof. Using Theorem (2.1) we see that we are in case (2.1.1). Indeed in case (2.1.2) we have $\nu < \tau - 2$ and $\tau + \nu \leq n$, which imply $\nu < (n-2)/2$. Thus we have $\nu = \tau - 2$. In this case we have

$$\dim\Phi(\ell) = 0, \quad \tau + \nu \geq n, \quad \text{and } \tau \geq n/2 + 1.$$

If $\Phi = \text{cont}_R$ then since $\dim\Phi(\ell) = 0$ it follows that $\ell \in R$. Since $L\cdot\ell = 1$ we see that ℓ satisfies the condition $-(n + 1) \leq K_X\cdot\ell \leq -1$ and hence ℓ is an extremal rational curve. The fact that $\text{Pic}(X) \cong \Phi^*\text{Pic}(Y) \oplus \mathbb{Z}[L]$ follows from the exact sequence (0.2.3), where the morphism $\text{Pic}(X) \xrightarrow{\cdot\ell} \mathbb{Z}$ is surjective since $L\cdot\ell = 1$.

Thus we can assume without loss of generality that Φ is not a Mori contraction and therefore that there is a contraction $\varphi = \text{cont}_R$ where R is an extremal ray contracted by Φ and $\ell \notin R$. Here Φ factors as π composed with φ. Since Φ is not a contraction it follows that $\rho(X) \geq 2$.

Let $\mu \in R$ be an extremal rational curve. Since $(K_X + \tau L)\cdot\mu = 0$ we conclude that $\text{length}(R) \geq \tau$. Let F be a fiber of φ. By (0.2.2.1), $\dim F \geq \text{length}(R) - 1 \geq \tau - 1$. As in the proof of (2.2) by combining (1.4.2) and (1.4.5) we get $\dim F \leq n - \nu - 1$. Thus from $\dim F \geq \tau - 1$ and $\tau + \nu \geq n$, we see that $\dim F = n - \nu - 1$. From (1.4.5) we conclude that $NE(X) = NE(F) + R_+[\ell]$. Since F is a positive dimensional fiber of a Mori contraction cont_R we conclude that $\overline{NE}(F) = R$ and thus that $\rho(X) = 2$. Thus $\rho(\varphi(X)) = 1$ and hence $\rho(Y) = 0$, i.e. Y is a point (see also Lemma (0.5)). From this we conclude that $K_X + \tau L$ is trivial. Using the Wiśniewski theorem [W3] we see that $X \cong \mathbb{P}^{n/2} \times \mathbb{P}^{n/2}$, $L \cong \mathcal{O}_X(1)$. Q.E.D.

As noted in the introduction the following application of the theorem above was at the start of the paper. Recall that if L is *very ample* on X, the discriminant locus, $\mathcal{D}$, of (X, L) is the set

$$\mathcal{D} := \{H \in |L|,\ H \text{ is singular}\}.$$

Note that $\mathcal{D}$ is irreducible since L is very ample. The defect def(X,L) = k is defined by

$$\mathrm{cod}_{\mathbf{P}(\Gamma(L))}\ \mathcal{D} = k+1.$$

See [BFS2] for applications of the following result to the dual variety. Recall also that n and k have the same parity [E1].

(2.4) Corollary. *Let* X *be a smooth connected* n*-fold,* L *a very ample line bundle on* X*. Let* def(X, L) = k *be positive. Then the morphism defined by the linear system* $|N(K_X+((n+k)/2+1)L|$, *for* $N >> 0$, *is the fiber type contraction of an extremal ray* R *and* $(n+k)/2+1$ *is the nef value of* L*. Furthermore the image of this contraction has dimension less than or equal to* $(n-k)/2$.

Proof. It is known that $k > 0$ implies that there is a line, ℓ, through every point $x \in X$ and $K_X \cdot \ell = -((n+k)/2+1)$ (see [E1]). Therefore $\nu : -K_X \cdot \ell - 2 = (n+k)/2-1 > n/2-1$ since $k > 0$. Thus by Theorem (2.3) the nef value τ of L is $\tau = (n+k)/2+1$ and $\Phi : X \to Y$ is the fiber type contraction of the extremal ray $R = R_+[\ell]$. Hence in particular $K_X + ((n+k)/2+1)L \cong \Phi^*\mathcal{L}$ for some ample line bundle $\mathcal{L}$ on Y.

The bound of the dimension of the image of the contraction is an immediate consequence of Lemma (1.4.4). Q.E.D.

We now push forward to the $\nu = (n-3)/2$ case .

(2.5) Theorem. *Let* X *be a smooth connected projective* n*-fold,* L *an ample line bundle on* X *and* ℓ *a line on* X *relative to* L*. Let* T *be a non-breaking family of rational curves which are deformations of* ℓ *and suppose that* (X, L), ℓ, T *satisfy the assumptions* (2.0)*. Let* τ *be the nef value of* L *and* $\Phi : X \to Y$ *the morphism associated to* $K_X+\tau L$*. Let* $\nu = -K_X \cdot \ell - 2$*. If* $\nu \geq (n-3)/2$*, then either* $\nu < \tau - 2$ *and* $X \cong \mathbf{P}^{\tau-1} \times \mathbf{P}^{\nu+1}$*, or* $\nu = \tau - 2$ *and one of the following is true:*

(2.5.1) $\dim Y \leq n - \nu - 1$, ℓ *is an extremal rational curve and* $\Phi = \mathrm{cont}_R$, *the fiber type contraction of the extremal ray* $R = R_+[\ell]$, *or*

(2.5.2) $\Phi : X \to Y$ *factors as* π *composed with* φ, $\Phi = \pi \circ \varphi$, *where* $\varphi : X \to W$ *is a* $\mathbf{P}^{\tau-1}$ *bundle over a smooth variety* W *of dimension* $\tau = (n+1)/2$*. Furthermore* Y *is a smooth curve. Denoting* $\mathcal{E} = \varphi_* L$, *there is a rank* τ *vector bundle* $\mathcal{F}$ *over* Y *such that* $W \cong \mathbf{P}(\mathcal{F})$ *and* $\mathcal{E}_\Delta \cong \oplus^\tau \mathcal{O}_{\mathbf{P}^{\tau-1}}(1)$ *for any fiber* Δ $(\cong \mathbf{P}^{\tau-1})$ *of* $\pi : W \to Y$,

(2.5.3) Y *is a point and* X *is a Fano manifold of index* $\tau = (n+1)/2$*, and Picard number* $\rho(X) \geq 2$*. Such* X *are classified in* [W4].

Proof. First let us assume that $\nu < \tau - 2$ so that $\nu + \tau > 2\nu + 2 \geq n - 1$. Let R be an extremal ray contracted by Φ and let μ be an extremal rational curve belonging to R and satisfying the condition $-(n+1) \leq K_X \cdot \mu \leq -1$. Since $K_X \cdot \mu = -\tau L \cdot \mu$ we conclude that

$$\tau(L \cdot \mu) \leq n + 1.$$

Since $\tau > \nu + 2 \geq (n+1)/2$ we conclude that $L \cdot \mu = 1$ and thus τ is the length of R. Thus $\nu + \tau \geq n$ since $\nu + \tau > n - 1$ and ν, τ are both integers. By (2.1.2) we get $\nu + \tau = n$ and by (2.2) we conclude that $X \cong \mathbf{P}^{\tau-1} \times \mathbf{P}^{\nu+1}$.

Thus we can assume that $\nu = \tau - 2$ so that Φ contracts ℓ. By (1.4.4), $\dim Y \leq n - \dim T_x - 1 \leq n - \nu - 1$. If $\Phi = \mathrm{cont}_R$, $R = R_+[\ell]$, then since $\dim \Phi(\ell) = 0$ it follows that $\ell \in R$. Since $L \cdot \ell = 1$ we see that ℓ is an extremal rational curve and we are in case (2.5.1).

Assume now that Φ is not the contraction of an extremal ray containing ℓ. We must show that either (2.5.2) or (2.5.3) hold. Let R be an extremal ray not containing ℓ that is contracted by Φ. Let μ be an extremal rational curve that is contained in R. By (2.3) we conclude that $\nu < (n-2)/2$ and hence $\nu = (n-3)/2$. Let $\varphi : X \to W$ denote the contraction associated to R and let F be any positive dimensional fiber of φ. Then by (1.4.5), $\dim F \leq n - \dim T_X - 1 \leq n - \nu - 1$ and if $\dim F = n - \nu - 1$ then $NE(X) = NE(F) + \mathbf{R}_+[\ell] = R + \mathbf{R}_+[\ell]$. In the latter case, X is a Fano manifold, Y is a point since $\rho(X) = 2$ (compare with the proof of (2.3)) and $\tau = \nu + 2 = (n+1)/2$. Since $L\cdot\ell = 1$, $K_X \approx -\tau L$ implies the index of X is τ and we are in the case (2.5.3).

Therefore we may assume that every fiber of the contraction φ is of dimension $\leq n - \nu - 2 = \tau - 1$. Since $(K_X + \tau L)\cdot \mu = 0$ we get

$$(L\cdot\mu)(n+1)/2 = \tau(L\cdot\mu) \leq n+1.$$

Thus $L\cdot\mu = 1$ or 2. In particular length(R) $\geq \tau$ and we conclude from (0.2.2.1) that φ is a contraction of fiber type and all fibers have dimension $\geq \tau - 1$. Thus by the above all fibres of φ have dimension $\tau - 1$. By the same argument as in the proof of (2.2), using the Kobayashi-Ochiai characterization of projective spaces, and Fujita's lemma ([Fu1],(2.12)), we see that $\varphi : X \to W$ is a $\mathbf{P}^{\tau-1}$ bundle. We know that Φ factors as $\pi : W \to Y$ composed with φ, $\Phi = \pi\circ\varphi$. Note if $\dim Y = 0$, then $K_X \approx -\tau L$ and $\rho(X) \geq 2$ since π is not an isomorphism. Thus we can assume that $\dim Y \geq 1$ since otherwise we are in case (2.5.3). From the canonical bundle formula for the bundle φ we have

$$K_X \otimes L^\tau \approx \varphi^*(K_W \otimes \det(\mathcal{E}))$$

where $\mathcal{E} = \varphi_* L$ is an ample vector bundle of rank τ. Moreover $K_W \otimes \det(\mathcal{E}) \approx \pi^* H$ where H is an ample line bundle on Y such that $K_X \otimes L^\tau \approx \Phi^* H$. Since $\tau = (n+1)/2$ note that $\dim W = n - (\tau - 1) = \tau$. Note also that $K_W \otimes \det(\mathcal{E})$ is nef but not ample (since we are assuming that π is not an isomorphism). Then Fujita's theorem, (0.6), applies to say that $(W,\mathcal{E})$ is one of the pairs listed in (0.6). In the first case, $\dim Y = 0$, and in the second case, $K_W \otimes \det(\mathcal{E})$ is not nef, and so neither occur. The third, fourth, and fifth cases are contained in case (2.5.3). This leaves the sixth case, i.e.

there is a rank τ vector bundle $\mathcal{F}$ over a smooth curve C such that $W \cong \mathbf{P}(\mathcal{F})$ and $\mathcal{E}_\Delta \cong \oplus^\tau \mathcal{O}_{\mathbf{P}^{\tau-1}}(1)$ for any fiber Δ $(\cong \mathbf{P}^{\tau-1})$ of β: $W \to C$.

In this case $K_X \otimes L^\tau \approx \varphi^*(K_W \otimes \det(\mathcal{E})) \approx \varphi^*\beta^*\mathcal{M}$ for some line bundle, $\mathcal{M}$, on C. Since $K_X \otimes L^\tau \approx \Phi^* H$ for some ample line bundle on Y where $\dim Y \geq 1$, we conclude that $Y \cong C$ and β is isomorphic to π. This gives (2.5.2). Q.E.D.

(2.6) Remark. More can be said in case (2.5.2). Since π is not an isomorphism there is an extremal ray R' on X different from R contracted by Φ. Let ℓ' be a line in a fiber of φ. Let T' be the family of deformations of ℓ'. Using R', T' in the roles of R, T in the above proof we get by the same reasoning a second and different $\mathbf{P}^{\tau-1}$ bundle over a manifold V which is a $\mathbf{P}^{\tau-1}$ bundle over Y. From this we conclude that X is actually a fiber product of two $\mathbf{P}^{\tau-1}$ bundles over Y.

§ 3. Applications.

In this section we discuss two main consequences of the results above.

(3.1) Relative version of a Mukai conjecture. Let X be a smooth complex connected projective variety of dimension n. Mukai conjectured the following (see [Mu])

if index(X) $> n/2 + 1$ *then* $\mathrm{Pic}(X) \cong \mathbf{Z}$.

Such a a conjecture has been proved by the third author in [W3]. The following result can be viewed as a "relative version" of the Mukai conjecture, which indeed can be deduced from it when the variety Y (from the theorem) is a point. Let us also point out that in the theorem below we don't assume *a priori* the existence of a non-breaking dominating family of lines.

(3.1.1) Theorem. *Let* X *be a smooth connected projective variety of dimension* n, L *an ample line bundle on* X. *Let* τ *be the nef value of* L *and let* $\Phi : X \to Y$ *be the morphism with connected fibers and normal image associated to* $|m(K_X+\tau L)|$ *for* $m >> 0$. *Assume that* Φ *is not birational.*

(3.1.1.1) *If* $\tau > (n - \dim Y + 1)/2$, *there exists a non-breaking dominating family,* T, *of lines relative to* L.

(3.1.1.2) *If* $\tau \geq n/2+1$, Φ *is a fiber type contraction of an extremal ray and* $\mathrm{Pic}(X) \cong \Phi^*\mathrm{Pic}(Y) \oplus \mathbf{Z}[L]$ *unless* $\tau = n/2 + 1$, $(X,L) \cong (\mathbf{P}^{n/2} \times \mathbf{P}^{n/2}, \mathcal{O}_{\mathbf{P}^{n/2} \times \mathbf{P}^{n/2}}(1))$, *and* $\dim Y = 0$.

Proof. Let F be a general fiber of Φ. Since $-K_F \approx \tau L_F$ is ample it is a well known general fact, which we have seen attributed to J. Kollár, that there exists an extremal rational curve through any general point of F (see e.g. [M3], [MM]). Furthermore for any such a curve, ℓ, in F we have by (0.2.1)

$$-K_X \cdot \ell = -K_F \cdot \ell \leq n - \dim Y + 1.$$

We claim that $L \cdot \ell = 1$. Indeed, if not, we get the numerical contradiction

$$n - \dim Y+1 = 2((n - \dim Y + 1)/2) < \tau L \cdot \ell = \tau L_F \cdot \ell = - K_F \cdot \ell \leq n - \dim Y+1.$$

So we know that through a general point of X there passes a curve ℓ which is contracted by Φ and has the intersection with L equal to 1. Thus, by finiteness of the number of the components of the Hilbert scheme parameterizing curves whose intersection with L is one, we can choose an irreducible variety, T, parameterizing such curves passing through a general point of X. Also we can assume that those curves are contracted by Φ. By the minimality of the intersection of L it follows that the family is non-breaking and dominates X in the sense of (2.0) (see also (1.4.2)).

To prove the second part of the statement, note that $(K_X+\tau L) \cdot \ell = (K_F+\tau L_F) \cdot \ell = 0$ and $L \cdot \ell = 1$ yield $K_X \cdot \ell = -\tau$ and hence $\nu := -K_X \cdot \ell - 2 = \tau - 2 \geq (n-2)/2$.

Thus if $\dim Y \neq 0$, (2.3) applies to say that ℓ is an extremal rational curve on X and $\Phi = \mathrm{cont}_R$ is the contraction of the extremal ray $R = \mathbf{R}_+[\ell]$.

To see that $\mathrm{Pic}(X) \cong \Phi^*\mathrm{Pic}(Y) \oplus \mathbf{Z}\,[L]$ use the exact sequence (0.2.3). Recall that the morphism $\mathrm{Pic}(X) \xrightarrow{\cdot \ell} \mathbf{Z}$ is surjective since $L \cdot \ell = 1$.

If $\dim\Phi(X) = 0$, then $\mathrm{index}(X) = \tau = n/2+1$ so that if $\mathrm{Pic}(X) \neq \mathbf{Z}$, then $(X,L) \cong (\mathbf{P}^{n/2} \times \mathbf{P}^{n/2}, \mathcal{O}(1))$ by [W3]. Q.E.D.

(3.2) Structure results for scrolls and quadric fibrations. Let X be a smooth connected projective variety of dimension n and L an *ample* line bundle on X. Assume that (X, L) is a *scroll*, $p : X \to Y$, onto a normal projective variety Y of dimension m. Then we have the following structure theorem for p.

(3.2.1) Theorem. *Let* (X, L) *be a scroll,* $p : X \to Y$, *as above. Let* $\mathbf{P}^d$, $d = n-m$, *be a general fiber of* p *and let* ℓ *be a line in* $\mathbf{P}^d$. *Assume* $n \geq 2m-1$. *Then* ℓ *is an extremal rational curve and* $p = \mathrm{cont}_R$ *is the fiber type contraction of* $R = \mathbf{R}_+[\ell]$. *In particular* p *has no divisorial fibers unless* $m = 1$.

Proof. We have $K_X+ (d+1)L \approx p^*\mathcal{L}$ for some ample line bundle $\mathcal{L}$ on Y. Then $K_X+ (d+1)L$

is nef but not ample and therefore $d+1 = \tau$, the nef value of L (see also [BFS1], (1.2)). Take a line ℓ in a smooth general fiber $\mathbf{P}^d$ of p. Hence $L\cdot\ell = 1$ so that one can construct a non-breaking family, T, of rational curves which are deformations of ℓ and which fill up X so that the assumptions (2.0) are satisfied (compare with (1.4.2)). Now compute

$$\nu = -K_X\cdot\ell - 2 = -K_{\mathbf{P}^d}\cdot\ell - 2 = d - 1.$$

Thus $\nu = \tau-2$, $\nu = n-m-1 \geq (n-3)/2$ and therefore theorem (2.5) applies to give the result. To see this note that in case (2.5.3) Y is a point, X is therefore a projective space, and the result is clear. In case (2.5.2), p is a composition of a $\mathbf{P}^{\tau-1}$ bundle projection, $\varphi : X \to W$, with a nontrivial morphism, $\pi : W \to Y$, with $\dim Y < \dim W$. In this case a general fibre of p is a nontrivial bundle and not a projective space. So we are in case (2.5.1).

To see that, if $m > 1$, p has no divisorial fibers note that since $p = \mathrm{cont}_R$, $R = \mathbf{R}_+[\ell]$ and $L\cdot\ell = 1$ we have the exact sequence (0.2.3). Any divisorial fiber F of p satisfies $F\cdot\ell = 0$ and hence $F \in p^*\mathrm{Pic}(Y)$, which is clearly not possible. Q.E.D.

(3.2.2) Remark. Note that the result above is sharp. Indeed in [BS2], (4.2) we produce an example of a (2n–2)-dimensional scroll over a n-fold with a divisorial fiber. ■

A consequence of Theorem (3.2.1) is the following result which states that, in a number of cases and if L is *very ample*, a scroll (X, L) is a $\mathbf{P}^d$ bundle.

(3.2.3) Proposition. *Let* (X, L) *be a* n-*dimensional scroll,* $p : X \to Y$, *with* L *very ample, over a normal projective variety* Y *of dimension* $m \leq 3$. *If* $n \geq 2m-1$, *then* p *is a* $\mathbf{P}^d$ *bundle,* $d = n-m$.

Proof. For m = 1, 2 this is shown in [S2], (3.3). So we can assume m = 3. Since $n \geq 2m-1 = 5$, Theorem (3.2.1) applies to say that p is the contraction of a numerically effective extremal ray $R = \mathbf{R}_+[\ell]$, ℓ a line in a general fiber $\mathbf{P}^d$ and p has no divisorial fibers. If all fibers are equal dimensional then it is easy to check (see e.g. [S2], (3.3) and also [Fu1], (2.12)) that $p : X \to Y$ is a bundle.

Thus it is enough to show that there are no fibers F of dimension n–2. Assume otherwise. By slicing with general hyperplane sections we immediately reduce to the n = 5 case where $\dim F = 3$.

Let V be the smooth 3-fold obtained as transversal intersection of 2 general members of $|L|$. Consider the restriction $\psi : V \to Y$ of p to V. Since $K_X \otimes L^3 \approx p^*\mathcal{L}$ for some ample line bundle $\mathcal{L}$ on Y we have $K_V \otimes L_V \approx \psi^*\mathcal{L}$ and, since Y is normal and the fibers of ψ are connected, ψ is the morphism associated to $|N(K_V+L_V)|$ for $N >> 0$. This morphism has been studied in [F] and [S1]. Let C be the curve obtained by transversal intersection of the fiber F of p with V, $C = F \cap V$. Then C is a 1-dimensional fiber of ψ. Thus by the results of [F] and [S1] we know that C is the fiber of a $\mathbf{P}^1$ bundle S that has $\mathcal{N}^V_{S|C} \cong \mathcal{O}_C(-1)$ and $L_C \cong \mathcal{O}_{\mathbf{P}^1}(1)$. Therefore $C \cong \mathbf{P}^1$ and $\mathcal{N}^V_C \cong \mathcal{O}_{\mathbf{P}^1} \oplus \mathcal{O}_{\mathbf{P}^1}(-1)$.

We claim that $(F, L_F) \cong (\mathbf{P}^3, \mathcal{O}_{\mathbf{P}^3}(1))$. To see this note that we could have chosen our two smooth sections giving V to pass through any point $x \in F$. Thus we have F intersected with two divisors is a smooth curve of F that contains x. Then x is a smooth point of F. Since x was an arbitrary point of F, F is smooth. Since C is a linear $\mathbf{P}^1$ we see that F is a linear $\mathbf{P}^3$. Indeed

$$(K_F+3L_F)\cdot L_F\cdot L_F = ((K_X+3L)_F + \det \mathcal{N}^X_F)\cdot L_F\cdot L_F = (\det \mathcal{N}^X_F)\cdot L_F\cdot L_F =$$
$$= \deg(\mathcal{N}^X_{F|C}) = \deg \mathcal{N}^V_C = -1.$$

It thus follows that K_F+3L_F has no sections and hence $(F, L) \cong (\mathbf{P}^3, \mathcal{O}_{\mathbf{P}^3}(1))$ (see e.g. [S2]).

Since the curve C is transversal intersection of F and V we have that $\mathcal{N}^X_{F|C} \cong \mathcal{O}_{\mathbf{P}^1} \oplus \mathcal{O}_{\mathbf{P}^1}(-1)$. Therefore the normal bundle $\mathcal{N}^X_F$ is a uniform rank 2 vector bundle on $F \cong \mathbf{P}^3$ and hence $\mathcal{N}^X_F \cong \mathcal{O}_{\mathbf{P}^3} \oplus \mathcal{O}_{\mathbf{P}^3}(-1)$ (see e.g. [OSS], I, § 3). Then $h^0(\mathcal{N}^X_F) = 1$, $h^1(\mathcal{N}^X_F) = 0$. It thus follows that there exists a 1-dimensional family of deformations of F. Furthermore since F is a fiber, the rigidity property of proper maps says that deformations of F are contained in nearby fibers Λ (hence in particular $\dim\Lambda \geq 3$) of p and by the semicontinuity of the dimension of fibers we have $3 = \dim F \geq \dim\Lambda$. Therefore $\dim\Lambda = 3$ so that one has in X a 1-dimensional family of 3-dimensional fibers of p. This family fills up a divisor, D, such that $D\cdot\ell = 0$ and hence $D \in p^*\mathrm{Pic}(Y)$ by (0.2.2.3). This is not possible since D maps down to a curve.

Thus we conclude that for any $n \geq 2m-1 = 5$ all fibers of p are of dimension $n-m = 2$. Since $L^2\cdot F = 1$ all fibers are generically reduced and irreducible. Since p is a morphism with equidimensional fibers it thus follows that all fibers are indeed reduced and isomorphic to $\mathbf{P}^2$. Therefore p is a $\mathbf{P}^d$ bundle, $d = n-m$, and (3.2.3) is proved. Q.E.D.

(3.2.4) Remark-Example (the case $n = 2m - 2$). Let (X,L) be a $(2n - 2)$-dimensional scroll, $p : X \to Y$, over a n-fold Y. Then the map p doesn't have to be a projective bundle even if it is a Mori contraction (so there are no divisorial fibers) and the statement of Proposition (3.2.3) is false in this case (compare with Remark (3.2.2)).

However, for $n = 3$, it turns out that the 2-dimensional fibers of p are isomorphic to $\mathbf{P}^2$ with normal bundle $\mathcal{T}^*_{\mathbf{P}^2}(1)$. To see this, let F be a 2-dimensional fiber of p. Then by looking over the proof of (3.2.3) we see that $F \cong \mathbf{P}^2$ and $\mathcal{N}^X_F$ is a uniform rank 2 vector bundle over $\mathbf{P}^2$ such that $\mathcal{N}^X_{F|\mathbf{P}^1} \cong \mathcal{O}_{\mathbf{P}^1} \oplus \mathcal{O}_{\mathbf{P}^1}(-1)$. Then either $\mathcal{N}^X_F \cong \mathcal{O}_{\mathbf{P}^2} \oplus \mathcal{O}_{\mathbf{P}^2}(-1)$ or $\mathcal{N}^X_F \cong \mathcal{T}_{\mathbf{P}^2}(a)$ for some integer a (see e.g. [OSS], p. 59). The first case leads to the same contradictions as in the proof of (3.2.3) while in the second we see that $a = -2$ since $(\det\mathcal{T}_{\mathbf{P}^2}(a))_{\mathbf{P}^1}$ $(\cong \mathcal{O}_{\mathbf{P}^1}(2a+3)) \cong \mathcal{O}_{\mathbf{P}^1}(-1)$. Note that $\mathcal{T}_{\mathbf{P}^2}(-2) \cong \mathcal{T}^*_{\mathbf{P}^2}(1)$.

Let us give explicit examples of two types of special fibers which can occur for $n \geq 3$. For $n = 3$ both have codimension 2.

(3.2.4.1) (Fibers of dimension $n - 1$ and codimension $n - 1$). Let Y be a smooth projective n-fold and $W := \mathbf{P}^{n-1} \times Y$. Let $q : W \to \mathbf{P}^{n-1}$, $p : W \to Y$ be the projections. Take $\mathcal{L} := q^*\mathcal{O}_{\mathbf{P}^{n-1}}(1) \otimes p^*\mathcal{M}$ for a very ample line bundle $\mathcal{M}$ on Y. Hence $\mathcal{L}$ is very ample and let X be a smooth divisor in W corresponding to a general member of $|\mathcal{L}|$. An easy check shows that $(X, \mathcal{O}_X(1))$ is a scroll, $p : X \to Y$, over Y with $c_1(\mathcal{M})^n$ $(n - 1)$-dimensional fibers F such that $\mathcal{N}^X_F \cong \mathcal{T}^*_{\mathbf{P}^{n-1}}(1)$. Note that such $(n - 1)$-dimensional fibers correspond to the points of Y which are zeroes of a general section of $p_*\mathcal{L} \cong \mathcal{M}^{\oplus n}$.

(3.2.4.2) (Fibers of dimension $2n - 4$ and codimension 2). Let $Y = \mathbf{P}(V^*)$ where V is an $n + 1$ dimensional complex vector space. Let us choose a point y_0 in Y which represents a one dimensional linear subspace, V_0, of V. Now let X be the incidence variety of projective 2-planes in Y meeting y_0:

$$X = \{(y,\Pi) \mid \text{the projective plane } \Pi \text{ in Y contains the points y and } y_0\}.$$

We have a projection $q : X \to \mathrm{Grass}(2, V/V_0)$, the Grassmannian of two dimensional vector subspaces of the quotient V/V_0. The map, q, makes X into a $\mathbf{P}^2$ bundle over $\mathrm{Grass}(2,V/V_0)$ so that X is smooth. On the other hand we have a projection $p : X \to Y$ which makes $(X, \mathcal{O}_X(1))$ a scroll over Y in the sense of (0.3). The special fiber over the point y_0 is then isomorphic to $\mathrm{Grass}(2,V/V_0)$. ■

Let us point out the following consequence of Proposition (3.2.3).

(3.2.5) Corollary. *Let* (X, L) *be a* n-*dimensional scroll*, $p : X \to Y$, *over a normal projective variety* Y *of dimension* $m \geq 3$ *and let* L *be very ample. Let* $\mathcal{Z} = \{y \in Y, \dim p^{-1}(y) > n-m\}$. *If the general fiber of* p *has dimension bigger or equal to 2, then* $\mathrm{cod}_Y \mathcal{Z} \geq 4$.

Proof. By slicing with general hyperplane sections on Y we can assume m = 3. Hence $\mathcal{Z} = \emptyset$ by (3.2.3) so are done. Q.E.D.

Assume now that (X,L) is a *quadric fibration*, $p: X \to Y$, over a normal projective variety Y of dimension m. Then we have the following structure theorem for p.

(3.2.6) Theorem. *Let* (X,L) *be a quadric fibration*, $p: X \to Y$, *as above with* L *ample line bundle on* X. *Let* ℓ *be a line in a smooth general fiber of* p. *Assume* $n-m \geq 3$ *and* $n \geq 2m+1$. *Then* ℓ *is an extremal rational curve and* p *is the fiber type contraction of* $R = R_+[\ell]$.

Proof. It runs parallel to that of (3.2.1). We have $K_X+dL \approx p^*\mathcal{L}$ for some ample line bundle $\mathcal{L}$ on Y, d = n–m. Then K_X+dL is nef but not ample and therefore $d = \tau$, the nef value of L. Take a line ℓ in a smooth fiber, Q, of p. Hence $L \cdot \ell = 1$ so that we can construct a non-breaking family T of rational curves which are deformations of ℓ and which fill up X so that the assumptions (2.0) are satisfied. Now compute $\nu = -K_X \cdot \ell - 2 = -K_Q \cdot \ell - 2 = d-2$. Thus $\nu = \tau - 2$, $\nu = n-m-2 \geq (n-3)/2$ and therefore (2.5) applies to give the result; neither of the cases (2.5.2) nor (2.5.3) is a quadric fibration over a nontrivial base. Q.E.D.

(3.2.7) Remark. Note that in [BS1], (2.8) we construct an example of a quadric fibration over a surface with a $\mathbf{P}^1 \times \mathbf{P}^1$ as a divisorial fiber. Indeed in [BS1] we prove that a n-dimensional quadric fibration (X, L) over a surface Y with $n \geq 4$ has equidimensional fibers and, if n = 3, the only divisorial fibers can occur are isomorphic to either $\mathbf{F}_0 = \mathbf{P}^1 \times \mathbf{P}^1$ or to the union $\mathbf{F}_0 \cup \mathbf{F}_1$, $\mathbf{F}_1 = \mathbf{P}(\mathcal{O}_{\mathbf{P}^1} \oplus \mathcal{O}_{\mathbf{P}^1}(-1))$.

APPENDIX

Applications to adjunction theory

by

Mauro C. Beltrametti, Andrew J. Sommese

The preceding results have immediate, important applications to adjunction theory. Let L^ be an ample and spanned line bundle on a smooth connected, projective n-fold X. General references for adjunction theory are [S2] and [BFS1].

If $K_{X^\wedge}+(n-1)L^\wedge$ is nef and big then the first reduction (X', L') of (X^, L^) exists. Recall that there is a birational morphism $r : X^\wedge \to X'$ with $L' := (r_* L^\wedge)^{**}$ ample, $X^\wedge \setminus r^{-1}(B) \cong X' \setminus B$ where B is a finite set of points and $L^\wedge \approx r^*L' - [r^{-1}(B)]$.

If $K_{X'}+(n-2)L'$ is nef and big then the second reduction $(X, \mathcal{K})$ exists, where $\mathcal{K}$ is an ample line bundle on X. Here there is a birational morphism $\varphi : X' \to X$ with $\varphi^*\mathcal{K} \approx K_{X'}+(n-2)L'$ and $X' \setminus \varphi^{-1}(Z) \cong X \setminus Z$ where Z is an algebraic subset of X and $\dim Z \leq 1$. From [S1] and [F] we know that X has terminal singularities and it is smooth outside of a finite set of isolated points $x \in Z$. Further (see [BFS1]), $L' \approx \varphi^*L - \mathcal{D}$ where $L = (\varphi_*L')^{**}$ and $\mathcal{D}$ is an effective 2-Cartier divisor with $\mathrm{supp}(\mathcal{D}) = \varphi^{-1}(Z)$. Recall also that $\mathcal{K} \approx K_X + (n-2)L$ where K_X and L are both 2-Cartier (see [BFS1], (0.2)).

We say that a rational curve ℓ on X^ (respectively on X' or X) is a *line* if $L\hat{}\cdot\ell = 1$ (respectively $L'\cdot\ell = 1$ or $L\cdot\ell = 1$). All the other notation are as in the previous sections.

(A.1) Theorem. *Notation as above. We have*

(A.1.1) *If* X^ *is not isomorphic to* X', *no lines on* X' *can meet* B,

(A.1.2) *If* X^ *is dominated by a non-breaking family* T^ *of lines then either* $X\hat{} \cong X'$ *or* $-K_{X\hat{}}\cdot\ell = 2$ *and the morphism* $r : X\hat{} \to X$ *blows up a single point,*

(A.1.3) *If* X' *is not isomorphic to* X, *no lines on* X *can meet* Z,

(A.1.4) *If* X' *is dominated by a non-breaking family,* T', *of lines then either* $X' \cong X$, *or* $-K_{X'}\cdot\ell = 3$ *and* X *is smooth, or* $-K_{X'}\cdot\ell = 2$. *In the latter case there is at most one divisorial fiber for* $\varphi : X' \to X$.

Proof. (A.1.1). Assume that there is a line ℓ on X' containing a point $x \in B$. Let γ be the proper transform of ℓ under $r : X\hat{} \to X'$. Then one has

$$L\hat{}\cdot \gamma \leq r^*L' \cdot \gamma - r^{-1}(x) \cdot \gamma = L'\cdot\ell - r^{-1}(x) \cdot \gamma = 0$$

which contradicts the ampleness of L^.

(A.1.2). Let X^ be filled out by a non-breaking family T^ of lines and assume that X^ is not isomorphic to X'. Then there is a linear $\mathbf{P}^{n-1} \subset X\hat{}$ which contracts to a point in X'. We can choose a line ℓ from the family T^ such that ℓ is not contained in $\mathbf{P}^{n-1}$ (since otherwise r would have a lower dimensional image) and $\ell \cap \mathbf{P}^{n-1} \neq \emptyset$. Let $x \in \ell$ be a smooth point of ℓ such that $x \notin \mathbf{P}^{n-1}$. Let $\hat{D}_x$ be the locus of the deformations of ℓ containing x. Recall that $\dim \hat{D}_x \geq -K_{X\hat{}}\cdot\ell - 1$ by (1.4.2). Then

$$\dim(\hat{D}_x \cap \mathbf{P}^{n-1}) \geq -K_{X\hat{}}\cdot\ell - 2.$$

If $-K_{X\hat{}}\cdot\ell > 2$ we get the usual contradiction as in § 2 by using the non-breaking Lemma (1.4.3), so that $-K_{X\hat{}}\cdot\ell \leq 2$. Since the lines fill out X^ we have that the normal bundle $\mathcal{N}_\ell$ of ℓ in X^ is spanned and by the adjunction formula $K_{X\hat{}|\ell} + \det\mathcal{N}_\ell \approx \mathcal{O}_\ell(-2)$. Then we see that $K_{X\hat{}}\cdot\ell \leq -2$ with the equality only if $\mathcal{N}_\ell$ is trivial. Thus $K_{X\hat{}}\cdot\ell = -2$ and $\mathcal{N}_\ell$ is trivial. Assume that B contains at least two points, b_1, b_2. Hence as before $\ell\cdot r^{-1}(b_1) = \ell\cdot r^{-1}(b_2) = 1$. By contracting $r^{-1}(b_1)$ and $r^{-1}(b_2)$ we get a family of non-breaking curves contradicting Lemma (1.4.3).

(A.1.3). Assume that there is a line ℓ on X containing a point $x \in Z$. Let γ be the proper transform of ℓ under $\varphi : X' \to X$. Then

$$L'\cdot \gamma = \varphi^*L\cdot \gamma - \mathcal{D} \cdot \gamma = 1 - \mathcal{D} \cdot \gamma < 1$$

which contradicts the ampleness of L'.

(A.1.4). Let X' be filled out by a non-breaking family T' of lines and assume that X' is not isomorphic to X. Then $Z \neq \emptyset$.

Assume that there exists an isolated point $x \in Z$. Therefore $\varphi^{-1}(x)$ belongs to the support of $\mathcal{D}$. The same argument as in the proof of (A.1.2) applies to give $X' \cong X$ if $-K_{X'}\cdot\ell \geq 4$ and X smooth if $-K_{X'}\cdot\ell = 3$. Recall that $-K_{X'}\cdot\ell = 2 + \det\mathcal{N}_\ell^X \cdot\ell \geq 2$.

Thus we can assume that Z consists of (smooth) curves. Therefore from [F], § 2, we know that X is smooth and there exist a smooth irreducible curve $C \subset Z$ and a $\mathbf{P}^{n-2}$ bundle $\mathcal{P} \subset X'$ with the restriction $p = \varphi_{\mathcal{P}} : \mathcal{P} \to C$ the bundle projection. Fix a fiber $\mathbf{P}^{n-2}$ of p. We can choose a line ℓ from the family T' such that $\ell \cap \mathbf{P}^{n-2} \neq \emptyset$ and ℓ is not contained in $\mathbf{P}^{n-2}$. Let x be a smooth point of ℓ, $x \notin \mathbf{P}^{n-2}$. Let D'_x be the locus of the deformations of ℓ containing x and recall that $\dim D'_x \geq -K_{X'} \cdot \ell - 1$. Then

$$\dim(D'_x \cap \mathbf{P}^{n-2}) \geq -K_{X'} \cdot \ell - 3.$$

If $-K_{X'} \cdot \ell > 3$ we contradict again Lemma (1.4.3), so that $-K_{X'} \cdot \ell \leq 3$. The argument is finished by reasoning analogous to that in the proof of (A.1.2). Q.E.D.

(A.2) Theorem. *Notation as above. Assume that* $K_{X^\wedge} + (n-1)L^\wedge$ *is nef and not big. Let* $\phi : X^\wedge \to Y$ *be the morphism with connected fibers onto a normal variety* Y *such that* $K_{X^\wedge} + (n-1)L^\wedge \approx \phi^* H$ *for an ample line bundle* H *on* Y. *Then* ϕ *is a Mori contraction if* $n \geq 4$ *unless* $X^\wedge \cong \mathbf{P}^2 \times \mathbf{P}^2$ *and* $L^\wedge \approx \mathcal{O}_{\mathbf{P}^2 \times \mathbf{P}^2}(1, 1)$.

Proof. Note that the nef value of $(X^\wedge, L^\wedge)$ is $\tau = n - 1$ (see e.g. [BFS1], (1.2)). First assume that Y is a point. Then if $n-1 \geq n/2+1$, i.e. $n \geq 4$, we have by [W3] that either $\mathrm{Pic}(X^\wedge) \cong \mathbf{Z}$ and ϕ is a Mori contraction or $X^\wedge \cong \mathbf{P}^{n/2} \times \mathbf{P}^{n/2}$ and $n-1 = n/2+1$. In this case $n = 4$. If Y is not a point then by (3.1.1.2), ϕ is a Mori contraction if $n-1 \geq n/2+1$, i.e. $n \geq 4$. Q.E.D.

Given a **Q**-Cartier divisor D on a projective variety W we let $\kappa(D, W)$ denote the *Kodaira dimension* of (W, D) or the D-*dimension* of W. This is the maximal dimensional image of W under the set of meromorphic maps given by $|mD|$ where mD is a Cartier divisor and $h^0(mD) > 0$.

Let D be a Cartier divisor such that $|mD|$ is basepoint free for $m >> 0$. In the following we refer to the morphism with connected fibers and normal image defined by $|mD|$, $m >> 0$, simply as the morphism associated to D.

(A.3) Theorem. *Notation as above. Assume that the first reduction* (X', L') *of* $(X^\wedge, L^\wedge)$ *exists (which is equivalent to* $K_{X^\wedge} + (n-1)L^\wedge$ *being nef and big). Let* τ *be the nef value of* (X', L'). *If*

$$2\tau \geq \dim X' - \kappa(K_{X'} + \tau L', X') + 3 \quad \textit{and} \quad \kappa(K_{X'} + \tau L', X') < \dim X'$$

then $(X^\wedge, L^\wedge) \cong (X', L')$.

Proof. By the Kawamata-Shokurov basepoint free theorem we know that there is a morphism $\phi : X' \to Y$ such that $N(K_{X'} + \tau L') \approx \phi^* H$ where H is an ample line bundle on Y and $N(K_{X'} + \tau L')$ is Cartier for some positive integer, N. Under the above assumptions a general fiber F of ϕ is smooth and positive dimensional and $K_F \approx -\tau L'_F$ where $\tau > \dim F/2+1$. Thus by (3.1.1.1), F and hence X' is covered by lines relative to L' and we are done by (A.1.1).

Q.E.D.

We get the following

(A.3.1) Corollary. *Assume that the first reduction* (X',L') *of* $(X^\wedge, L^\wedge)$ *exists and* $K_{X'} + (n-2)L'$ *is nef but not big. If* $n \geq 7$, *then* $(X', L') \cong (X^\wedge, L^\wedge)$ *and the morphism,* Φ, *associated to* $K_{X'} + (n-2)L'$ *is a Mori contraction.*

Proof. Let τ be the nef value of (X',L'). Since $K_{X'} + (n-2)L'$ is nef but not big one has $\tau = n - 2$. Then by (A.3) we have $(X', L') \cong (X^\wedge, L^\wedge)$ if

$$2(n-2) \geq n - \kappa(K_{X'} + (n-2)L', X') + 3,$$

that is if $n \geq 7 - \kappa(K_{X'} + (n-2)L', X')$.

From (3.1.1.2) we know that Φ is a Mori contraction if $n-2 > n/2+1$, i.e. $n \geq 7$.

Q.E.D.

(A.4) Theorem. *Notation as above. Assume that the second reduction* $(X, \mathcal{K})$ *exists (which is equivalent to* $\kappa(K_{X^\wedge}+(n-2)L^\wedge, X^\wedge) = n$) *and* K_X *is not nef. There is a positive rational number* t *such that* K_X+tL *is nef and not ample and such that there is a morphism* $\phi : X \to Y$ *with connected fibers and normal image* Y *and* $N(K_X+tL) \approx \phi^*H$ *where* H *is an ample line bundle on* Y *and* N *is some positive integer such that* $N(K_X+tL)$ *is a Cartier divisor. Assume that* $\dim Y < \dim X$. *Then if* $t \geq n/2+1$ *either* $\dim Y = 0$ *or* ϕ *is a Mori contraction of fiber type and* $(X,L) \cong (X',L') \cong (X^\wedge,L^\wedge)$, *where* $\mathcal{K} \approx K_X+(n-2)L$ *on* X.

Proof. By the assumption of existence on the 2nd reduction $(X, \mathcal{K})$, we have that $\mathcal{K} \approx K_X+(n-2)L$ is ample on X. Let σ be the rational number such that $K_X+\sigma\mathcal{K}$ is nef and not ample. Thus

$$K_X+(\sigma(n-2)/(1+\sigma))L = (1/(1+\sigma)(K_X+\sigma\mathcal{K}))$$

is nef and not ample. Furthermore there is a morphism, $\phi : X \to Y$, and a positive integer m such that $m(K_X+\sigma\mathcal{K}) \approx \phi^*H$ where H is ample and $m(K_X+\sigma\mathcal{K})$ is Cartier. Note that $t = \sigma(n-2)/(1+\sigma)$, and that N can be taken to be any positive integer such that N divides m and $N/(1+m)$ is integral.

If $\dim\phi(X) \neq 0$, then note that the general fiber, F, of ϕ is smooth, since X has isolated singularities, and $K_F \approx -tL_F$. Thus $t \geq n/2+1$ implies that $t > \dim F/2+1$. Also L_F is ample on F since $K_F+\sigma\mathcal{K}_F$ is trivial where $\mathcal{K}_F$ is the restriction of $\mathcal{K}$ to F and $\sigma = t/(n-2-t)$. Therefore by the argument of (3.1.1.1) we see that F and hence X is covered by lines relative to L. Thus $(X, L) \cong (X',L') \cong (X^\wedge,L^\wedge)$ by combining (A.1.3) and (A.1.1). Q.E.D.

(A.4.1) Remark. We can still say something if $\dim Y = 0$ in the above theorem. We know that $|2\mathcal{K}|$ is basepoint free whenever the second reduction exists ([S3], § 2). Since Sing(X) is a finite set of points, we can choose a smooth $A \in |2\mathcal{K}|$ which doesn't meet Sing(X). The assumption $\dim Y = 0$ implies that $(K_X+tL)_A \sim \mathcal{O}_A$. Therefore, $K_A+(3t-2(n-2))L_A \sim \mathcal{O}_A$. So if $3t > 2(n-2)$ then $K_A+(3t-2(n-2))L_A \approx \mathcal{O}_A$ (see e.g. [BFS1], (0.7)).

If A is covered by lines relative to L, then so is X. Note that by (3.1.1.1), A is covered by lines if $3t-2(n-2) > (n-1)/2+1$ or, equivalently, $t \geq 5n/6-1$. So this inequality implies that $(X, L) \cong (X', L') \cong (X^\wedge, L^\wedge)$ by Theorem (A.1) as well as $K_X \approx -tL$ and $\mathrm{Pic}(X) \cong \mathbf{Z}$ by (3.1.1.2).

(A.4.2) Corollary. *Assume* $L^\wedge$ *is very ample. Assume also that the* 2nd *reduction* $(X,\mathcal{K})$ *of* $(X^\wedge,L^\wedge)$ *exists and* $n \geq 8$. *Then either* $K_X+(n-3)L$ *is nef and big,* $K_X+(n-3)L \approx \mathcal{O}_X$ *or* $(X^\wedge,L^\wedge) \cong (X',L') \cong (X,L)$ *and the morphism* ϕ *associated to* $K_X+(n-3)L$ *is a Mori contraction.*

Proof. By Theorem (3.1) of [BSF1] we know that $K_X+(n-3)L$ is nef when $n \geq 7$ and if not big then L is a line bundle and X is Gorenstein. Furthermore $K_X+(n-3)L \approx \mathcal{O}_X$ if the morphism ϕ associated to $K_X+(n-3)L$ has 0-dimensional image. Thus we can assume that $K_X+(n-3)L$ is nef and ϕ has a lower positive dimensional image. Therefore by Theorem (A.4) we are done if $n-3 \geq n/2+1$, i.e. $n \geq 8$. Q.E.D.

(A.4.3) Remark. Note that if $\dim\phi(X) = 0$ in (A.4.2), then by the Remark (A.4.1) we have $\mathrm{Pic}(X) \cong \mathbf{Z}$ and $(X,L) \cong (X',L') \cong (X^\wedge,L^\wedge)$ if $n-3 \geq 5n/6-1$, or $n \geq 12$. ■

The above results suggest the following conjecture.

(A.5) ***Conjecture.*** Notation as above. Assume $L^\wedge$ is very ample on $X^\wedge$. If

$$t > n/2+1 \quad \text{and} \quad 0 \le \kappa(K_{X^\wedge}+tL^\wedge, X^\wedge) < n$$

then $K_{X^\wedge}+tL^\wedge$ is nef and the associated morphism $\phi : X \to Y$ with $K_{X^\wedge}+tL^\wedge \approx \phi^*H$ and H an ample Q-Cartier divisor, is a Mori contraction of fiber type. In particular $\kappa(K_{X^\wedge}+tL^\wedge, X^\wedge) = \dim Y \le n - t + 1$.

(A.5.1) Remark. True (and reckless) optimists would conjecture further:
(A.5.1.1) $K_{X^\wedge}+tL^\wedge$ is spanned under the above conditions with H very ample except for "obvious" exceptions.
(A.5.1.2) In fact the conjecture (A.5) is true if

$$t > (n - \kappa(K_{X^\wedge}+ tL^\wedge, X^\wedge))/2 + 1 \quad \text{and} \quad 0 \le \kappa(K_{X^\wedge}+ tL^\wedge, X^\wedge) < n.$$

REFERENCES

[BFS1] M.Beltrametti, M.L.Fania, A.J.Sommese, *On the adjunction theoretic classification of projective varieties*, Math. Ann., **290** (1991), 31–62.

[BFS2] M.Beltrametti, M.L.Fania, A.J.Sommese, *On the discriminant variety of a projective manifold*, preprint.

[BS1] M.Beltrametti, A.J.Sommese, *New properties of special varieties arising from adjunction theory*, to appear in J. Math. Soc. Japan.

[BS2] M.Beltrametti, A.J.Sommese, *Comparing the classical and the adjunction theoretic definition of scrolls*, to appear in the Proceedings of the 1990 Cetraro Conference "Geometry of Complex Projective Varieties".

[E1] L.Ein, *Varieties with small dual varieties*, I, Invent. Math., **86** (1986), 63–74.

[E2] L.Ein, *Varieties with small dual varieties*, II, Duke Math. Jour., **52** (1985), 895–907.

[F] M.L.Fania, *Configurations of −2 rational curves on sectional surfaces of* n*-folds*, Math. Ann., **275** (1986), 317–325.

[Fu1] T. Fujita, *On polarized manifolds whose adjoint bundles are not semipositive*, Algebraic Geometry, Sendai, 1985, Advanced Studies in Pure Math. **10** (1987), 167–178.

[Fu2] T.Fujita, *On adjoint bundles of ample vector bundles*, preprint.

[I] P.Ionescu, *Generalized adjunction and applications*, Math. Proc. Cambridge, Phil. Soc., **99** (1986), 457–472.

[KMM] Y.Kawamata, K.Matsuda, K.Matsuki, *Introduction to the minimal model problem*, Algebraic Geometry, Sendai, 1985, Advanced Studies in Pure Math., **10** (1987), 283–360.

[L] R. Lazarsfeld, *Some applications to the theory of positive vector bundles*, "Complete Intersections", Proceedings Acireale (1983), ed. by S. Greco and R. Strano, Lecture Notes in Math., **1092**, Springer-Verlag (1984), 29–61.

[M1] S.Mori, *Projective manifolds with ample tangent bundle*, Ann. Math., **110** (1979), 593–606.

[M2] S.Mori, *Threefolds whose canonical bundles are not numerically effective*, Ann. Math., **116** (1982), 133–176.

[M3] S.Mori, *Hartshorne conjecture and extremal ray*, Sugaku Exp., **0** (1988), 15–37.

[MM] S.Mori, Y.Miyaoka, *A numerical criterion for uniruledness*, Ann. Math., **124** (1986), 65–69.

[Mu] S.Mukai, *Birational geometry of algebraic varieties. Open problems.* The 23rd International Symposium of the Division of Mathematics of the Taniguchi Foundation, Katata, August 1988.

[OSS] C.Okonek, M.Schneider, H.Spindler, *Vector Bundles on Complex Projective Spaces*, Progress in Math., **3** (1980), Birkhäuser.

[S1] A.J.Sommese, *Configuration of −2 rational curves on hyperplane sections of projective threefolds*, Classification of Algebraic and Analytic Manifolds, (ed. K.Ueno), Progress in Mathematics, **39** (1983), Birkhäuser.

[S2] A.J.Sommese, *On the adjunction theoretic structure of projective varieties*, Complex Analysis and Algebraic Geometry, Proceedings Göttingen 1985, ed. by H. Grauert, Lecture Notes in Math., **1194**, Springer-Verlag (1986), 175–213.

[S3] A.J.Sommese, *On the nonemptiness of the adjoint linear system of a hyperplane section of a threefold*, Jour. für die reine und angew. Math., **402** (1989), 211–220; erratum, Jour. für die reine und angew. Math., **411** (1990), 122–123.

[W1] J.A.Wiśniewski, *Length of extremal rays and generalized adjunction*, Math. Z., **200** (1989), 409–427.

[W2] J.A.Wiśniewski, *On contractions of extremal rays of Fano manifolds*, to appear in Jour. für die reine und angew. Math.

[W3] J.A.Wiśniewski, *On a conjecture of Mukai*, manuscripta math., **68** (1990), 135–141.

[W4] J.A.Wiśniewski, *On Fano manifolds of large index*, manuscripta math., **70** (1991), 145–152.

Mauro C.Beltrametti
Dipartimento di Matematica
Università degli Studi di Genova
Via L.B.Alberti, 4, I-16132 GENOVA (ITALY)
MBELTRA@IGECUNIV.bitnet

Andrew J. Sommese
Department of Mathematics
University of Notre Dame
NOTRE DAME, Indiana 46556, U.S.A.
SOMMESE@IRISHMVS.bitnet

Jarosław A.Wiśniewski
Institute of Mathematics
Warsaw University
Palac Kultury 9 p.
00-901 WARSZAWA, POLAND

THE STABILITY OF CERTAIN VECTOR BUNDLES ON $\mathbb{P}^n$

Guntram Bohnhorst and Heinz Spindler

FB Mathematik/Informatik, Albrechtstr.28, D-4500 Osnabrück

The starting point of this work was the question raised by G.Ottaviani on the Bayreuth Conference, whether the Schwarzenberger bundles on $\mathbb{P}^n$ ([S], [S-T]) are stable. Indeed, this is true.

More general, we show that the stability of rank n vector bundles $\mathcal{E}$ on $\mathbb{P}^n$ with minimal resolution

$$0 \longrightarrow \bigoplus_{i=1}^{k} \mathcal{O}(a_i) \longrightarrow \bigoplus_{j=1}^{n+k} \mathcal{O}(b_j) \longrightarrow \mathcal{E} \longrightarrow 0$$

can be characterized by a simple inequality in the a_i and b_j only (Theorem 2.7). The proof of this mainly consists of a spectral sequence argument applied to the exterior powers of the sequence above.

We end this paper with some remarks on moduli spaces of these vector bundles.

After having finished the main part of this work, G. Trautmann informed us that the special case of Schwarzenberger bundles also has been treated by H. Völlinger ([Vö]). His method however is completely different from ours. Recently Ancona and Ottaviani proved the stability of the Schwarzenberger bundles independently in a similar way ([A-O]). But it seems that their method doesn't give the general result.

1 Preliminaries

Throughout this paper let K be an algebraically closed field of characteristic 0. $\mathbb{P}^n$ denotes the n-dimensional projective space over K and $S = K[z_0, \ldots, z_n]$ its homogeneus coordinate ring. The notion stable/semistable always means μ-stable/μ-semistable in the sense of Maruyama ([Ma]). For a coherent sheaf $\mathcal{E}$ on $\mathbb{P}^n$ the graded S-modul $\bigoplus_{m\in\mathbb{Z}} \mathrm{H}^q(\mathcal{E}(m))$ is denoted by $\mathrm{H}^q_*(\mathcal{E})$. For example, $\mathrm{H}^0_*(\mathcal{O}(m))$ is the twisted modul $S(m)$ with $S(m)_d = S_{m+d}$. [1]

We have the following useful proposition (compare [H])

[1]Acknowledgment: We thank M. Schneider for pointing out a mistake in the first version of this paper.

1.1 Proposition. Let $\mathcal{E}$ be a vector bundle on $\mathbb{P}^n$ of rank r. If $H^0(\mathbb{P}^n, \Lambda^l \mathcal{E}(m)) = 0$ for all $m \leq -l\,\mu(\mathcal{E})$ (resp. $m < -l\,\mu(\mathcal{E})$) and for all $1 \leq l < r$, then $\mathcal{E}$ is stable (resp. semistable). Here and in the rest of this paper $\Lambda^l \mathcal{E}(m)$ means $(\Lambda^l \mathcal{E})(m)$. Also $\mu(\mathcal{E}) = c_1(\mathcal{E})/\operatorname{rk}(\mathcal{E})$ as usual.-

1.2 Example. For $\mathcal{E} = \Omega^1_{\mathbb{P}^n}$ the cohomology $H^q_*(\Lambda^l \mathcal{E})$ is known completely (comp. [O-S-S], p. 8). It follows from this that $l+1$ is the smallest twist m such that $H^0(\Lambda^l \mathcal{E}(m)) \neq 0$. On the other hand $c_1(\mathcal{E}) = -n-1$ and therefore $-l\,\mu(\mathcal{E}) = l + \frac{l}{n}$. By 1.1 $\Omega^1_{\mathbb{P}^n}$ is stable. This is an easy argument for the stability of the tangent bundle $T\mathbb{P}^n$.-

We remark that for stable bundles the converse of 1.1 is not true, as the example of the nullcorrelation bundles shows. Let $\mathcal{E}$ be a stable bundle on $\mathbb{P}^n$ and $0 < l < \operatorname{rk}\mathcal{E}$. Then one knows that $\Lambda^l \mathcal{E}$ is a direct sum of stable bundles $\mathcal{E}_1, \ldots, \mathcal{E}_\rho$ with $\mu(\mathcal{E}_i) = l\,\mu(\mathcal{E})$ for $i = 1, \ldots, \rho$ ([D], [U-Y], [Ko], [L] and the Lefshetz principle). If $l\mu(\mathcal{E})$ is an integer, say $m = l\,\mu(\mathcal{E})$, then the line bundle $\mathcal{O}(m)$, which of course is stable, may occur in this decomposition. Therefore there is a well defined number $s \geq 0$ such that the trivial bundle $\mathcal{O}^s$ is a direct summand of $\Lambda^l \mathcal{E}(-m)$ and $H^0(\Lambda^l \mathcal{E}(-m)) \cong H^0(\mathcal{O}^s)$. Recently Ancona and Ottaviani have shown that the symplectic special instanton bundles on $\mathbb{P}^{2n+1}$ are stable although their second exterior powers admit $\mathcal{O}$ as a direct summand ([A-C]). However, if $\mathcal{E}$ is semistable, then $\Lambda^l \mathcal{E}$ is semistable, too. Therefore we have $H^0(\mathbb{P}^n, \Lambda^l \mathcal{E}(m)) = 0$ for any $m < -l\,\mu(\mathcal{E})$ in this case.

We now state some more or less known facts about resolutions of vector bundles.

1.3 Definition. Let $\mathcal{E}$ be a vector bundle on $\mathbb{P}^n$. A resolution of $\mathcal{E}$ is an exact sequence $0 \to \mathcal{F}_d \to \mathcal{F}_{d-1} \to \cdots \to \mathcal{F}_1 \to \mathcal{F}_0 \to \mathcal{E} \to 0$, where every $\mathcal{F}_i$ splits as a direct sum of line bundles. We call the minimal number of such resolutions the homological dimension of $\mathcal{E}$ and denote it by $\operatorname{hd}(\mathcal{E})$.-

By applying the functor $M \mapsto \tilde{M}$ (see [Ha]) every free resolution of the graded S-module $M = H^0_*(\mathcal{E})$ induces a resolution of $\mathcal{E}$ in the above sense, but not conversely. It follows that $\operatorname{hd}(\mathcal{E}) \leq \operatorname{hd}(M) \leq n$.

1.4 Proposition. Let $\mathcal{E}$ be a vector bundle on $\mathbb{P}^n$. Then

$$\operatorname{hd}(\mathcal{E}) \leq d \iff H^q_*(\mathcal{E}) = 0 \quad \forall\ 1 \leq q \leq n-d-1.$$

For the proof we need

1.5 Lemma. Let $\mathcal{E}$ be a vector bundle on $\mathbb{P}^n$. Then there exist a splitting vector bundle $\mathcal{F} = \bigoplus_{i=1}^k \mathcal{O}(a_i)$ and an epimorphism $\phi : \mathcal{F} \to \mathcal{E}$ such that

$$\mathrm{H}^0(\phi)(m) : \mathrm{H}^0(\mathcal{F}(m)) \longrightarrow \mathrm{H}^0(\mathcal{E}(m))$$

is surjective for all $m \in \mathbb{Z}$.-

Proof. Since $\mathcal{E}$ is locally free, there is an integer m_0 such that $\mathrm{H}^0(\mathcal{E}(m)) = 0$ for all $m \leq m_0$. Therefore $\mathrm{H}^0_*(\mathcal{E}) = \bigoplus_{m \geq m_0} \mathrm{H}^0(\mathcal{E}(m))$ is a finitely generated S-modul and thus there is a surjective homomorphism $\bigoplus_{i=1}^k S(a_i) \to \mathrm{H}^0_*(\mathcal{E})$, which induces an epimorphism of the desired form.-

Proof of 1.4. We proof both directions by induction on d. For $d = 0$ the assertion is Horrocks splitting theorem. So we assume $d \geq 1$.

"$\Rightarrow$": Let $0 \to \mathcal{F}_d \to \mathcal{F}_{d-1} \to \cdots \to \mathcal{F}_1 \to \mathcal{F}_0 \xrightarrow{\epsilon} \mathcal{E} \to 0$ be a resolution of $\mathcal{E}$. Set $\tilde{\mathcal{E}} = \ker \epsilon$. Then $\mathrm{hd}(\tilde{\mathcal{E}}) \leq d-1$. By induction hypothesis we have $\mathrm{H}^q_*(\tilde{\mathcal{E}}) = 0 \quad \forall\ 1 \leq q \leq n-d$. The claim now follows from the exact sequence $0 \to \tilde{\mathcal{E}} \to \mathcal{F}_0 \to \mathcal{E} \to 0$.

"$\Leftarrow$": Choose an epimorphism $\epsilon : \mathcal{F}_0 \to \mathcal{E}$ as in 1.5 and set $\tilde{\mathcal{E}} = \ker \epsilon$. The same short exact sequence as above gives $\mathrm{H}^q_*(\tilde{\mathcal{E}}) = 0 \quad \forall\ 2 \leq q \leq n-d$. By the construction of $\mathcal{F}_0$ also $\mathrm{H}^1_*(\tilde{\mathcal{E}}) = 0$. By induction hypothesis $\mathrm{hd}(\tilde{\mathcal{E}}) \leq d-1$ and thus $\mathrm{hd}(\mathcal{E}) \leq d$.-

1.6 Corollary. Let $\mathcal{E}$ be a vector bundle on $\mathbb{P}^n$ and $M = \mathrm{H}^0_*(\mathcal{E})$. Then $\mathrm{hd}(\mathcal{E})$ coincides with the homological dimension of the graded S-module M.

Proof. Set $d = \mathrm{hd}(\mathcal{E})$. We have to show that $\mathrm{hd}(M) \leq d$. If $\mathcal{F}_.$ is a resolution of $\mathcal{E}$ then in general $\mathrm{H}^0_*(\mathcal{F}_.)$ is not a free resolution of M. But the proof of 1.4 shows, that we always can construct a resolution $\mathcal{F}_.$ of length d such that this is true.-

From 1.4 and the splitting criterion of Griffith and Evans ([E-G], Thm. 2.4) we get

1.7 Corollary. Let $\mathcal{E}$ be a non splitting vector bundle on $\mathbb{P}^n$. Then

$$\mathrm{rk}(\mathcal{E}) \geq n + 1 - \mathrm{hd}(\mathcal{E}).$$

We denote by $\mathcal{M}$ the moduli space $\mathcal{M}_{\mathbb{P}^n}(r; c_1, \ldots, c_n)$ of stable vector bundles with rank r and Chern classes $c_1, \ldots, c_n$ and by $\mathcal{M}^d$ the subspace $\{\ [\mathcal{E}] \in \mathcal{M} \mid \mathrm{hd}(\mathcal{E}) \leq d\ \}$. From 1.4 and the semicontinuity theorem one gets immediately

1.8 Theorem. $\mathcal{M}^d$ is a Zariski open subset of $\mathcal{M}$.

2 Vector Bundles on $\mathbb{P}^n$ of Homological Dimension 1

We now consider vector bundles $\mathcal{E}$ on $\mathbb{P}^n$ with $\mathrm{hd}(\mathcal{E}) = 1$ and the smallest possible rank, i.e. $\mathrm{rk}(\mathcal{E}) = n$. In particular $n \geq 2$. So the following situation is assumed

2.1 There is a resolution

$$\mathcal{F}_. : \quad 0 \longrightarrow \mathcal{F}_1 \xrightarrow{\phi} \mathcal{F}_0 \longrightarrow \mathcal{E} \longrightarrow 0$$

of $\mathcal{E}$ with $\mathcal{F}_1 = \bigoplus_{i=1}^{k} \mathcal{O}(a_i)$ and $\mathcal{F}_0 = \bigoplus_{j=1}^{n+k} \mathcal{O}(b_j)$. It follows that

$$c_1(\mathcal{E}) = \sum_{j=1}^{n+k} b_j - \sum_{i=1}^{k} a_i \quad \text{and} \quad \mu(\mathcal{E}) = \frac{1}{n}\, c_1(\mathcal{E}).$$

Œ $a_1 \geq a_2 \geq \ldots \geq a_k, \quad b_1 \geq b_2 \geq \ldots \geq b_{n+k}$.
The homomorphism ϕ is given as a matrix $\phi = (\phi_{ji})$, where $\phi_{ji} \in S$ is a homogeneous polynomial of degree $b_j - a_i$.-

The resolutions of this kind form a category. If $\tilde{\mathcal{F}}_. : \quad 0 \longrightarrow \tilde{\mathcal{F}}_1 \xrightarrow{\tilde{\phi}} \tilde{\mathcal{F}}_0 \longrightarrow \tilde{\mathcal{E}} \longrightarrow 0$ is another resolution, a homomorphism $\rho : \tilde{\mathcal{F}}_. \to \mathcal{F}_.$ is a pair $\rho = (\alpha, \beta) \in \mathrm{Hom}(\tilde{\mathcal{F}}_1, \mathcal{F}_1) \times \mathrm{Hom}(\tilde{\mathcal{F}}_0, \mathcal{F}_0)$ with $\phi\alpha = \beta\tilde{\phi}$. ρ induces a unique homomorphism $\overline{\rho} : \tilde{\mathcal{E}} \to \mathcal{E}$ such that the diagram

$$\begin{array}{ccccccccc}
0 & \longrightarrow & \tilde{\mathcal{F}}_1 & \xrightarrow{\tilde{\phi}} & \tilde{\mathcal{F}}_0 & \longrightarrow & \tilde{\mathcal{E}} & \longrightarrow & 0 \\
 & & \downarrow \alpha & & \downarrow \beta & & \downarrow \overline{\rho} & & \\
0 & \longrightarrow & \mathcal{F}_1 & \xrightarrow{\phi} & \mathcal{F}_0 & \longrightarrow & \mathcal{E} & \longrightarrow & 0
\end{array}$$

commutes. It is not difficult to show that every homomorphism from $\tilde{\mathcal{E}}$ to $\mathcal{E}$ is obtained in this way.

2.2 Example. The Schwarzenberger bundle $\mathcal{S}_k$ is defined by the exact sequence

$$0 \longrightarrow \mathcal{O}(-1)^{k+1} \xrightarrow{\phi} \mathcal{O}^{n+k+1} \longrightarrow \mathcal{S}_k \longrightarrow 0,$$

where ϕ is given by the $(n+k+1) \times (k+1)$ – matrix

$$\begin{pmatrix} z_0 & \cdots & 0 \\ \vdots & \ddots & \vdots \\ z_n & & z_0 \\ \vdots & \ddots & \vdots \\ 0 & \cdots & z_n \end{pmatrix} .-$$

Let $a = (a_1, \ldots, a_k)$ and $b = (b_1, \ldots, b_{n+k})$ with $a_1 \geq \ldots \geq a_k$, $b_1 \geq \ldots \geq b_{n+k}$. The pair (a, b) is called **admissible** if $a_1 < b_{n+1}, a_2 < b_{n+2}, \ldots, a_k < b_{n+k}$. The resolution is called **minimal**, if there is no commutative diagram

$$\begin{array}{ccccccccc} 0 & \longrightarrow & \tilde{\mathcal{F}}_1 & \xrightarrow{\tilde{\phi}} & \tilde{\mathcal{F}}_0 & \longrightarrow & \mathcal{E} & \longrightarrow & 0 \\ & & \cap & & \cap & & \| & & \\ 0 & \longrightarrow & \mathcal{F}_1 & \xrightarrow{\phi} & \mathcal{F}_0 & \longrightarrow & \mathcal{E} & \longrightarrow & 0 \end{array}$$

such that $\tilde{\mathcal{F}}_i$ is a non trivial summand of $\mathcal{F}_i$ for $i = 0, 1$ and also $0 \to \tilde{\mathcal{F}}_1 \to \tilde{\mathcal{F}}_0 \to \mathcal{E} \to 0$ is a resolution. Obviously any resolution can be reduced to a minimal one.

2.3 Proposition. The resolution 2.1 is minimal if and only if the following conditions hold

(a) (a, b) is admissible.

(b) $\phi_{ji} = 0$ for all i, j with $b_j = a_i$.

Furthermore, if the resolution $\mathcal{F}_.$ is minimal, then k is minimal and $\mathcal{F}_.$ is unique up to isomorphism. Especially (a, b) is uniquely determined by $\mathcal{E}$ in this case and called the **type** of $\mathcal{E}$.

Proof. " $\Longrightarrow$ " : We may transform the matrix ϕ by elementary column or row transformations. If therefore ϕ_{ji} is a nonzero constant (especially $a_i = b_j$), we may assume that

$$\phi = \begin{pmatrix} \phi_{11} & \cdots & 0 & \cdots & \phi_{1k} \\ \vdots & & \vdots & & \vdots \\ 0 & \cdots & 1 & \cdots & 0 \\ \vdots & & \vdots & & \vdots \\ \phi_{n+k,1} & \cdots & 0 & \cdots & \phi_{n+k,k} \end{pmatrix} \begin{matrix} \\ \\ - j\text{-th} \\ \\ \\ \end{matrix}$$

$$\quad\quad\quad\quad i\text{-th}$$

But then we can cancell the i -th column and the j -th row of ϕ, i.e. the resolution was not minimal. This proves (b).

For the proof of (a) consider some $1 \leq l \leq k$. Let s be the maximal number such that $(\phi_{s1}, \ldots, \phi_{sl})$ is nonzero. Let $\psi : \bigoplus_{i=1}^{l} \mathcal{O}(a_i) \to \bigoplus_{j=1}^{s} \mathcal{O}(b_j)$ be the homomorphism $\psi = (\phi_{ji})_{\substack{1 \leq i \leq l \\ 1 \leq j \leq s}}$. With ϕ also ψ is fiberwise injective and therefore defines a vector bundle $\tilde{\mathcal{E}} :=$ coker ψ. Then $\mathrm{hd}(\tilde{\mathcal{E}}) \leq 1$. In the case $\mathrm{hd}(\tilde{\mathcal{E}}) = 1$ $\mathrm{rk}\,\tilde{\mathcal{E}} = s - l \geq n$ by 1.7. By construction there is some $i \leq l$ with $\phi_{si} \neq 0$. It follows that $a_l \leq a_i < b_s \leq b_{n+l}$.

Now assume that $\tilde{\mathcal{E}}$ splits as a direct sum of line bundles. Then the sequence $0 \to \bigoplus_{i=1}^{l} \mathcal{O}(a_i) \xrightarrow{\psi} \bigoplus_{j=1}^{s} \mathcal{O}(b_j) \to \tilde{\mathcal{E}} \to 0$ splits and thus ψ can be extended to a non-singular matrix $A = (\psi, \tilde{\psi})$, i.e. $\det A \in K \setminus \{0\}$. Expanding out this determinant gives a contradiction to (b).

" $\Longleftarrow$ " : Let $\mathcal{F}_.$: $0 \longrightarrow \mathcal{F}_1 \xrightarrow{\phi} \mathcal{F}_0 \longrightarrow \mathcal{E} \longrightarrow 0$ be a resolution with (a), (b). We compare this with a minimal resolution $\tilde{\mathcal{F}}_.$: $0 \longrightarrow \tilde{\mathcal{F}}_1 \xrightarrow{\tilde{\phi}} \tilde{\mathcal{F}}_0 \longrightarrow \tilde{\mathcal{E}} \longrightarrow 0$. There are homomorphisms $\rho : \mathcal{F}_. \to \tilde{\mathcal{F}}_.$, $\tilde{\rho} : \tilde{\mathcal{F}}_. \to \mathcal{F}_.$, which induce the identity on $\mathcal{E}$. The composition $\tilde{\rho}\rho$ gives a commutative diagram

$$\begin{array}{ccccccccc} 0 & \longrightarrow & \mathcal{F}_1 & \xrightarrow{\phi} & \mathcal{F}_0 & \longrightarrow & \mathcal{E} & \longrightarrow & 0 \\ & & \downarrow \alpha & & \downarrow \beta & & \| & & \\ 0 & \longrightarrow & \mathcal{F}_1 & \xrightarrow{\phi} & \mathcal{F}_0 & \longrightarrow & \mathcal{E} & \longrightarrow & 0 \end{array}$$

Then $\mathrm{Im}(\beta - \mathrm{id}_{\mathcal{F}_0}) \subset \mathcal{F}_1$ and therefore $\beta = \mathrm{id}_{\mathcal{F}_0} + \phi\psi$ for some $\psi \in \mathrm{Hom}(\mathcal{F}_0, \mathcal{F}_1)$.

We claim that $\phi\psi$ is nilpotent. For the proof we choose $i \geq j$ and $1 \leq l \leq k$. If $\phi_{il} \neq 0$ then $b_i - a_l = \deg \phi_{il} \geq 1$. Therefore $a_l - b_j \leq a_l - b_i < 0$ and thus $\psi_{lj} = 0$. It follows that $(\phi\psi)_{ij} = 0$ for all $i \geq j$. So $\phi\psi$ is nilpotent. It follows that β is an isomorphism.

We conclude that $\tilde{\rho}\rho$ and by the same reasoning $\rho\tilde{\rho}$ are isomorphisms. But then, already ρ is an isomorphism. So $\mathcal{F}_.$ is minimal.

Moreover, the proof shows that up to isomorphism there is only one minimal resolution.-

2.4 Remark. If (a, b) is admissible then there exists a rank n vector bundle $\mathcal{E}$ on $\mathbb{P}^n$ with $\mathrm{hd}(\mathcal{E}) = 1$, which is of type (a, b).

Proof. Define $\phi : \mathcal{F}_1 \to \mathcal{F}_0$ by the matrix

$$\begin{pmatrix} z_0^{b_1 - a_1} & \dots & 0 \\ \vdots & & \vdots \\ \vdots & & z_0^{b_k - a_k} \\ z_n^{b_{n+1} - a_1} & & \vdots \\ \vdots & & \vdots \\ 0 & \dots & z_n^{b_{n+k} - a_k} \end{pmatrix}$$

Then ϕ is fiberwise injective and thus defines a vector bundle $\mathcal{E} = \mathrm{coker}\,\phi$ of the required type. Every constant entry of ϕ is zero. Therefore $\mathcal{E}$ is of homological dimension 1 and of type (a, b) by (the proof of) 2.3.-

2.5 Proposition. For every vector bundle $\mathcal{E}$ as in 2.1 the following hold.

(a) $\quad \mathrm{H}^q_*(\Lambda^l \mathcal{E}) = 0 \quad \forall\, 1 \leq q \leq n - l - 1.$

(b)

$$\mathrm{h}^0(\Lambda^l \mathcal{E}\,(m)) = \sum_{1 \leq j_1 < \ldots < j_l \leq n+k} \binom{n + m + b_{j_1} + \ldots + b_{j_l}}{n}' - \sum_{p=1}^{l} \sum_{1 \leq i_1 < \ldots < i_p \leq k} \mathrm{h}^0(\Lambda^{l-p} \mathcal{E}\,(m + a_{i_1} + \cdots + a_{i_p}))$$

for $0 \leq l \leq n - 1$.

(c)

$$h^0(\mathcal{E}(m)) = \sum_{j=1}^{n+k} \binom{n+m+b_j}{n}' - \sum_{i=1}^{k} \binom{n+m+a_i}{n}'.$$

Here $\binom{m}{n}'$ means $\binom{m}{n}' = \binom{m}{n}$ if $m \geq n$ and $\binom{m}{n}' = 0$ if $m < n$.-

For the proof of this we need

2.6 Lemma. Let $0 \to \mathcal{E}' \to \mathcal{E} \to \mathcal{E}'' \to 0$ be an exact sequence of vector bundles on a scheme X. Then for any $l \leq \operatorname{rk} \mathcal{E}$ there is a spectral sequence

$$E_1^{pq} = H^{p+q}(\Lambda^p \mathcal{E}' \otimes \Lambda^{l-p} \mathcal{E}''(m)) \implies E^{p+q} = H^{p+q}(\Lambda^l \mathcal{E}(m))$$

Proof. The assertion follows from the fact that there is a filtration $\Lambda^l \mathcal{E} = \mathcal{F}_0 \supset \mathcal{F}_1 \supset \ldots \supset \mathcal{F}_l \supset \mathcal{F}_{l+1} = 0$ with $\mathcal{F}_p/\mathcal{F}_{p+1} \cong \Lambda^p \mathcal{E}' \otimes \Lambda^{l-p} \mathcal{E}''$. This is excercise 5.16(d) in [Ha]. For the proof of it we fix some $x \in X$ and put $R := \mathcal{O}_{X,x}$, $M := \mathcal{E}_x$, $M' := \mathcal{E}'_x$, $M'' := \mathcal{E}''_x$. We consider M' as a submodule of M and have the exact sequence $0 \to M' \xrightarrow{\iota} M \xrightarrow{\tau} M'' \to 0$. This induces the sequence $0 \to \Lambda^l M' \xrightarrow{\Lambda^l \iota} \Lambda^l M \xrightarrow{\Lambda^l \tau} \Lambda^l M'' \to 0$, but for $l \geq 2$ this is no longer exact in the middle, whereas the exactness at the edge terms is preserved. Setting $K_l := \ker \Lambda^l \tau$ we get $\Lambda^l M'' = \Lambda^l M/K_l$. Now define F_p to be the submodule of $\Lambda^l M$ generated by all elements of the form $x_1 \wedge \ldots \wedge x_l$ with $x_1, \ldots, x_p \in M'$. Especially $F_1 = K_l$. Then $\Lambda^l M = F_0 \supset F_1 \supset \ldots \supset F_l \supset F_{l+1} = 0$ and by definition there is a surjective homomorphism $\Lambda^p M' \otimes \Lambda^{l-p} M \to F_p$. By taking suitable bases one computes that the kernel of $\Lambda^p M' \otimes \Lambda^{l-p} M \to F_p/F_{p+1}$ is just $\Lambda^p M' \otimes K_{l-p}$. This gives a canonical isomorphism $\Lambda^p M' \otimes \Lambda^{l-p} M'' \to F_p/F_{p+1}$. If we now set $(\mathcal{F}_p)_x := F_p$ the claim follows very easily.-

Proof of 2.5. (a) We proof this by induction on l. The case $l = 0$ is clear. Therefore we may assume that $l \geq 1$ and the assertion is true for $l' < l$. By 2.6 and 2.1 there is a spectral sequence

$$E_1^{pq} = \bigoplus_{1 \leq i_1 < \ldots < i_p \leq k} H^{p+q}(\Lambda^{l-p} \mathcal{E}(m + a_{i_1} + \cdots + a_{i_p}))$$

which converges to

$$E^q = \bigoplus_{1 \leq j_1 < \ldots < j_l \leq n+k} H^q(\mathcal{O}(m + b_{j_1} + \cdots + b_{j_l})).$$

By induction hypothesis $E_1^{pq} = 0$ for $l - p < l$ and $1 \leq p + q \leq n - (l - p) - 1$ i. e. for $p \geq 1$ and $1 - p \leq q \leq n - l - 1$. Therefore $E_1^{0q} = E_\infty^{0q} = E^q$ for $1 \leq q \leq n - l - 1$. But $E^q = 0$ for $1 \leq q \leq n - 1$. It follows that $H^q(\Lambda^l \mathcal{E}(m)) = E_1^{0q} = 0$ for $1 \leq q \leq n - l - 1$.

(b) The same spectral sequence as in (a) gives us $E_1^{p,-p} = E_\infty^{p,-p}$ for $l \leq n - 1$. Therefore $\dim E^0 = \sum_{p=0}^{l} \dim E_1^{p,-p}$. From this the claimed equation follows.

(c) This follows at once from 2.1.-

2.7 Theorem. Let $\mathcal{E}$ be a rank n vector bundle on $\mathbb{P}^n$ with $\operatorname{hd}\mathcal{E} = 1$ and admissible resolution

$$0 \longrightarrow \bigoplus_{i=1}^{k} \mathcal{O}(a_i) \longrightarrow \bigoplus_{j=1}^{n+k} \mathcal{O}(b_j) \longrightarrow \mathcal{E} \longrightarrow 0.$$

Then the following conditions are equivalent

(a) $\mathcal{E}$ is stable (resp. semistable).

(b) $\mathrm{H}^0(\mathcal{E}(m)) = 0$ for $m \le$ (resp. $<$) $-\mu(\mathcal{E})$.

(c) $b_1 <$ (resp. $\le$) $\mu(\mathcal{E}) = \frac{1}{n}(\sum_{j=1}^{n+k} b_j - \sum_{i=1}^{k} a_i)$

(d) $\mathrm{H}^0(\Lambda^l \mathcal{E}\,(m)) = 0 \quad \forall\, m \le$ (resp. $<$) $-l\,\mu(\mathcal{E}), \quad \forall\, 1 \le l \le n-1$.

If $b_1 = \cdots = b_n$, then $\mathcal{E}$ is stable in any case. In particular all Schwarzenberger bundles $\mathcal{S}_k$ are stable.

Proof. By m_l we denote the smallest twist m such that $\mathrm{H}^0(\Lambda^l \mathcal{E}\,(m)) \ne 0$. By 2.5(c) and (b) we have $m_1 = -b_1$ and $m_l \ge -(b_1 + \cdots + b_l)$ for $1 \le l \le n-1$.

(a) $\Rightarrow$ (b) is known (and true for every vector bundle).

(b) $\Rightarrow$ (c) is evident.

(c) $\Rightarrow$ (d) follows from $m_l \ge -(b_1 + \cdots + b_l) \ge -l\,b_1 \ge$ (resp. $>$) $-\,l\,\mu(\mathcal{E})$.

(d) $\Rightarrow$ (a) is a consequence of 1.1.-

3 Some Remarks on Moduli Spaces

We close this note with some remarks on moduli spaces. We are interested in the open part $\mathcal{M}^1 \subset \mathcal{M} = \mathcal{M}_{\mathbb{P}^n}(n; c_1, \ldots, c_n)$ of stable bundles with homological dimension 1 (comp. 1.8). It is easy to see that the type (a, b) defines a stratification of $\mathcal{M}^1$ by constructible subsets $\mathcal{M}^1(a, b)$, where $\mathcal{M}^1(a, b)$ is the set of isomorphism classes $[\mathcal{E}] \in \mathcal{M}^1$ of bundles $\mathcal{E}$ of type (a, b).

3.1 The pair (a, b) belongs to the finite set $I = I(n; c_1, \ldots, c_n)$ of pairs (a, b) with $a = (a_1, \ldots, a_k)$, $b = (b_1, \ldots, b_{n+k})$ such that

(a) $a_1 \geq \ldots \geq a_k, \quad b_1 \geq \ldots \geq b_{n+k}$.

(b) (a, b) is admissible.

(c) $1 + c_1 t + \cdots + c_n t^n = \prod_{j=1}^{n+k} (1 + b_j t) \prod_{i=1}^{k} (1 + a_i t)^{-1}$.

(d) (a, b) is 'stable', i.e. $b_1 < \frac{c_1}{n}$.

We remark that there is an estimation $1 \leq k \leq k_{max}$, where k_{max} is a bound, which depends on the Chern classes $c_1, \ldots, c_n$ only.-

Let us fix such a pair (a, b) and define $\mathcal{F}_1, \mathcal{F}_0$ as in 2.1. We consider the linear subspace V of $\mathrm{Hom}(\mathcal{F}_1, \mathcal{F}_0)$ consisting of those homomorhisms $\phi : \mathcal{F}_1 \to \mathcal{F}_0$ satisfying condition 2.3(b). $X = \{ \phi \in V \mid \phi \text{ is injective, } \operatorname{coker} \phi \text{ is locally free} \}$ is a Zariski open subspace of V. The cokernel of the universal homomorphism

$$\Phi : \mathrm{pr}_2^* \mathcal{F}_1 \longrightarrow \mathrm{pr}_2^* \mathcal{F}_0 \quad \text{on} \quad X \times \mathbb{P}^n$$

induces a morphism

$$\tau : X \longrightarrow \mathcal{M}.$$

By definition $\mathcal{M}^1(a, b) = \tau(X)$. Especially $\mathcal{M}^1(a, b)$ is a constructible subset of $\mathcal{M}$.

The group $G = \mathrm{Aut}(\mathcal{F}_1) \times \mathrm{Aut}(\mathcal{F}_0)$, which in general is not reductive, operates on X via

$$(\alpha, \beta)\, \phi = \beta\, \phi\, \alpha^{-1}.$$

At least set theoretically τ induces a bijection of the orbit set ${}^{X}\!/_{G}$ onto $\mathcal{M}^1(a, b)$. The argument for this is similar to that in the proof of 2.3. To compute the dimension of the fibers of τ we need information about the isotropy groups G_ϕ at points $\phi \in X$.

3.2 Lemma. The isotropy group G_ϕ of $\phi \in X$ is

$$G_\phi = \{ (\lambda \,\mathrm{id} + \psi\phi, \lambda\, id + \phi\psi) \mid \lambda \in K^*,\ \psi \in \mathrm{Hom}(\mathcal{F}_0, \mathcal{F}_1) \}.$$

Moreover, $\psi\phi$ and $\phi\psi$ are nilpotent for all $\phi \in V$, $\psi \in \mathrm{Hom}(\mathcal{F}_0, \mathcal{F}_1)$.

Proof. The inclusion " $\supset$ " is easy. For the converse inclusion, let $(\alpha,\beta) \in G$. Then we have the commutative diagram

$$\begin{array}{ccccccccc} 0 & \longrightarrow & \mathcal{F}_1 & \xrightarrow{\phi} & \mathcal{F}_0 & \longrightarrow & \mathcal{E} & \longrightarrow & 0 \\ & & \downarrow \alpha & & \downarrow \beta & & \| \wr \overline{\beta} & & \\ 0 & \longrightarrow & \mathcal{F}_1 & \xrightarrow{\phi} & \mathcal{F}_0 & \longrightarrow & \mathcal{E} & \longrightarrow & 0 \end{array}$$

The stability of $\mathcal{E}$ implies that $\overline{\beta} = \lambda$ id for some $\lambda \in K^*$. Then $\operatorname{Im}(\beta - \lambda \operatorname{id}_{\mathcal{F}_0}) \subset \operatorname{Im}\phi$ and therefore there exists a $\psi \in \operatorname{Hom}(\mathcal{F}_0,\mathcal{F}_1)$ with $\beta = \lambda \operatorname{id}_{\mathcal{F}_0} + \phi\psi$. It follows that $\phi\alpha = \beta\phi = \lambda\phi + \phi\psi\phi = \phi(\lambda \operatorname{id}_{\mathcal{F}_1} + \psi\phi)$ i.e. $\alpha = \lambda \operatorname{id}_{\mathcal{F}_1} + \psi\phi$.

The last part follows as in the proof of 2.3.-

Especially all isotropy groups G_ϕ have the same dimension $\dim \operatorname{Hom}(\mathcal{F}_0,\mathcal{F}_1) + 1$ and therefore τ has constant fiber dimension $\dim \operatorname{End}\mathcal{F}_1 + \dim \operatorname{End}\mathcal{F}_0 - \dim \operatorname{Hom}(\mathcal{F}_0,\mathcal{F}_1) - 1$. Since $\tau : X \longrightarrow \overline{\mathcal{M}^1(a,b)}$ is a dominant morphism we finally get

3.3 Proposition. $\overline{\mathcal{M}^1(a,b)}$ is an irreducible algebraic subset of $\mathcal{M}^1$ of dimension

$\dim \operatorname{Hom}(\mathcal{F}_1,\mathcal{F}_0) + \dim \operatorname{Hom}(\mathcal{F}_0,\mathcal{F}_1) - \dim \operatorname{End}(\mathcal{F}_1) - \dim \operatorname{End}(\mathcal{F}_0) + 1 - \#\{(i,j) \mid a_i = b_j\}$.–

We introduce a partial order on I, namely $(\tilde{a},\tilde{b}) \leq (a,b)$ iff there are $\alpha_1,\ldots,\alpha_l \in \mathbb{Z}$ such that up to order

$$\tilde{a} = (a_1,\ldots,a_k,\alpha_1,\ldots,\alpha_l) \quad \text{and} \quad \tilde{b} = (b_1,\ldots,b_{n+k},\alpha_1,\ldots,\alpha_l).$$

The pair $(a,b) \in I$ is called **maximal** iff it is maximal with respect to this order. This is equivalent to $\{a_i\} \cap \{b_j\} = \emptyset$. For every $(\tilde{a},\tilde{b}) \in I$ there is exactly one maximal $(a,b) \in I$ with $(\tilde{a},\tilde{b}) \leq (a,b)$.

3.4 Lemma. Let $(\tilde{a},\tilde{b})$, $(a,b) \in I$. Then we have

$$(\tilde{a},\tilde{b}) \leq (a,b) \quad \Longrightarrow \quad \mathcal{M}^1(\tilde{a},\tilde{b}) \subset \overline{\mathcal{M}^1(a,b)}.$$

Proof. We may assume that $(\tilde{a},\tilde{b})$ is obtained from (a,b) by adding one number α to a and b, i.e. $\tilde{a} = (a_1,\ldots,a_k,\alpha)$, $\tilde{b} = (b_1,\ldots,b_{n+k},\alpha)$ up to order. Let $\mathcal{E}_0 \in \mathcal{M}^1(\tilde{a},\tilde{b})$ be given with minimal resolution

$$0 \longrightarrow \mathcal{F}_1 \oplus \mathcal{O}(\alpha) \xrightarrow{\phi_0} \mathcal{F}_0 \oplus \mathcal{O}(\alpha) \longrightarrow \mathcal{E}_0 \longrightarrow 0$$

and $\phi_0 = \begin{pmatrix} \tilde{\phi} & * \\ * & c \end{pmatrix}$. Proposition 2.3 implies c = 0. Then the family $\phi_t = \begin{pmatrix} \tilde{\phi} & * \\ * & t \end{pmatrix}$ defines a vector bundle $\mathcal{E}$ in a suitable neighborhood T of 0 in the affine line over K. We have $\mathcal{E}_0 \in \mathcal{M}^1(\tilde{a},\tilde{b})$ but $\mathcal{E}_t \in \mathcal{M}^1(a,b)$ for $t \in T \setminus 0$. This proves our assertion.-

3.5 Theorem. Assume that $n \geq 3$.

(a) If $[\mathcal{E}] \in \mathcal{M}^1(a,b)$, then $\dim \overline{\mathcal{M}^1(a,b)} = \mathrm{h}^1(End(\mathcal{E})) - \#\{(i,j) \mid a_i = b_j\}$.

(b) The irreducible components of $\mathcal{M}^1$ are precisely the sets $\overline{\mathcal{M}^1(a,b)}$ corresponding to maximal elements $(a,b) \in I$, and $\mathcal{M}^1$ is smooth at every point $[\mathcal{E}] \in \mathcal{M}^1(a,b)$, (a,b) maximal.

(c) If $n \geq 4$, then $\mathcal{M}^1$ is smooth and $\mathcal{M}^1(\tilde{a},\tilde{b}) \cap \mathcal{M}^1(a,b) = \emptyset$ for $(\tilde{a},\tilde{b}) \not\leq (a,b)$.

Proof. We take a vector bundle $\mathcal{E} \in \mathcal{M}^1(a,b)$ with minimal resolution 2.1. We tensor the sequence dual to 2.1 with $\mathcal{E}$ and get the exact sequence

$$0 \longrightarrow End(\mathcal{E}) \longrightarrow \mathcal{E} \otimes \mathcal{F}_0^* \longrightarrow \mathcal{E} \otimes \mathcal{F}_1^* \longrightarrow 0.$$

From $n \geq 3$ we conclude $\mathrm{H}^1_*(\mathcal{E}) = 0$ and this gives

$$\mathrm{h}^1(End(\mathcal{E})) = \mathrm{h}^0(\mathcal{E} \otimes \mathcal{F}_1^*) - \mathrm{h}^0(\mathcal{E} \otimes \mathcal{F}_0^*) + 1.$$

By tensoring the sequence in 2.1 with $\mathcal{F}_0^*$ or by a direct computation with 2.5(c) we get

$$\mathrm{h}^0(\mathcal{E} \otimes \mathcal{F}_0^*) = \dim \mathrm{End}(\mathcal{F}_0) - \dim \mathrm{Hom}(\mathcal{F}_0, \mathcal{F}_1)$$

and analogous

$$\mathrm{h}^0(\mathcal{E} \otimes \mathcal{F}_1^*) = \dim \mathrm{Hom}(\mathcal{F}_1, \mathcal{F}_0) - \dim \mathrm{End}(\mathcal{F}_1).$$

Combining these equations with 3.3 gives us (a).

(b) is now obvious since $\mathrm{H}^1(End(\mathcal{E}))$ is the Zariski tangent space of $\mathcal{M}^1$ at the point $[\mathcal{E}]$.

(c): The same exact sequence as in (a) gives us $\mathrm{H}^2(End(\mathcal{E})) = 0$ for $n \geq 4$. Therefore $\mathcal{M}^1$ is smooth. Especially the different $\overline{\mathcal{M}^1(a,b)}$ with (a,b) maximal don't intersect.-

For $n = 3$ it may possibly happen that $\mathcal{M}^1(\tilde{a},\tilde{b}) \cap \overline{\mathcal{M}^1(a,b)} \neq \emptyset$ for some $(\tilde{a},\tilde{b})$, $(a,b) \in I$ with $(\tilde{a},\tilde{b}) \not\leq (a,b)$. This phenomenon definitely appears in the case $n = 2$. In this case $\mathcal{M} = \mathcal{M}^1$ is smooth and connected of dimension $4\,c_2 - c_1^2 - 3$. So there is exactly one (a,b) with $\mathcal{M} = \overline{\mathcal{M}^1(a,b)}$ and all other strata lie in the boundary of $\mathcal{M}^1(a,b)$. Of course, this 'generic' (a,b) is maximal, but in general $\mathcal{M} = \mathcal{M}^1$ contains more than one maximal $\mathcal{M}^1(a,b)$. For example, if we consider $\mathcal{M} = \mathcal{M}_{\mathbb{P}^2}(2;0,5)$ with $\dim \mathcal{M} = 17$, then the stratification of $\mathcal{M}$ is given by

a	b		codim $\mathcal{M}^1(a,b)$
(-3,-3,-3)	(-1,-2,-2,-2,-2)	maximal	0
(-4)	(-1,-1,-2)	maximal	2
(-3,-4)	(-1,-1,-2,-3)	not maximal	3
(-2,-5)	(-1,-1,-1,-4)	maximal	5

It is not easy to understand how this happens. We shall study these questions in a subsequent paper.

References

[A-O] Ancona, V.; Ottaviani G.: On the stability of special instanton bundles on $\mathbb{P}^{2n+1}$. preprint 1990

[D] Donaldson, S.K.: Infinite determinants, stable bundles and curvature. Duke Math. J. 54 (1987), 231-247

[E-G] Evans, E.G.; Griffith, P.: The syzygy problem. Ann. Math. 114 (1981), 323-333

[Ha] Hartshorne R.: Algebraic Geometry. Springer 1977

[H] Hoppe, H. J.: Generischer Spaltungstyp und zweite Chernklasse stabiler Vektorraumbündel vom Rang 4 auf $\mathbb{P}^4$. Math.Zeitschr. 187 (1984), 345-360

[Ko] Kobayashi, S.: Differential geometry of complex vector bundles. Princeton Univ. Press 1987.

[L] Lübke, M.: Stability of Einstein-Hermite vector bundles. Manuscr. Math. 42 (1983), 245-257

[Ma] Maruyama, M.: The theorem of Grauer-Mülich-Spindler. Math. Ann. 255 (1981), 317-333

[O-S-S] Okonek, C.; Schneider, M.; Spindler, H.: Vector bundles on complex projective spaces. Progr. in Math. 3, Birkhäuser Boston 1980

[S] Schwarzenberger, R.L.E.: Vector bundles on the projective plane. Proc. London Math. Soc. 11 (1961), 623-640

[S-T] Spindler, H.; Trautmann, G.: Special instanton bundles on $\mathbb{P}_{2N+1}$, their geometry and their moduli. Math. Ann. 286 (1990), 559-592

[U-Y] Uhlenbeck, K.K.; Yau, S.T.: On the existence of Hermite -Yang-Mills connections in stable vector bundles. Comm. Pure Appl. Math. 39 (1986), 257-293

[Vö] Völlinger, H.: Moduli von Kernbündeln auf $\mathbb{P}_N(\mathbb{C})$ ... Dissertation Kaiserslautern 1988

VECTOR BUNDLES, LINEAR SYSTEMS AND EXTENSIONS OF π_1

F. Catanese - F. Tovena
Dipartimento di Matematica, Universita' di Pisa
Via Buonarroti 2, 56127 Pisa (Italy)

§0 Introduction

I. Reider ([Rei]) introduced a new method to prove that certain linear systems on algebraic surfaces are free from base points (respectively very ample).

He uses a construction due to Schwarzenberger, ([Sch1,2], [GH]) producing, if Z is a 0-cycle not imposing independent conditions on the linear system $|K+L|$, a certain rank 2 bundle occurring as an extension

$$0 \rightarrow \mathcal{O}_S \rightarrow \mathcal{E} \rightarrow \mathcal{I}_Z L \rightarrow 0$$

and then derives a contradiction if the vector bundle $\mathcal{E}$ is numerically unstable according to Bogomolov ([Bog]).

As pointed out by D. Kotschick ([Kot]), if the numerical inequality $c_1^2(\mathcal{E}) - 4c_2(\mathcal{E}) \geq 0$ becomes an equality, i.e. $L^2 - 4 \deg Z = 0$, and the Chern class L is divisible by 2, then $\mathcal{E}$ is the twist (by a line bundle) of a vector bundle with trivial Chern classes; hence, by a deep theorem of Donaldson ([Do]), if $\mathcal{E}$ is Mumford-stable with respect to an ample divisor H, then this vector bundle arises from an irreducible SU(2)-representation of the fundamental group $\pi_1(S)$ of the surface.

In fact, cf. [Ko], when one has equality, and one can prove that the bundle $\mathcal{E}$ is stable for some ample divisor H, then the associated projective bundle $P(\mathcal{E})$ arises from a PU(2)-representation of the fundamental group $\pi_1(S)$ of the surface.
We thus get a central extension Γ of $\pi_1(S)$ by a cyclic group of order 2, whose extension class measures the obstruction to lifting the PU(2) to a SU(2)-representation.

We apply this method to the study of bicanonical systems on surfaces S with $K_S^2 = 4$, where $L = K_S$: in this case one cannot have a SU(2)-irreducible representation since the numerical class of $L = K_S$ is not divisible by 2 . Whence, we get in §1 the result that $|2K_S|$ is free from base points if $H^2(\pi_1(S), \mathbb{Z}/2\mathbb{Z})=0$.

This partial result is of some interest in view of the open problem (cf. §1) whether $|2K_S|$ has base points only when $K_S^2 = 1$, and $p_g = 0$ (the only cases which are left open being the ones where $p_g = 0$, $K_S^2 = 2,3,4$).

After the results of §2, we obtain a sharper theorem (Thm. 3.2) implying in particular that $|2K_S|$ is base point free if the pull back of K_S to the universal covering is not 2-divisible, or it gives rise to a trivial extension.

The second section is devoted to the geometrical analysis of the possible central extension of the fundamental group. This problem is treated in a greater generality, by considering the standard m^{th} root extraction covering trick (cf. [Mi2]), under which the pull-back of a divisor L becomes m-divisible; we first show (cf. Lemma 2.1) that in this situation the fundamental group changes up to a central extension by a cyclic group of order dividing m (this argument is essential for the main result of [Mi2]). Later on, we give a complete description of the extension which occurs, in terms of the divisibility properties of the pull-back of L to the universal covering of S.

Our first example where the first homology group would not change, but the fundamental group would, was the case of a $(\mathbb{Z}/2\mathbb{Z})^2$-Galois cover of an Abelian surface: here L gives a polarization of type (1,4) and the fundamental group of the cover is the classical Heisenberg central extension of the fundamental group of the Abelian surface associated to the mod 2 reduction of the alternating form given by the Chern class of L.

We appealed again to Donaldson's theorem in order to calculate the fundamental group of the cover, just by providing the existence of some stable bundle with trivial Chern classes (the ideas here were influenced by the article [BLvS], whose results by the way can also be reproved using the above ideas (cf. work in progress by the second author)).

Later on, guided by the conjecture raised by this nice example, we worked out completely the general case, where we essentially investigate the spectral sequence describing the cohomology of the quotient S in terms of group cohomology.

The main result of the paper is the following

Main Theorem: Let $Y \to X$ be the $(\mathbb{Z}/m\mathbb{Z})^2$-Galois covering given by the mth root extraction of the divisor D.

Then we have a central extension

$$0 \to \mathbb{Z}/r\mathbb{Z} \to \pi_1(Y) \to \pi_1(X) \to 0$$

where, if $\pi: \tilde{X} \to X$ is the universal covering of X, $\tilde{D} = \pi^*(D)$, d the divisibility index of $\tilde{D}$, then r = G.C.D.(m,d) and the extension class in $H^2(G,\mathbb{Z}/r\mathbb{Z})$, $(G = \pi_1(X))$, is given by the Chern class of (–D) modulo r, via the exact sequence

$$0 \to H^2(G,\mathbb{Z}/r\mathbb{Z}) \to H^2(X,\mathbb{Z}/r\mathbb{Z}) \to H^2(\tilde{X},\mathbb{Z}/r\mathbb{Z})^G \quad .$$

Acknowledgements : both authors would like to acknowledge support from M.U.R.S.T. ; the first author would like to acknowledge the warm hospitality of the Max Planck Institute in October '90, where the final part of the paper was prepared.

We want to thank Igor Reider and Ingrid Bauer Kosarew for some useful conversation.

Mathias Kreck showed us kindly how one could avoid the use of Lefschetz duality in 2.18.

Added in proof: we would like to call the reader's attention to related partial results, concerning base points of the bicanonical system, by Weng Lin ([We]).

§1 Bicanonical systems on surfaces of general type.

Let S be a smooth (complete) minimal surface of general type, and consider the bicanonical linear system $|2K_S|$, where K_S is a canonical divisor on S .

Through work of several people (Moishezon [Moi], Kodaira [Kod], Bombieri [Bo1,2], Francia [Fr], Reider [Rei] and others, e.g. [Ca-Ci], [Mi1])) it is known that the above linear system has no base points if p_g is ≥ 1, and also in the case $p_g = 0$, provided $K^2 \geq 5$, and particular cases when $K^2 = 2$ (cf. [Pet2], [Xi] Thm 5.5 page 77).

Since $|2K_S|$ is a pencil exactly when $K_S^2 = 1$, one may ask about the remaining cases $p_g = 0$, $K_S^2 = 3,4$.

There are no known examples of surfaces with invariants $p_g = 0$, $K^2 = 3,4$ such that the bicanonical linear system has base points.

In this paragraph we shall give some sufficient conditions, concerning the fundamental group $\pi_1(S)$, which ensure that $|2K_S|$ be (base point) free.

On the other hand, by looking at some examples of surfaces with the above invariants constructed by Burniat and Keum ([Bu], [Pet], [Ke]), we shall see however that those conditions are not necessary (but, using the results of the next paragraph one gets some weaker sufficient conditions).

As we remarked in the introduction, it remains an interesting question to know whether the case $p_g = 0$ and $K_S^2 = 1$ is indeed the only exception to $|2K_S|$ being (base point) free.

Theorem 1.1 Let S be a minimal surface with $p_g = 0$, $K_S^2 = 4$, and such that no nontrivial central extension Γ

$$1 \to \mathbb{Z}/2\mathbb{Z} \to \Gamma \to \pi_1(S) \to 0$$

of $\pi_1(S)$ has an irreducible SU(2)-representation.
Then the bicanonical linear system $|2K_S|$ is (base point) free.

Corollary 1.2 In particular, the theorem holds if $H^2(\pi_1(S),\mathbb{Z}/2\mathbb{Z}) = 0$ (e.g. if $\pi_1(S)$ is cyclic, or it has odd order).

Proof of Theorem 1.1 The proof of the theorem will be divided in two different cases and will involve some assertions that will be justified at a later time.

Assume that the bicanonical linear system $|2K_S|$ has a base point x. Then (cf. [GH], [Rei]) there exists a vector bundle $\mathcal{E}$ of rank 2 on S occurring as an extension

(1.3) $$0 \to \mathcal{O}_S \to \mathcal{E} \to \mathcal{I}_x(K_S) \to 0$$

where $\mathcal{I}_x$ is the ideal sheaf of the given point x. We will show that (with no assumption on $\pi_1(S)$) the vector bundle $\mathcal{E}$ is K-stable (prop. 1.4) and, furthermore, it is stable with respect to a suitable ample line bundle H on S (see prop. 1.9).

Then, (cf. e.g. [Ko], thm. 10.19, page 236, and thm. 4.7, page 114) $\mathcal{E}$ admits a Hermite-Einstein structure, and, since $c_1^2(\mathcal{E}) - 4\,c_2(\mathcal{E}) = 0$, it is projectively flat, i.e., it comes from an irreducible PU(2)-representation of $\pi_1(S)$. We can lift thus the central extension

$$1 \to \mathbb{Z}/2\mathbb{Z} \to SU(2) \to PU(2) \to 0$$

to obtain

$$1 \to \mathbb{Z}/2\mathbb{Z} \to \Gamma \to \pi_1(S) \to 0 .$$

If this last extension were split, then the bundle $\mathcal{E}$ would arise from an irreducible SU(2)-representation; hence its Stiefel-Whitney class w_2 would vanish. Then, since w_2 is the mod 2 reduction of the first Chern class of $\mathcal{E}$, K_S would be 2-divisible.

But we claim that in fact K_S is not 2-divisible even in Num(S) = $H^2(S,\mathbb{Z})$/torsion . In fact, if K_S is numerically equivalent to 2L, then $L^2 = 1$ and $K_S \cdot L = 2$: by the genus formula we have then a contradiction.

Q.E.D. for theorem 1.1

Proposition 1.4 The vector bundle $\mathcal{E}$ is stable with respect to the canonical bundle K_S.

Proof Otherwise there exists an invertible and saturated subsheaf N of $\mathcal{E}$ which satisfies the inequality $N \cdot K_S \geq (1/2)\, K_S^2$ and gives a diagram of exact arrows of the following form

$$\begin{array}{ccccc} & & 0 & & \\ & & \downarrow & & \\ & & N & & \\ & & \downarrow & & \\ 0 \to \mathcal{O}_S \to & & \mathcal{E} & \to \mathcal{I}_x(K_S) \to & 0 \\ & & \downarrow & & \\ & & \mathcal{I}_Z(M) = \mathcal{I}_Z(K_S - N) & & \\ & & \downarrow & & \\ & & 0 & & \end{array} \tag{1.5}$$

where Z is a 0 dimensional subscheme of S and $M = K_S - N$ is then a line bundle on S with the property

(1.6) $\quad M \cdot K_S \leq 2$.

We can compare the two expressions of the Euler characteristic of the bundle $\mathcal{E}$ obtained from the two above exact sequences:

$$\begin{aligned}\chi(\mathcal{E}) &= \chi(\mathcal{O}_S) + \chi(\mathcal{I}_x(K_S)) = 2\chi(\mathcal{O}_S) - 1 \\ &= \chi(N) + \chi(\mathcal{I}_Z(M)) = 2\chi(M) - \deg Z .\end{aligned}$$

Hence we have the equality $1 + \deg Z = 2\chi(M)$ and therefore, applying the Riemann-Roch theorem to M and the fact that $M \cdot (K_S - M)$ is even, we infer that M satisfies the further inequality:

(1.7) $\quad M^2 \geq M \cdot K_S$.

By the Index theorem, we have

(1.8) $\quad M^2 K_S{}^2 - (M \cdot K_S)^2 \leq 0$

and so, by 1.6 and 1.7, $M^2 = M \cdot K_S = 0$. But, again by the Index theorem, this can only happen if $M = 0$.

Diagram 1.5 then gives a contradiction, since $N = K_S$. In fact there is no non zero morphism $K_S \to \mathcal{O}_S$ nor $K_S \to \mathcal{I}_x(K_S)$. Q.E.D.

Proposition 1.9 There exists an ample line bundle H on S such that the vector bundle $\mathcal{E}$ is H-stable.

Proof If K_S is ample, it is enough to take $H = K_S$. Otherwise, let $E_1, \ldots, E_\mu$ be the finitely many curves ($\cong \mathbb{P}^1$) on S such that $K_S \cdot E_i = 0$ for each $i = 1, \ldots, \mu$. We recall that the intersection matrix $(E_i \cdot E_j)$ is negative definite.

One can easily construct an effective divisor W on S of the form $W = \Sigma_i n_i E_i$ $(n_i \in \mathbb{N})$ such that $W \cdot E_i < 0$ for each i: if D is a divisor on S such that $\dim(\operatorname{supp} D \cap \operatorname{supp} E_i) = 0$ and $D \cdot E_i > 0$ for each i, and $\pi\colon S \to X$ is the blow down of the E_i's, the divisor W can be defined by the condition $\pi^*(\pi_* D) = W + D$.

By the Nakai-Moishezon criterion the divisor $H_t = K_S - tW$ is then ample for $0 < t << 1$: in fact we can assume that D is ample and this implies

$$\begin{aligned}H_t &= K_S - tW = K_S - t\pi^*(\pi_* D) + tD = \\ &= \pi^*(K_X - t\pi_* D) + tD ;\end{aligned} \tag{1.10}$$

moreover, $K_X - t\pi_* D$ is ample for $t << 1$ because K_X is ample on X.

For each effective divisor M on S we decompose M as the sum $M = M' + M''$ of two effective divisors, where $M'' \in \langle E_1, \ldots, E_\mu \rangle$ and $\dim(M' \cap E_i) = 0$ for each i .

If, for each t, there exists a line bundle N_t destabilizing $\mathcal{E}$ with respect to H_t, then for each t we have a diagram

$$
\begin{array}{ccccccccc}
 & & & & 0 & & & & \\
 & & & & \downarrow & & & & \\
 & & & & N_t & & & & \\
 & & & & \downarrow & & & & \\
(1.11) \quad & 0 & \to & \mathcal{O}_S & \to & \mathcal{E} & \to & \mathcal{J}_x(K_S) \to & 0 \\
 & & & & \downarrow & & & & \\
 & & & & \mathcal{J}_{Z_t}(M_t) = \mathcal{J}_{Z_t}(K_S - N_t) & & & & \\
 & & & & \downarrow & & & & \\
 & & & & 0 & & & & .
\end{array}
$$

If the "oblique arrow" $\mathcal{O}_S \to \mathcal{J}_{Z_t}(M_t)$ were zero, then $\mathcal{O}_S \subset N_t$, and, since both are saturated, $\mathcal{O}_S \cong N_t$ contradicting $N_t \cdot H_t > 0$.

It follows that M_t is an effective divisor for each t and, moreover, that

(1.12) $\quad M_t \cdot H_t \le (1/2)\, K_S \cdot H_t = (1/2)\, K_S^2 - t\, W \cdot K_S =$
$= (1/2)\, K_S^2 = 2$.

But $M_t \cdot H_t = M_t \cdot K_S - t\, M_t'' \cdot W - t\, M_t' \cdot W$: the inequality $M_t \cdot K_S \le 2$ would contradict the K_S-stability of $\mathcal{E}$ and so we have $M_t \cdot K_S \; (= M_t' \cdot K_S) \ge 3$ and $M_t \cdot H_t \le 2$ for each t. It follows that

(1.13) $\quad t\, (M_t \cdot W) \ge 1$

and then $0 < M_t \cdot W = M_t' \cdot W + M_t'' \cdot W$, where $M_t'' \cdot W < 0$ by the choice of W.

Let us fix now an index t_o such that H_{t_o} is ample. Then

$$M_t \cdot H_{t_o} = M_t \cdot (K_S - t_o W) = M_t \cdot (K_S - t\, W - (t_o - t)\, W) =$$
$$= M_t \cdot H_t - (t_o - t)\, M_t \cdot W < 2 \qquad \text{for } 0 < t < t_o .$$

This implies that $\{M_t\}$ is a bounded family and so the set $\{W \cdot M_t\}$ is bounded too: but this is absurd, because then the inequality $W \cdot M_t \ge t^{-1}$ is impossible. Q.E.D.

We will now consider two examples of surfaces S of general type with $p_g = 0$, $K_S^2 = 4$, which have a base point free bicanonical system but fail to satisfy the hypotheses of Theorem 1.1.

<u>Example 1.14</u> The Burniat surface B(2) has fundamental group $\pi_1(B(2))$ isomorphic to $\mathbb{H} \oplus (\mathbb{Z}/2\mathbb{Z})^2$, where $\mathbb{H}$ is the quaternion group of order 8 (cf. table 13 of [BPV], [Bu], [Pet]): so $\pi_1(B(2))$ has a non trivial central extension given by $\mathbb{H}^2$, which clearly admits an irreducible SU(2)-representation, since the group $\mathbb{H}$ admits the following irreducible SU(2)-representation, given by

$$i \to \begin{pmatrix} -\sqrt{(-1)} & 0 \\ 0 & -\sqrt{(-1)} \end{pmatrix}$$

$$j \to \begin{pmatrix} 0 & -1 \\ 1 & 0 \end{pmatrix}$$

$$k \to \begin{pmatrix} 0 & -\sqrt{(-1)} \\ -\sqrt{(-1)} & 0 \end{pmatrix}.$$

<u>Proposition 1.15</u> The bicanonical system is free from base points for the Burniat surface with K^2=4.

<u>Proof</u> By lemma 3.3. of ([Pet], (ii) page 118) it follows that, S being a $(\mathbb{Z}/2\mathbb{Z})^2$-Galois cover

of the blow up Y of the plane in 5 points of which 3 lie on a line, all the sections of $H^0(S,2K_S)$ are Galois invariant and are pull-backs of rational tensor 2-forms with simple poles on the branch divisor. Hence the linear system $|2K_S|$ factors through the $(\mathbb{Z}/2\mathbb{Z})^2$-Galois cover and indeed through the anticanonical mapping of Y which is a birational morphism onto a quartic weak del Pezzo surface with a node corresponding to the line containing the 3 collinear points. Q.E.D.

Example 1.16 In [Ke], J.H. Keum gives an example of a surface S of general type with $p_g = 0$, $K_S^2 = 4$ and with fundamental group $\pi_1(S) \cong \mathbb{Z}^4 \rtimes (\mathbb{Z}/2\mathbb{Z})^2$. Also here there exists a non split central extension $\tilde{\Gamma}$ of Γ by $\mathbb{Z}/2\mathbb{Z}$ which admits an irreducible SU(2)-representation, since there is a surjective map ψ obtained as a composition

$$\psi\colon \Gamma \to \mathbb{Z}^4 \rtimes (\mathbb{Z}/2\mathbb{Z})^3 \to (\mathbb{Z}/2\mathbb{Z})^2$$

and as above one can pull-back by ψ the extension $1 \to \mathbb{Z}/2\mathbb{Z} \to \mathbb{H} \to (\mathbb{Z}/2\mathbb{Z})^2 \to 0$.

Proposition 1.17 If S is the surface in Keum's example, the bicanonical system $|2K_S|$ is base point free.

Proof We recall the notation of the quoted paper. Let $A = E_1 \times E_2$ be a product of two elliptic curves $E_i = \mathbb{C}/\mathbb{Z}+\tau_i\mathbb{Z}$ and e_i for i=1,2 be a nonzero 2-torsion point of E_i. Then the endomorphism θ: A → A defined by

$$\theta(z_1,z_2) = (-z_1+ e_1, z_2+ e_2)$$

induces a fixed-point-free involution on the Kummer surface K of A and the quotient surface $Y = K/\theta$ is an Enriques surface. Keum's surface S is the minimal model of the canonical resolution of singularities $\overline{X} \to X$ of a ramified double covering X of Y; this covering X is determined by a square root of a reduced divisor B of Y with at most simple singularities and B is defined as follows. Let q : A → K and p: K → Y be the natural maps; we denote by R_1, R_5 (resp. R_3, R_7) the images under the composition map pq of the subsets of A of the form (a 2-torsion point of E_1) $\times E_2$ (resp. of the form $E_1 \times$ (a 2-torsion point of E_2)) and by R_2, R_9, R_4, R_{11}, R_6, R_{12}, R_8, R_{10} the remaining 16 exceptional lines corresponding to the 2-torsion points of A; the lines intersect as in the picture below:

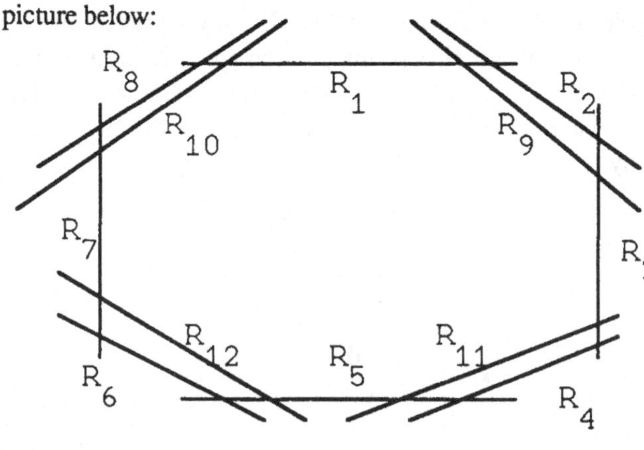

The branch divisor B is defined by

$$B = R_2+R_4+R_6+R_8+R_9+R_{10}+R_{11}+R_{12}+F+G$$

where F, G are smooth elliptic curves belonging to the elliptic pencils $|2R_3+R_2+R_4+R_9+R_{11}|$ and $|2R_1+R_2+R_8+R_9+R_{10}|$ respectively. In particular, it holds

$$(pq)^*(F) = a \times E_2 + (-a) \times E_2 + (a+e_1) \times E_2 + (-a+e_1) \times E_2$$
$$(pq)^*(G) = E_1 \times b + E_1 \times (-b) + E_1 \times (b+e_2) + E_1 \times (-b+e_2)$$

for some $a \in E_1$, $b \in E_2$.

To prove the proposition, it is enough to check that the bicanonical system $|2K_{\overline{X}}|$ of the surface $\overline{X}$ has base locus consisting entirely of the exceptional curves of the first kind counted with multiplicity one. Moreover, we can reduce the problem to the inspection of the linear system $|F+G|$ on Y, via the isomorphisms $H^0(2K_{\overline{X}}) \cong H^0(\mathcal{O}_Y(B)) \cong H^0(\mathcal{O}_Y(F+G))$. This follows since both $|F|$ and $|G|$ are base point free pencils. Q.E.D.

§2 m^{th} root extraction trick and change of fundamental group

In this paragraph we are going to sharpen the result of theorem 1.1, showing that the extension appearing there can be realized by the fundamental group of a $(\mathbb{Z}/2\mathbb{Z})^2$-Galois cover Y of S . By pulling back $\mathcal{E}$ to Y, and showing that the pull back still remains stable (and not only semistable) we shall be able to apply Donaldson's theorem to a stable vector bundle on Y with trivial Chern classes ([Do], [Ko]).

In order to do so , we recall the mth root extraction trick, which will produce the desired Y in the case m=2.

In the rest of the paragraph we shall explicitly describe how the fundamental group of Y can be computed, later on we shall apply this recipe in some concrete examples.

Lemma 2.1 (m^{th} root extraction trick) Let S be a smooth algebraic surface and $f : S \to \mathbb{P}^2$ be a holomorphic map associated to a base point free subsystem of a linear system $|D|$. Let Y be obtained as the fibre product $Y = S \times_{\mathbb{P}^2} \mathbb{P}^2$ of the previous $f : S \to \mathbb{P}^2$, and, for a general choice of coordinates, of the mth power map $g : \mathbb{P}^2 \to \mathbb{P}^2$ (i.e., $g(y_0, y_1, y_2) = (y_0^m, y_1^m, y_2^m)$).

Then, if F is the natural morphism $F: Y \to S$, $F_*: \pi_1(Y) \to \pi_1(S)$ is surjective, and its kernel K is cyclic of order dividing m; moreover K is contained in the centre of $\pi_1(Y)$.

Proof We denote for simplicity by H_i (i=0,1,2) the coordinate lines on $\mathbb{P}^2$, by D_i the inverse image of H_i under f, by C_i the inverse image of D_i under F, by H' the union of the H_i's, and similarly we define D' and C'. By the genericity of the H_i's, all the above divisors have global normal crossings, Y is smooth, F is a Galois $(\mathbb{Z}/m\mathbb{Z})^2$-cover branched on D' and totally ramified at the singular points of D'. In particular $\pi_1(Y-C') \to \pi_1(S-D')$ is a normal subgroup with quotient group $(\mathbb{Z}/m\mathbb{Z})^2$.

We have the following diagram of sequences

$$\begin{array}{ccccccc}
 & & 0 & & 0 & & \\
 & & \uparrow & & \uparrow & & \\
 & (\mathbb{Z}/m\mathbb{Z})^3 \to & \mathcal{K}/\mathcal{H} & \to & (\mathbb{Z}/m\mathbb{Z})^2 & \to 0 & \\
 & & \uparrow & & \uparrow & & \\
(2.2) \quad 0 & \to & \mathcal{K} & \to & \pi_1(S-D') & \to & \pi_1(S) \to 0 \\
 & & \uparrow & & \uparrow F_* & & \uparrow F_* \\
0 & \to & \mathcal{H} & \to & \pi_1(Y-C') & \to & \pi_1(Y) \to 0 \\
 & & \uparrow & & \uparrow & & \\
 & & 0 & & 0 & &
\end{array}$$

which are exact, with the exception of the first row. Here exactness of the middle column was already mentioned, whereas exactness of the second and third row is standard, and the inclusion $\mathcal{H} \subset \mathcal{K}$ is obvious.

Claim 2.3 $\mathcal{K}, \mathcal{H}$ are central and with respective generators $\gamma_0, \gamma_1, \gamma_2$ for $\mathcal{K}$, $\gamma_0{}^m, \gamma_1{}^m, \gamma_2{}^m$ for $\mathcal{H}$.

Proof of the Claim 2.3 : let γ_i, for i=0,1,2, be a loop consisting of the conjugate (under a path in S–D') of a small circle in the normal space to a smooth point of D_i. Argueing as in theorem 1.6 of [Ca1], one shows that $\mathcal{K}$ is generated by conjugates of the γ_i's, and that the γ_i's lie in the centre of $\pi_1(S-D')$.
An entirely similar argument applies to $\mathcal{H}$, since $\gamma_i{}^m$ is obtained in Y–C' by the same procedure by which γ_i is gotten. Q.E.D. for the claim

It is worthwhile to notice that, since γ_i, γ_j can be chosen to be local generators of $\pi_1(U-D')$ for a suitable neighbourhood of $x \in D_i \cap D_j$, by looking at the local monodromy of F, we obtain that they map to 2 generators of $(\mathbb{Z}/m\mathbb{Z})^2$.

By diagram chasing, the surjectivity of F_* follows from the surjectivity of the map $\mathcal{K}/\mathcal{H} \to (\mathbb{Z}/m\mathbb{Z})^2$.

Clearly $K = \ker(F_*\colon \pi_1(Y) \to \pi_1(S)) = (\mathcal{K} \cap \pi_1(Y-C'))/\mathcal{H}$, is contained in $\mathcal{K}/\mathcal{H}$. But, as we mentioned above, we have two surjective maps $(\mathbb{Z}/m\mathbb{Z})^3 \to \mathcal{K}/\mathcal{H} \to (\mathbb{Z}/m\mathbb{Z})^2$ such that any 2 of the 3 standard generators of $(\mathbb{Z}/m\mathbb{Z})^3$ map to 2 generators of $(\mathbb{Z}/m\mathbb{Z})^2$. Hence K is cyclic of order dividing m.

We can be more precise since in fact K is isomorphic to the kernel of $\mathcal{K}/\mathcal{H} \to (\mathbb{Z}/m\mathbb{Z})^2$, therefore K is the image of the cyclic group $(\mathbb{Z}/m\mathbb{Z}) = \ker\big((\mathbb{Z}/m\mathbb{Z})^3 \to (\mathbb{Z}/m\mathbb{Z})^2\big)$ inside $\pi_1(Y)$. Q.E.D.

Corollary 2.4 If S is simply connected, $\pi_1(Y)$ is cyclic of order r where r = G.C.D. (m,d), and d is the divisibility index of the divisor D (i.e., d is the order of the cyclic group $(\mathbb{Q}D \cap H^2(S,\mathbb{Z}))/\mathbb{Z}D)$). More generally, the kernel K always admits a surjective homomorphism onto a cyclic group of order r.

Proof Hypotheses of proposition 1.8 of [Ca1] are satisfied. Applying this proposition, one can see that $\pi_1(Y)$ is the quotient of $\ker((\mathbb{Z}/m\mathbb{Z})^3 \to (\mathbb{Z}/m\mathbb{Z})^2)$ by the image of the map obtained as the composition of $(H^2(S,\mathbb{Z}) \to H^2(D,\mathbb{Z}) \cong \mathbb{Z}^3 \to (\mathbb{Z}/m\mathbb{Z})^3)$. By Poincare' duality, this image consists of the elements divisible by d and our assertion is proven.
In the general case, the above quotient represents exactly the natural image of K in $H_1(S-D',\mathbb{Z})$ (cf. the proof of cor. 1.7, prop. 1.8 ibidem). Q.E.D.

We can indeed prove a much more precise statement (see Th. 2.16 below), which is our main result. If m is a fixed positive integer, the assertion in lemma 2.1 shows that (it suffices, in general, to add to the divisor D m times a suitably very ample divisor H, such that $|D+mH|$ yields a finite morphism to $\mathbb{P}^2$) to each divisor D on a smooth surface S we can associate a central extension of the fundamental group of the surface by a cyclic group $\mathbb{Z}/r'\mathbb{Z}$, where r' is divisible by r = G.C.D. (m,d) and d is the divisibility index of D: this is the extension describing the fundamental group of a $(\mathbb{Z}/m\mathbb{Z})^2$-Galois covering of S under which the divisor D becomes m-divisible.

Looking now at S as a quotient of its universal covering space $\widetilde{S}$, we will see that, by general properties of the cohomology of good quotient spaces, to each divisor D and to each integer m is uniquely associated another central extension of $\pi_1(S)$ by a cyclic group: we shall call the latter the "algebraic construction" of the extension.

The key fact is that this algebraic construction yields the same extension that we obtain by the geometrical construction of extracting the mth root of D, as it will be shown in Th. 2.16.

We firstly recall some basic facts and notation concerning the spectral sequence for the cohomology of a quotient $X = \widetilde{X}/G$ by a properly discontinuous group G (see [Mu], appendix to section 2, and [G], ch. 5).

If $\mathfrak{F}$ is a G-linearized sheaf on $\widetilde{X}$ and $\pi\colon \widetilde{X}\to X$ is the quotient map, one can describe the functor $\Gamma(\widetilde{X},\mathfrak{F})^G$ as a composition in two different ways:

(2.5) $\quad \mathfrak{F} \to \Gamma(\widetilde{X},\mathfrak{F}) \to \Gamma(\widetilde{X},\mathfrak{F})^G$

(2.6) $\quad \mathfrak{F} \to \pi_*(\mathfrak{F})^G \to \Gamma(X,\pi_*(\mathfrak{F})^G)$.

The derived cohomology functors are then given with two filtrations whose associated gradings are the limits of two spectral sequences

(2.7) $\quad H^p(G,H^q(\widetilde{X},\mathfrak{F})) = E_2^{p,q}$

(2.8) $\quad H^i(X,\mathcal{R}_G{}^j\,\pi_*(\mathfrak{F})) = E'_2{}^{i,j}$

where $H^p(G,-)$ is the group cohomology and $\mathcal{R}_G{}^j\,\pi_*(-)$ denote the derived cohomology functors of $\pi_*(-)^G$. If the action of G is free and gives indeed a covering space, then the functor $\mathfrak{F}\to(\pi_*\mathfrak{F})^G$ is exact and consequently $\mathcal{R}_G{}^j\pi_*(\mathfrak{F}) = 0$ for each $j \geq 0$. So the spectral sequence degenerates at E'_2 and the first spectral sequence converges to $H^i(X, (\pi_*\mathfrak{F})^G)$:

(2.9) $\quad H^p(G,H^q(\widetilde{X},\mathfrak{F})) \Rightarrow H^*(X,(\pi_*\mathfrak{F})^G)$.

Assume now that $\mathfrak{F} = \mathbb{Z}_{\widetilde{X}}$. Then $(\pi_*\mathbb{Z}_{\widetilde{X}})^G = \mathbb{Z}_X$, thus

(2.10) $\quad H^p(G,H^q(\widetilde{X},\mathbb{Z}_{\widetilde{X}})) \Rightarrow H^*(X,\mathbb{Z}_X)$.

If we assume moreover that $H^1(\widetilde{X},\mathbb{Z}_{\widetilde{X}}) = 0$, e.g. if $\widetilde{X}$ is the universal cover of X, the $E_2^{p,1}$-term in the spectral sequence 2.7 vanishes and so the differential $d_2 : E_2^{p,q}\to E_2^{p+2,q-1}$ is zero for each $q \leq 2$. So we can say that $E_2^{p,q} = E_3^{p,q}$ for each $q \leq 2$ and, finally, that

(2.11) $\quad d_3 : H^p(G, H^2(\widetilde{X},\mathbb{Z}_{\widetilde{X}})) \to H^{p+3}(G, H^0(\widetilde{X},\mathbb{Z}_{\widetilde{X}}))$

is the only non zero map for $q \leq 2$. The edge-morphisms give then an isomorphism

(2.12) $\quad H^1(X,\mathbb{Z}_X) \cong H^1(G,\mathbb{Z})$

and an exact sequence

(2.13) $\quad 0 \to H^2(G,\mathbb{Z}) \to H^2(X,\mathbb{Z}_X) \xrightarrow{\pi^*} H^2(\widetilde{X},\mathbb{Z}_{\widetilde{X}})^G \to H^3(G,\mathbb{Z})$

where the arrow $H^2(X,\mathbb{Z}_X) \to H^2(\widetilde{X},\mathbb{Z}_{\widetilde{X}})^G$ is given by the inverse image π^*. We similarly have an analogous sequence for any system $\mathbb{F}$ of coefficients ($\mathbb{F} = \mathbb{Z}/m\mathbb{Z}$, $\mathbb{Q}$, ...) and in particular, associated to the exact sequence $0 \to m\mathbb{Z} \to \mathbb{Z} \to \mathbb{Z}/m\mathbb{Z} \to 0$ we have a commutative diagram

$$
\begin{array}{ccccccccc}
0 \to H^2(G,\mathbb{Z}) & \to & H^2(X,\mathbb{Z}_X) & \xrightarrow{\pi^*} & H^2(\widetilde{X},\mathbb{Z}_{\widetilde{X}})^G & \to & H^3(G,\mathbb{Z}) \\
m\downarrow & & m\downarrow & & m\downarrow & & m\downarrow \\
0 \to H^2(G,\mathbb{Z}) & \to & H^2(X,\mathbb{Z}_X) & \xrightarrow{\pi^*} & H^2(\widetilde{X},\mathbb{Z}_{\widetilde{X}})^G & \to & H^3(G,\mathbb{Z}) \\
\downarrow & & \downarrow & & \downarrow & & \downarrow \\
0 \to H^2(G,\mathbb{Z}/m\mathbb{Z}) & \to & H^2(X,\mathbb{Z}_X/m\mathbb{Z}_X) & \to & H^2(\widetilde{X},\mathbb{Z}_{\widetilde{X}}/m\mathbb{Z}_{\widetilde{X}})^G & \to & H^3(G,\mathbb{Z}/m\mathbb{Z})
\end{array}
\tag{2.14}
$$

which is exact in the rows and in the columns.

Algebraic construction Now let $\pi: \tilde{X} \to X$ be the universal covering, let G be the foundamental group of X and m be a fixed positive integer. By the universal coefficients formula, the group $H^2(\tilde{X}, \mathbb{Z}_{\tilde{X}})$ is torsion free, hence, if d is the divisibility index of a class $\tilde{D}$ in $H^2(\tilde{X}, \mathbb{Z}_{\tilde{X}})^G$, then the image $\tilde{\delta} \in H^2(\tilde{X}, \mathbb{Z}/m\mathbb{Z}_{\tilde{X}})^G$ under the map of diagram 2.14 has period exactly m/r, where r = G.C.D.(m,d). In particular, the image $\hat{\delta}$ in $H^2(\tilde{X}, \mathbb{Z}/r\mathbb{Z}_{\tilde{X}})^G$ is 0. Hence if $\tilde{D} \in H^2(\tilde{X}, \mathbb{Z}_{\tilde{X}})^G$ equals $\pi^*(D)$ for $D \in H^2(X, \mathbb{Z}_X)$, the class D maps to a cohomology class $\delta \in H^2(X, \mathbb{Z}/r\mathbb{Z}_X)$ coming from a cohomology class $\delta \in H^2(G, \mathbb{Z}/r\mathbb{Z})$, since we must have $\hat{\delta} = \pi^*(\delta) = 0$ in $H^2(\tilde{X}, \mathbb{Z}/r\mathbb{Z}_{\tilde{X}})^G$. So we get:

Construction To each class $D \in H^2(X, \mathbb{Z}_X)$ we can associate $\tilde{D}$, r, δ as above, hence a unique (up to isomorphism) central extension

$$0 \to \mathbb{Z}/r\mathbb{Z} \to G(\delta) \to G \to 0 \quad . \tag{2.15}$$

Theorem 2.16 If Y is obtained from X via the mth root extraction trick associated to $|D|$, then we have that $\pi_1(Y) \cong G(-\delta)$, i.e., as an extension of $G \cong \pi_1(X)$, $\pi_1(Y)$ is the extension associated to $-D$ by the "algebraic construction" described before.

Proof Let us take the fibre product Y' of the universal covering $\pi: \tilde{X} \to X$ and the mth root extraction $(\mathbb{Z}/m\mathbb{Z})^2$-Galois covering $F: Y \to X$; we have the following diagram

$$\begin{array}{ccccc} \tilde{Y} & \to & Y' & \xrightarrow{q} & \tilde{X} \\ & & p \downarrow & & \downarrow \pi \\ & & Y & \xrightarrow{F} & X \\ & & \psi \downarrow & & \downarrow f \\ & & \mathbb{P}^2 & \xrightarrow{(y_i^m)} & \mathbb{P}^2 \end{array} \tag{2.17}$$

where

i) $0 \to \mathbb{Z}/r'\mathbb{Z} \to \pi_1(Y) \to \pi_1(X) = G \to 0$ is a central extension and r' divides m, as we know from lemma 2.1;

ii) p is the covering associated to $\mathbb{Z}/r'\mathbb{Z}$, so $\pi_1(Y') \cong \mathbb{Z}/r'\mathbb{Z}$;

iii) $f^* \mathcal{O}(1) \equiv D$;

iv) $(\mathbb{Z}/m\mathbb{Z})^2 \times \Gamma$ operates on Y'. The covering $\tilde{Y} \to Y'$ is an etale $(\mathbb{Z}/r'\mathbb{Z})$-Galois cover induced by the universal covering $\tilde{Y} \to Y$, since Γ operates freely.

Let $D' = D_1 \cup D_2 \cup D_3$ and $\tilde{D}' = \pi^{-1}(D')$, where the D_i's are the inverse image under the map f of the coordinate lines on $\mathbb{P}^2$.

Then $\pi_1(X - \tilde{D}') = \ker(\pi_1(X - D') \to \pi_1(X)) = \mathcal{K}$ is abelian and is generated by $\gamma_0, \gamma_1, \gamma_2$.

Step I We show that $r = r'$.

By Lefschetz's theorem $\pi_1(D_i)$ surjects onto $\pi_1(X)$, hence $\pi^{-1}(D_i) = \tilde{D}_i$ is connected and smooth.

The map $Y' \to \tilde{X}$ is an abelian $(\mathbb{Z}/m\mathbb{Z})^2$-cover which is unramified on $\tilde{X} - \tilde{D}'$. Since $\pi_1(Y')$ is cyclic, it suffices to calculate the first homology group $H_1(Y', \mathbb{Z})$.

We proceed as in [Ca1], sequel to cor. 1.7., and prop. 1.8. Here we have to apply Lefschetz's duality as in [Dol] prop. 7.14 page 297, by which it follows that, for a manifold M with boundary $L = \partial M$, and of dimension n, $H_{n-i}(X) \cong H_{n-i}(X - \partial X) \cong H_c^i(X, \partial X)$ the last group denoting cohomology with compact supports.

Hence, argueing as in loc. cit., $H_1(\tilde{X} - \tilde{D}') = H_c^3(\tilde{X}, \tilde{D}')$, which, by the exact sequence

(2.18) $$H_c^2(\widetilde{X}) \xrightarrow{\rho} H_c^2(\widetilde{D}') \to H_c^3(\widetilde{X},\widetilde{D}') \to 0$$
is isomorphic to coker (ρ), $\widetilde{X}$ being simply connected.
Now, by Poincare' duality $H_c^2(\widetilde{X}) \cong H_2(\widetilde{X})$ is a free $\mathbb{Z}$-module, whereas by Mayer-Vietoris and again Poincare' duality $H_c^2(\widetilde{D}') \cong \oplus_i H_c^2(\widetilde{D}'_i) \cong \oplus_i H_0(\widetilde{D}'_i) \cong \mathbb{Z}^3$, the $\widetilde{D}'_i$'s being as we mentioned smooth and connected. The map ρ is given by geometrical intersection and its dual sends $\mathbb{Z}^3 \to H^2(\widetilde{X},\mathbb{Z})$ by mapping each generator e_1, e_2, e_3 to the class of $\widetilde{D}_i$.

Argueing as in loc. cit. and as in claim 2.3, we obtain that, if d is the divisibility index of $\widetilde{D}$, then firstly coker (ρ) is isomorphic to $\mathbb{Z}^3/\mathbb{Z}(de_1+de_2+de_3)$, hence secondly $\pi_1(Y') \cong (\mathbb{Z}/m\mathbb{Z})/(d) = \mathbb{Z}/r\mathbb{Z}$ if r = G.C.D.(m,d) as in our notation.

Step II We can reduce the proof of the theorem to the case where r = m.

In fact, we can factor F: Y → X as $Y \xrightarrow{F''} Z \xrightarrow{F'} X$ where F' is obtained by extracting the rth root of D, hence $F'^*(D) \equiv rL$. We shall show in Step III that the divisibility of the pull back $\widetilde{L}$ of L to the universal cover $\widetilde{Z}$ of Z is precisely d/r.

Since F" is obtained by extracting the $(m/r)^{th}$ root of L, it follows by step I that $\pi_1(Y) \cong \pi_1(Z)$, thereby reducing the proof of the theorem to the case r = m.

Proof of step II (Proof of the theorem in the special case r=m)

We first introduce some notation to describe explicitly the cocycles on X and $\widetilde{X}$.
Let $\{U_\alpha\}$ be a sufficiently fine cover of X, such that, for each U_α, $\pi^{-1}(U_\alpha) = \cup_{g\in G} g(V_\alpha)$ where the union is disjoint and we have made a non canonical choice of V_α, a connected component of $\pi^{-1}(U_\alpha)$. We shall also write, for further use,

(2.19) $g(V_\alpha) = g \cdot V_\alpha = V_{(\alpha,g)}$ (so $V_{(\alpha,1)} = V_\alpha$)

and we let G act on the left.

One can observe the following facts:

a) For each (α,β) such that $U_\alpha \cap U_\beta \neq \emptyset$, there exists a unique element h(α,β) of G such that

(2.20) $V_{(\alpha,1)} \cap V_{(\alpha,h(\alpha,\beta))} \neq \emptyset$.

b) If $U_\alpha \cap U_\beta \cap U_\gamma \neq \emptyset$, since π is a local homeomorphism

(2.21) $\emptyset \neq V_{(\gamma,h(\alpha,\gamma))} \cap V_{(\beta,h(\alpha,\beta))}$ $(= h(\alpha,\beta) \cdot (V_{(\gamma,h(\beta,\gamma))} \cap V_{(\beta,1)}))$

Hence, if G acts on the left, we have the relation

(2.22) $h(\alpha,\gamma) = h(\alpha,\beta)\, h(\beta,\gamma)$ for each $U_\alpha \cap U_\beta \cap U_\gamma \neq \emptyset$.

In particular $h(\beta,\alpha) = h(\alpha,\beta)^{-1}$.

c) If $U_\alpha \cap U_\beta \neq \emptyset$, $V_{(\alpha,g)}$ intersects exactly $V_{(\beta,g\,h(\alpha,\beta))}$.

Therefore, if $(f_{\alpha\beta})$ is a cocycle for $L = \mathcal{O}_X(D)$ relative to the covering $\{U_\alpha\}$ on X, there exists, for the line bundle $\widetilde{L}$ on $\widetilde{X}$ such that $(\widetilde{L}^{\otimes r}) \cong \pi^*(\mathcal{O}_X(D))$ (whose existence is guaranteed by our assumption), a cocycle $(\widetilde{f}_{(\alpha,g)(\beta,\, g\,h(\alpha,\beta))})$ such that $(\widetilde{f}_{(\alpha,g)(\beta,\, g\,h(\alpha,\beta))})^r = f_{\alpha\beta}$.
For short, we write $(\widetilde{f}_{(\alpha,g)(\beta,g')})$ but we recall that $g' = g\,h(\alpha,\beta)$.

We write $z_{(\alpha,g)}$ for a local generator of on $V_{(\alpha,g)}$, so that

(2.23) $z_{(\alpha,g)} = \widetilde{f}_{(\alpha,g)(\beta,g')}\, z_{(\beta,g')}$.

Using the isomorphism $\widetilde{L}^{\otimes r} \cong \pi^*(L)$ we can assume that G acts on $(\widetilde{L}^{\otimes r})$ by sending the local generators $(z_{(\alpha,g)})^r$ of $(\widetilde{L}^{\otimes r})$ one to another, i.e. for each $\bar{g} \in G$

(2.24) $\bar{g}$: $(z_{(\alpha,g)})^r \to (z_{(\alpha,\bar{g}g)})^r$.

Since $\widetilde{L} \in H^1(\widetilde{X}, \mathcal{O}^*_{\widetilde{X}})^G$, for each $\bar{g} \in G$ there is an isomorphism $\widetilde{L} \to \bar{g}_*(\widetilde{L})$, which we still denote by $\bar{g}$ and which induces the above action on $(\widetilde{L}^{\otimes r})$ (by which $(L^{\otimes r}) \cong \pi^*(L)$). Hence $\bar{g}$ acts on $\widetilde{L}$ by

$$(2.25) \qquad \bar{g}: \ z_{(\alpha,g)} \ \to c^{\bar{g}}_{(\alpha,g)} \ z_{(\alpha,\bar{g}g)}$$

where $c^{\bar{g}}_{(\alpha,g)}$ satisfies the identity $(c^{\bar{g}}_{(\alpha,g)})^r = 1$. The constants $c^{\bar{g}}_{(\alpha,g)}$ must satisfy some compatibility condition, since

$$(2.26) \qquad \begin{array}{ccc} z_{(\alpha,g)} = \tilde{f}_{(\alpha,g)(\beta,g')} \ z_{(\beta,g')} & \xrightarrow{\bar{g}} & \tilde{f}_{(\alpha,g)(\beta,g')} \, c^{\bar{g}}_{(\beta,g')} \ z_{(\beta,\bar{g}g')} \\ \downarrow \bar{g} & & \\ c^{\bar{g}}_{(\alpha,g)} \ z_{(\alpha,\bar{g}g)} = c^{\bar{g}}_{(\alpha,g)} \ \tilde{f}_{(\alpha,\bar{g}g)(\beta,\bar{g}g')} \ z_{(\beta,\bar{g}g')} & & \end{array}$$

where $g' = g\,h(\alpha,\beta)$ as before. Hence:

$$(2.27) \qquad c^{\bar{g}}_{(\alpha,g)} \ \tilde{f}_{(\alpha,\bar{g}g)(\beta,\bar{g}\,gh(\alpha,\beta))} = \tilde{f}_{(\alpha,g)(\beta,gh(\alpha,\beta))} \ c^{\bar{g}}_{(\beta,gh(\alpha,\beta))} .$$

The above formula shows that $c^{\bar{g}}_{(\beta,gh(\alpha,\beta))}$ is completely determined by $c^{\bar{g}}_{(\alpha,g)} \in \mu_r$, where μ_r is the group of the rth roots of the unity.

Since $\widetilde{X}$ is connected, the $c^{\bar{g}}_{(\alpha,g)}$ are completely determined by one of them. Moreover, once fixed a local generator $z_{(\alpha,1)}$ for the bundle L such that $(z_{(\alpha,g)})^r$ is G-invariant as before, for each $\bar{g} \in G$ one can also choose the root $z_{(\alpha,\bar{g})}$ such that the action is given by

$$\bar{g}: z_{(\alpha,1)} \to z_{(\alpha,\bar{g})} .$$

In other words, we may assume:

$$(2.28) \qquad c^{g}_{(\alpha,1)} = 1 \qquad \text{for each } g \in G .$$

We can now check how the composite action of $(g_1 g_2)^{-1} \cdot g_1 \cdot g_2$ fails to act as the identity:

$$(2.29) \qquad \begin{array}{l} z_{(\alpha,g)} \xrightarrow{g_2} c^{g_2}_{(\alpha,g)} \ z_{(\alpha,g_2g)} \xrightarrow{g_1} c^{g_2}_{(\alpha,g)} \ c^{g_1}_{(\alpha,g_2g)} \ z_{(\alpha,g_1g_2g)} \\ \qquad\qquad\qquad\qquad \downarrow (g_1 g_2)^{-1} \\ \qquad c^{g_2}_{(\alpha,g)} \ c^{g_1}_{(\alpha,g_2g)} \ c^{(g_1g_2)^{-1}}_{(\alpha,g_1g_2g)} \ z_{(\alpha,g)} = \\ \qquad = c^{g_2}_{(\alpha,g)} c^{g_1}_{(\alpha,g_2g)} \left(c^{(g_1g_2)}_{(\alpha,g)}\right)^{-1} z_{(\alpha,g)} \end{array}$$

where the last equality follows by observing that we can assume:

$$(2.30) \qquad c^{\bar{g}^{-1}}_{(\alpha,g)} = \left(c^{\bar{g}}_{(\alpha,(\bar{g})^{-1}g)} \right)^{-1} .$$

By 2.27 and the connectedness of $\widetilde{X}$ we get then:

$$(2.31) \qquad c^{g_2}_{(\alpha,g)} c^{g_1}_{(\alpha,g_2g)} \left(c^{(g_1g_2)}_{(\alpha,g)}\right)^{-1} \text{ is independent of } (\alpha,g).$$

This is in fact an element of $H^2(G,\mu_r)$ classifying the extension of groups ("theta group" extending G): by the assumption 2.28, we know that for each α:

(2.32) $c^{g_2}_{(\alpha,g)} c^{g_1}_{(\alpha,g_2g)} (c^{(g_1g_2)}_{(\alpha,g_1g_2g)})^{-1} = c^{g_1}_{(\alpha,g_2)} = c(g_1, g_2)$

and we found that:

(2.33) $c^{g_1}_{(\alpha,g_2)} = c^{g_1}_{g_2} = c\,(g_1, g_2)$ is independent of α.

We must then explicitly write the cocycle $c_{\alpha\beta\gamma} \in H^2(X,\mu_r)$ which is associated to the cocycle $c(g_1, g_2)$ given in 2.32. We use the description proven in ([Mu], page 23) of the image of $c(g_1, g_2)$ which yields the following formula:

(2.34) $c_{\alpha\beta\gamma} = c^{h(\alpha,\beta)}_{h(\beta,\gamma)}$.

We want now to show that, taking Chern classes modulo r, the inverse $(c_{\alpha\beta\gamma})^{-1}$ gives exactly the Chern class of L.

We start by describing more explicitly this class, in terms of the chosen cocycles for L and $\tilde{L}$. We will use the exact sequence

(2.35) $0 \to H^2(G,\mu_r) \to H^2(X,\mu_r) \xrightarrow{\pi^*} H^2(\tilde{X},\mu_r)^G \to H^3(G,\mu_r)$

obtained as in 2.13.

We can apply the theory of spectral sequences for G-linearized sheaves discussed above to the case $\mathfrak{F} = \mathcal{O}_{\tilde{X}}, \mathcal{O}^*_{\tilde{X}}$: specializing 2.9, we get

(2.36) $H^p(G,H^q(\tilde{X},\mathcal{O}_{\tilde{X}})) \Rightarrow H^*(X,\mathcal{O}_X)$

(2.37) $H^p(G, H^q(\tilde{X},\mathcal{O}^*_{\tilde{X}})) \Rightarrow H^*(X,\mathcal{O}^*_X)$ (since $\pi_*(\mathcal{O}_{\tilde{X}}) = \mathcal{O}_X$ and $\pi_*(\mathcal{O}^*_{\tilde{X}}) = \mathcal{O}^*_X$) and an exact sequence

(2.38) $0 \to H^1(G,H^0(\mathcal{O}^*_{\tilde{X}})) \to H^1(X,\mathcal{O}^*_X) \xrightarrow{\pi^*} H^1(\tilde{X},\mathcal{O}^*_{\tilde{X}})^G \xrightarrow{\partial} H^2(G,H^0(\mathcal{O}^*_{\tilde{X}}))$

In this sequence, to each G-linearized line bundle $\tilde{\mathfrak{L}}$ on $\tilde{X}$ is associated the "theta group" $\mathcal{G}(\tilde{\mathfrak{L}})$: this is a central extension

(2.39) $0 \to \mathrm{Aut}(\tilde{\mathfrak{L}}) = H^0(\mathcal{O}^*_{\tilde{X}}) \to \mathcal{G}(\tilde{\mathfrak{L}}) \to G \to 0$

classified by $\partial(\tilde{\mathfrak{L}}) \in H^2(G, H^0(\mathcal{O}^*_{\tilde{X}}))$. In particular this applies when $\tilde{\mathfrak{L}}$ is a pull back bundle from X.

We will firstly consider the $(2\pi i)$-twisted exponential sequence

(2.40) $0 \to \mathbf{Z} \to \mathcal{O}_X \xrightarrow{e} \mathcal{O}^*_X \to 0$

on X (and the corresponding on $\tilde{X}$) that gives rise to the diagram

$$\begin{array}{ccccccccc}
0 \to & H^1(G,H^0(\mathcal{O}_{\tilde{X}})) & \to & H^1(X,\mathcal{O}_X) & \xrightarrow{\pi^*} & H^1(\tilde{X},\mathcal{O}_{\tilde{X}})^G & \to & H^2(G,H^0(\mathcal{O}_{\tilde{X}})) \\
& e\downarrow & & e\downarrow & & \downarrow & & \downarrow \\
0 \to & H^1(G,H^0(\mathcal{O}^*_{\tilde{X}})) & \to & H^1(X,\mathcal{O}^*_X) & \xrightarrow{\pi^*} & H^1(\tilde{X},\mathcal{O}^*_{\tilde{X}})^G & \xrightarrow{\partial} & H^2(G,H^0(\mathcal{O}^*_{\tilde{X}})) \\
& \downarrow & & \delta\downarrow & & \delta_G\downarrow & & \downarrow \\
0 \to & H^2(G,\mathbf{Z}) & \to & H^2(X,\mathbf{Z}) & \xrightarrow{\pi^*} & H^2(\tilde{X},\mathbf{Z})^G & \to & H^3(G,\mathbf{Z})
\end{array} \tag{2.41}$$

in which rows and columns are exact and the map δ is the first Chern class: so Im δ is the Neron Severi group NS(X) of X and Im δ_G is NS^G .

On the other hand, we can consider the Kummer sequence

(2.42) $0 \to \mu_r \xrightarrow{r} \mathcal{O}^*_X \xrightarrow{r} \mathcal{O}^*_X \to 0$,

where by $\mathcal{O}^*_X \xrightarrow{r} \mathcal{O}^*_X$ we denote the rth power (and the corresponding sequence on $\tilde{X}$): the exact sequence of cohomology groups gives then

(2.43) $H^1(X,\mathcal{O}^*_X) \xrightarrow{r} H^1(X,\mathcal{O}^*_X) \to H^2(X,\mu_r)$

and, again, we have a diagram with exact rows and colums

$$
\begin{array}{lccccccc}
 & 0 \to H^1(G,H^0(\mathcal{O}^*_{\tilde X})) & \to & H^1(X,\mathcal{O}^*_X) & \xrightarrow{\pi^*} & H^1(\tilde X,\mathcal{O}_{\tilde X})^G & \to & H^2(G,H^0(\mathcal{O}^*_{\tilde X})) \\
 & r\downarrow & & r\downarrow & & \downarrow & & \downarrow \\
(2.44) & 0 \to H^1(G,H^0(\mathcal{O}^*_{\tilde X})) & \to & H^1(X,\mathcal{O}^*_X) & \xrightarrow{\pi^*} & H^1(\tilde X,\mathcal{O}^*_{\tilde X})^G & \xrightarrow{\partial} & H^2(G,H^0(\mathcal{O}^*_{\tilde X})) \\
 & \downarrow & & \downarrow & & \downarrow & & \downarrow \\
 & 0 \to H^2(G,\mu_r) & \to & H^2(X,\mu_r) & \xrightarrow{\pi^*} & H^2(\tilde X,\mu_r)^G & \to & H^3(G,\mu_r) .
\end{array}
$$

By diagram chasing, using 2.44, we check that the Chern class $\hat c_1(L)$ (this is the first Chern class $c_1(L)$ modulo r) of L is given by

$$(2.45)\quad \hat c_1(L) = f_{\alpha\beta}^{\ 1/r}\, f_{\beta\gamma}^{\ 1/r}\, f_{\alpha\gamma}^{\ -1/r}$$

$$= \tilde f_{(\alpha,1)(\beta,h(\alpha,\beta))}\, \tilde f_{(\beta,1)(\gamma,h(\beta,\gamma))}\, (\tilde f_{(\alpha,1)(\gamma,h(\alpha,\gamma))})^{-1}.$$

(we can take the cocycle $(1/2\pi i)[\log f_{\alpha\beta} + \log f_{\beta\gamma} + \log f_{\gamma\alpha}]$ as a representative for $c_1(L)$).

Since $f_{\alpha\beta} = (\tilde f_{(\alpha,g)(\beta,gh(\alpha,\beta))})^r$ the class is zero in $H^2(\tilde X,\mu_r)$ as follows from the equality

$$(2.46)\quad \tilde f_{(\alpha,g)(\beta,gh(\alpha,\beta))}\, \tilde f_{(\beta,gh(\alpha,\beta))(\gamma,gh(\alpha,\gamma))} = \tilde f_{(\alpha,g)(\gamma,gh(\alpha,\gamma))}.$$

Hence $\hat c_1(L)$ is cohomologous to a class coming from $H^2(G,\mu_r)$ in 2.35, and our claim is that this class is the inverse $(c_{\alpha\beta\gamma})^{-1}$ of the class $c_{\alpha\beta\gamma}$ described in 2.34.

So we have to show that

$$(2.47)\quad \hat c = \hat c_1(L) = \tilde f_{(\alpha,1)(\beta,h(\alpha,\beta))}\, \tilde f_{(\beta,1)(\gamma,h(\beta,\gamma))}(\tilde f_{(\alpha,1)(\gamma,h(\alpha,\gamma))})^{-1}$$

is cohomologous (in $H^2(X,\mu_r)$) to

$$(2.48)\quad (c_{\alpha\beta\gamma})^{-1} = (c^{h(\alpha,\beta)}_{h(\beta,\gamma)})^{-1} = c(h(\alpha,\beta),h(\beta,\gamma))^{-1},$$

where, for each g, $\bar g$, α,β it holds (notice that $c^{\bar g}_{(\alpha,g)} \in \mu_r$)

$$(2.49)\quad c^{\bar g}_{g}\ \tilde f_{(\alpha,\bar g g)(\beta,\bar g g h(\alpha,\beta))} = c^{\bar g}_{gh(\alpha,\beta)}\ \tilde f_{(\alpha,g)(\beta,gh(\alpha,\beta))}\ .$$

In particular

$$(2.50)\quad \tilde f_{(\alpha,\bar g)(\beta,\bar g h(\alpha,\beta))} = c^{\bar g}_{h(\alpha,\beta)}\ \tilde f_{(\alpha,1)(\beta,h(\alpha,\beta))}\ .$$

We shall use again the cocycle condition

$$(2.51)\quad 1 = \tilde f_{(\alpha,1)(\beta,h(\alpha,\beta))}\ \tilde f_{(\beta,h(\alpha,\beta))(\gamma,h(\alpha,\gamma))}\ (\tilde f_{(\alpha,1)(\gamma,h(\alpha,\gamma))})^{-1}\ .$$

$$(2.52)\quad \hat c = \frac{\hat c}{1} = \frac{\tilde f_{(\beta,1)(\gamma,h(\beta,\gamma))}}{\tilde f_{(\beta,h(\alpha,\beta))(\gamma,h(\alpha,\gamma))}}$$

which, using 2.50, gives the desired equality:

$$(2.53)\quad \hat c = \frac{1}{c^{h(\alpha,\beta)}_{h(\beta,\gamma)}}\ .$$

<u>Step III</u> We must now prove that, if $F': Z \to X$ is obtained by extracting the rth root of D, and we thus have $(F')^*(D) \equiv rL$, then the pull back $\tilde L$ of L to the universal cover $\tilde Z$ of Z is exactly (d/r)-divisible, if the pull back $\tilde D$ of D to the universal cover $\tilde X$ is exactly d-divisible.

We have thus $\tilde M$ such that $\tilde D \equiv d\tilde M$, and we remark that in the previous steps we have proved the following:

Fact: the map $f: \tilde{Z} \to \tilde{X}$ is a $(\mathbb{Z}/r\mathbb{Z})^3$-Galois cover, obtained as the fibre product of three elementary cyclic covers $f_i: \tilde{Y}_i \to \tilde{X}$ (hence f is the composition of three elementary cyclic covers). Each f_i is gotten by taking the rth root of a smooth and connected divisor $\tilde{D}_i$, linearly equivalent to $\tilde{D}$, inside the line bundle associated to the divisor $(d/r)\tilde{M}$.

Since $\tilde{L} = f^*((d/r)\tilde{M})$, the desired result shall follow by iterated application of the following:

Proposition 2.55 Let $f: Y \to X$ be an elementary cyclic covering of connected and simply connected complex manifolds, i.e. there is an effective smooth irreducible divisor D given by a section σ of $\mathcal{L}^{\otimes r}$, and Y is the submanifold of the total space of the line bundle associated to $\mathcal{L}$, obtained by extracting the rth root of D. Then, NS(X) denoting the Neron Severi group of X, the map f^* induces an isomorphism between NS(X) and $NS(Y)^{\mathbb{Z}/r\mathbb{Z}}$, and in particular the divisibility index of the class of a divisor M equals the one of $f^*(M)$.

Proof: We shall argue as in lemma 4 of [Ca2], recalling that $NS(X) \subset H^2(X,\mathbb{Z}_X)$ is given by the Chern classes of invertible sheaves, for which, though, we shall use the notation as for divisors (by real abuse of notation).

First of all, $f_*f^*: NS(X) \to NS(X)$ is given by multiplication by r, hence clearly, $H^2(X,\mathbb{Z}_X)$ being free, f^* is injective.

Moreover, if $f^*(M) = kN$, then first of all $N \in NS(Y)^{\mathbb{Z}/r\mathbb{Z}}$. In fact, $N = (1/k) f^*(M)$, hence $N \in H^2(S,\mathbb{Z}) \cap \mathbb{Q}NS(Y)^{\mathbb{Z}/r\mathbb{Z}} = NS(Y)^{\mathbb{Z}/r\mathbb{Z}}$. In view of the exact sequence

(2.56) $\quad 0 \to H^1(Y,\mathcal{O}_Y) \to H^1(Y,\mathcal{O}^*_Y) \to NS(Y) \to 0$,

we can assume that, if $f^*(M) = kN$, and M is an invertible sheaf on X, we have an invertible sheaf N such that $f^*(M) \equiv N^{\otimes k}$. Since $H^1(\mathbb{Z}/r\mathbb{Z}, H^1(Y,\mathcal{O}_Y)) = 0$ (these are homomorphisms of $\mathbb{Z}/r\mathbb{Z}$ into a $\mathbb{C}$-vector space), we can achieve that $N \in H^1(Y,\mathcal{O}^*_Y)^{\mathbb{Z}/r\mathbb{Z}}$.

To N we associate the theta group of automorphisms of the line bundle associated to N which cover the action of $\mathbb{Z}/r\mathbb{Z}$ on Y, we have thus a (non central) extension

(2.57) $\quad 0 \to H^0(Y,\mathcal{O}^*_Y) \to \mathcal{G}(N) \to \mathbb{Z}/r\mathbb{Z} \to 0$.

Claim: The sequence 2.57 splits.

Assuming the claim, we obtain that there is an action of $(\mathbb{Z}/r\mathbb{Z})$ on the invertible sheaf N, and it suffices then to show that the invariant direct image sheaf $\mathcal{N} = f_*(N)^{\mathbb{Z}/r\mathbb{Z}}$ is invertible, since then $N \cong f^*(\mathcal{N})$. The sheaf $\mathcal{N}$ is clearly invertible outside the branch divisor D, whereas, locally at D, $f_*\mathcal{O}_Y = \{\Sigma_{i=0}^{r-1} f_i(x,\sigma) z^i\}$, where $z^r = \sigma$, $\sigma = 0$ being the local equation of D, and (x,σ) is a local coordinate vector for X at a point of D.

Locally a generator g of $\mathbb{Z}/r\mathbb{Z}$ acts on Y by $z \to \varepsilon z$, $\varepsilon = \exp(2\pi i/r)$, and, if N is locally trivialized with w a fibre variable, by $(z, w) \to (\varepsilon z, \varphi(x,z)w)$: here $\Pi_{i=0}^{r-1}\varphi(x,\varepsilon^i z) = 1$, since $g^r = 1$.

Writing $\varphi(x,z) = \Sigma_{i=0}^{r-1} \varphi_i(x) z^i$, we obtain $\varphi_0(x)^r = 1$, hence there exists h such that $0 \le h \le r-1$ and

(2.58) $\quad \varphi(x,z) = \varepsilon^h (1+\Sigma_{i\ge1} \varphi_i(x)z^i) = \varepsilon^h \exp(\Sigma_{i\ge1} \psi_i(x) z^i)$.

Changing the local trivialization of N by

(2.59) $\quad w \to \exp(\Sigma_{j\ge1} a_j(x) z^j)$

we replace $\varphi(x,z)$ by $\varepsilon^h \exp(\Sigma_{j\geq 1} z^j(\psi_j(x) - a_j(x) + \varepsilon^j a_j(x)))$, hence we may assume $\psi_i(x) \equiv 0$ for i not divisible by r.

Finally, since $\varphi = \varphi(x,\sigma)$, the equality $\Pi_{i=0}^{r-1} \varphi(x,\varepsilon^i z) = 1$ implies that $\varphi = \varepsilon^h$. But then the locally invariant sections are given by functions $\zeta(x,z)$ such that $\zeta(x,\varepsilon z) = \varepsilon^h \zeta(x,z)$, hence we can write $\zeta(x,z) = z^h \xi(x,z)$, thereby proving that $f_*(N)^{\mathbb{Z}/r\mathbb{Z}}$ is invertible.

There remains to prove the claim.

Proof of the claim: We choose an acyclic cover $\{U_\alpha\}$ of X such that each inverse image $\varphi^{-1}(U_\alpha) = V_\alpha$ is also acyclic. Then V_α is defined by local equations in $U_\alpha \times \mathbb{C}$, $z_\alpha^r = \sigma_\alpha$, with $z_\alpha = g_{\alpha\beta} z_\beta$, $g_{\alpha\beta}$ being a cocycle in $H^1(X, \mathcal{O}_X^*)$ for $\mathcal{L}$.

Then $H^0(\mathcal{O}_Y) \cong H^0(\mathcal{O}_X) \oplus (\oplus_{i=1}^{r-1} H^0(L^{-i}))$, and we have just written the eigenspace decomposition for the action of $\mathbb{Z}/r\mathbb{Z}$ on Y.

Let $(n_{\alpha\beta})$ be a cocycle for N relative to the cover $\{V_\alpha\}$ of Y: saying that $N \in H^1(Y, \mathcal{O}_Y^*)^{\mathbb{Z}/r\mathbb{Z}}$ means that, if $n_{\alpha\beta} = \Sigma_{i=0}^{r-1} n_{i,\alpha\beta} z_\alpha^i$, $\varepsilon = \exp(2\pi i/r)$, then the cocycle $\hat{n}_{\alpha\beta} = \Sigma_{i=0}^{r-1} n_{i,\alpha\beta} \varepsilon^i z_\alpha^i$ is cohomologous to $n_{\alpha\beta}$. I.e.,

(2.60) $\quad n_{\alpha\beta} = \hat{n}_{\alpha\beta} (\Sigma_{i=0}^{r-1} \psi_{i,\alpha} z_\alpha^i)^{-1} (\Sigma_{i=0}^{r-1} \psi_{i,\beta} (g_{\alpha\beta} z_\alpha)^i)$.

The equation 2.60 is indeed equivalent to the assertion: if $w_\alpha = n_{\alpha\beta} w_\beta$ is a fibre coordinate for the line bundle associated to N, then $w_\alpha \to \psi_\alpha w_\alpha$ ($\psi_\alpha = \Sigma_{i=0}^{r-1} \psi_{i,\alpha} z_\alpha^i$) lifts the action of the generator g of $\mathbb{Z}/r\mathbb{Z}$ from Y to N.

One can lift this action in a different way, just by multiplying ψ_α by a global invertible function ν on Y, and what we have to show amounts to prove that we can choose ν in such a way that this action has period r. In other words, we want

(2.60) $\quad \Pi_{i=0}^{r-1} g^i(\psi_\alpha \nu) = 1$.

We notice that $\xi_\alpha = \Pi_{i=0}^{r-1} g^i(\psi_\alpha)$ is an invariant invertible function on V_α, hence $\xi_\alpha \in \mathcal{O}_X^*(U_\alpha)$; moreover, by the previous equation 2.60, $\xi_\alpha = \xi_\beta$ and we have $\xi \in H^0(X, \mathcal{O}_X^*)$. Since X is simply connected, we can choose ν to be the inverse of an rth root of ξ, whence $\Pi_{i=0}^{r-1} g^i(\psi_\alpha \nu) = \nu^r \cdot \xi = 1$. Q.E.D.

§3 Back to stable bundles and linear systems

In order to apply the previous theorem, let now $\pi: Y \to S$ be a $(\mathbb{Z}/2\mathbb{Z})^2$-Galois cover of S as in Lemma 2.1, such that $\pi^*(K_S) \equiv 2L$ for a line bundle L on Y and $\pi_*(\mathcal{O}_Y) = \mathcal{O}_S \oplus (\mathcal{O}_S(-3K_S))^3$.

Proposition 3.1 The pullback $\pi^*(\mathcal{E})$ is $\pi^*(H)$-stable if H is an ample line bundle on S such that $\mathcal{E}$ is H-stable.

Proof By pullback under π we have the exact sequence on Y

(*) $\quad 0 \to \mathcal{O}_Y \to \pi^*\mathcal{E} \to \mathcal{I}_{\pi^{-1}(x)}(\pi^* K_S) \to 0$.

Let $0 \to N' \to \pi^*\mathcal{E} \to \mathcal{I}_Z(M') \to 0$ be a $H'(=\pi^*H)$-semi-destabilizing sequence for $\pi^*\mathcal{E}$ on Y. The Galois group $G = (\mathbb{Z}/2\mathbb{Z})^2$ acts on Y and $\pi^*\mathcal{E}$ has a natural G-linearization. There are two different cases

i) $g^*N' = N'$ for each $g \in G$;
ii) $g^*N' \neq N'$ for some $g \in G$.

In case i) the line bundle inherits from $\pi^*\mathcal{E}$ a G-linearization and then there exists an invertible subsheaf N of $\mathcal{E}$ on S such that $N' = \pi^*N$. But then $H \cdot N = (1/4)\, H' \cdot N' \geq (1/8)\, \pi^*K_S \cdot H' = (1/2)\, K_S \cdot H$ and $\mathcal{E}$ is not H-stable, which is absurd.

In case ii), let us set $g^*N' = N''$. The line bundle N" is still a subsheaf of $\mathcal{E}$ and satisfies also the equalities $(N')^2 = (N'')^2$, $H' \cdot N' = H' \cdot N''$.

By hypothesis N' and N" are distinct, so the map obtained as a composition $\beta: N'' \to \pi^*\mathcal{E} \to \mathcal{I}_S(M')$ gives a non zero element of $H^0(\mathcal{I}_Z(M'-N''))$.

But $\mathrm{div}(\beta) \cdot H' = H' \cdot M' - H' \cdot N'' \leq 0$ implies $N' \cdot H' = M' \cdot H'$, $N'' \equiv M'$, $Z = \emptyset$. So the bundle $\pi^*\mathcal{E}$ splits as a direct sum $\pi^*\mathcal{E} \cong N' \oplus N''$ and there exists two respective global holomorphic sections of N' and N" such that the sequence (*) has the following form

$$0 \to \mathcal{O}_Y \xrightarrow{{}^t(n',n'')} \pi^*\mathcal{E} \xrightarrow{(-n'',n')} \mathcal{I}_{(\pi^{-1}(x))}(\pi^*K_S) \to 0 .$$

In particular, $(n'=0) \cap (n''=0) = \pi^{-1}(x)$ and $N' \cdot N'' = 4$. But the long exact sequence of cohomology associated to the sequence (*) gives

$$0 \neq h^0(\mathcal{I}_{(\pi^{-1}(x))}(\pi^*K_S)) \leq h^0(\pi_*\pi^*K_S) = h^0(K_S \oplus \mathcal{O}_Y(-2K_S)^3) = 0.$$

Q.E.D.

Since, as we saw in theorem 1.1, the canonical bundle is not 2-divisible, we consider the $(\mathbb{Z}/2\mathbb{Z})^2$-Galois cover $\pi: Y \to S$ of S described in Lemma 2.1 and associated to the linear system $|D| = |3K_S|$. Then there exists a line bundle L on Y such that $\pi^*(K_S) \equiv 2L$ and $\pi_*(\mathcal{O}_Y) = \mathcal{O}_S \oplus (\mathcal{O}_S(-3K_S))^3$. By Prop. 3.1 the pullback $\pi^*(\mathcal{E})$ is $\pi^*(H)$-stable for any ample line bundle H on S such that $\mathcal{E}$ is H-stable. Moreover, $\pi^*(\mathcal{E})(-L)$ has trivial Chern classes, hence it gives rise to an irreducible SU(2)-representation of $\pi_1(Y)$ not induced by $\pi_1(Y) \to \pi_1(S)$.

We get thus the theorem

<u>Theorem 3.2</u> Let S be a minimal surface with $p_g = 0$, $K_S^2 = 4$. Then $|2K_S|$ is base point free if, $\pi: \tilde{S} \to S$ being the universal cover, either
i) $\pi^*(K_S)$ is not 2-divisible,
ii) $\pi^*(K_S)$ is 2-divisible and either its Chern class modulo 2 is trivial or it gives a central extension

$$1 \to \mathbb{Z}/2\mathbb{Z} \to \Gamma \to \pi_1(S) \to 0$$

not associated to any irreducible PU(2)-representation of $\pi_1(S)$.

We give now an alternative proof (using Donaldson's theorem) of the existence of an example where $m = 2$, D is not 2-divisible, but $\pi_1(Y) \neq \pi_1(S)$.

Let A be a simple minimal abelian surface admitting a polarization L of type (1,4). Consider now a $G=(\mathbb{Z}/2\mathbb{Z})^2$-Galois ramified cover $\pi: Y \to A$ over A associated to $|L|$ as in Lemma 2.1. The pullback of L under π is 2-divisible and $\pi_*(\mathcal{O}_Y) = \mathcal{O}_S \oplus (\mathcal{O}_S(-L))^3$. The map associated to the linear system $|L|$ is a well defined birational but not injective morphism $\varphi = \varphi_{|L|}$ on A (cf. [BLvS]); let $Z = \{x_1, x_2\}$ be a subset of A such that $\varphi(x_1) = \varphi(x_2)$. There exists then a vector bundle $\mathcal{E}$ of rank 2 on S occurring as an extension

(3.3) $$0 \to \mathcal{O}_A \to \mathcal{E} \to \mathcal{I}_Z(L) \to 0 .$$

Proposition 3.4 The vector bundle $\mathcal{E}$ is L-stable (since A is simple).

Proof Otherwise there exists an invertible and saturated subsheaf N of $\mathcal{E}$ which satisfies the inequality $N \cdot L \geq (1/2) L^2 = 4$ and gives a diagram of exact arrows of the following form

$$\begin{array}{ccccccccc} & & & & 0 & & & & \\ & & & & \downarrow & & & & \\ & & & & N & & & & \\ & & & & \downarrow & & & & \\ (3.5) \quad 0 & \to & \mathcal{O}_A & \to & \mathcal{E} & \to & \mathcal{J}_Z(L) & \to & 0 \\ & & & & \downarrow & & & & \\ & & & & \mathcal{J}_W(M) = \mathcal{J}_W(L-N) & & & & \\ & & & & \downarrow & & & & \\ & & & & 0 & & & & \end{array}$$

for a suitable 0-dimensional subset of A.

The composition of maps $\alpha: N \to \mathcal{E} \to \mathcal{J}_Z(L)$ is non-zero and so M is an effective non-zero divisor. We have that $N^2 \geq M^2$, because of the inequality

(3.6) $\quad 0 \leq (N - M) \cdot L = (N - M) \cdot (N + M) = N^2 - M^2$,

and by computing the first Chern class of $\mathcal{E}$, we get the equality $M \cdot N + \deg W = 2$ which implies that $M \cdot N \leq 2$.

Then we get $M^2 \leq 2$ from the inequality $M \cdot L \leq 4$.

By the hypothesis that A is simple, since M^2 is even, the only possibility is $M^2 = 2$. But then by the Index theorem $M^2 N^2 \leq 4 \Rightarrow N^2 \leq 2 \Rightarrow M^2 = N^2 = 2$, $M \cdot N = 2$, and again by the Index theorem M and N differ by a topologically trivial line bundle on A: but this is impossible, because the line bundle L is not 2-divisible on A. Q.E.D.

Theorem 3.7 The pullback bundle $\pi^*(\mathcal{E})$ on Y is $\pi^*(L)$-stable.

Corollary 3.8 The corresponding surjective map $\pi_1(Y) \to \pi_1(A)$ between fundamental groups has a kernel isomorphic to $\mathbb{Z}/2\mathbb{Z}$.

Proof of corollary 3.9 Otherwise, Lemma 2.1 would imply that $\pi_1(Y) \cong \pi_1(A) \cong \mathbb{Z}^4$ and, by Donaldson's Theorem, Y does not admit any stable rank two vector bundle with trivial Chern classes because every SU(2)-representation of $\pi_1(Y)$ is reducible. But $\pi^*(L) \equiv 2L''$ for some line bundle L'' on Y and the vector bundle $\pi^*(\mathcal{E}) \otimes (L'')^{-1}$ cannot be stable, contradicting the theorem. Q.E.D. for corollary 3.8

Proof of theorem 3.8 As in Prop. 3.1, we can assume otherwise that the bundle $\pi^*(\mathcal{E})$ splits as a direct sum $\pi^*\mathcal{E} \cong N' \oplus N''$ of line bundles on Y such that $(N')^2 = (N'')^2 = N' \cdot N'' = 8$ and so by the Index theorem N' is homologous to N''. In particular, by pullback under π, the sequence 3.3 gives rise to an exact sequence on Y of the form

$$(3.10) \qquad 0 \to \mathcal{O}_Y \xrightarrow{{}^t(n',n'')} \pi^*\mathcal{E} \xrightarrow{(-n'',n')} \mathcal{J}_{(\pi^{-1}(x))}(\pi^* K_S) \to 0.$$

where n' and n'' are respective holomorphic sections of N' and N'' whose divisors have no common components and meet (tranversally) in 8 points.

Since in any case $H^1(A,\mathbb{Z}) \cong H^1(Y,\mathbb{Z})$, then $\mathrm{Pic}^0(Y) \cong \mathrm{Pic}^0(A)$ and there exists a topologically trivial line bundle M such that $N'' \cong N' \otimes \pi^*(M)$.

It follows that $(\pi_*(N'\oplus N'')) \cong \pi_*(\pi^*\mathcal{E}) \cong \mathcal{E}\otimes\pi_*\mathcal{O}_Y \cong \mathcal{E}\oplus\mathcal{E}(-L)^3$. This is absurd: in fact, being also $(\pi_*(N'\oplus N'')) \cong \pi_*(N')\oplus(\pi_*(N')\otimes M))$, for each stable quotient $\mathcal{F}$ in the Harder-Narasimhan filtration of $(\pi_*(N'\oplus N''))$, a corresponding quotient $\mathcal{F}\otimes M$ must also appear (cf. [Ko], ch. 5), whereas $\mathcal{E}$ and $\mathcal{E}(-L)$ are stable.

Q.E.D. for Theorem 3.9

REFERENCES

[BLvS] Ch. Birkenhake, H. Lange, D. Van Straten, Abelian surfaces of type (1,4), Math. Ann., 285 (1989), 625-646

[Bo1] E. Bombieri, The pluricanonical map of a complex surface, Lecture Notes in Math., 155 (1971), Springer Verlag, 35-87

[Bo2] E. Bombieri, Canonical models of surfaces of general type, Inst. Hautes Etudes Sci. Publ. Math. , 42 (1973), 171-219

[Bog] F.A. Bogomolov, Holomorphic tensors and vector bundles on projective varieties, Math. USSR Izvestija, 13 (1979), 499-555

[BPV] W. Barth, C. Peters and A. Van de Ven, Compact complex surfaces, Ergebnisse der Math., Springer Verlag, 1984

[Bu] P. Burniat, Sur les surfaces de genre $P_{12} > 0$, Ann. Math. Pura Appl. (4), 71 (1966), 1-24

[Ca1] F. Catanese, On the moduli spaces of surfaces of general type, J. Differential Geometry, 19 (1984),483-515

[Ca2] F. Catanese, Connected components of moduli spaces, J. Differential Geometry, 24 (1986), 395-399

[Ca-Ci] F. Catanese and C. Ciliberto, Surfaces with $p_g = q = 1$, in "Problems in the theory of surfaces and their classification", Cortona, Italy, Oct. 1988, Symposia Math. 32, Academic Press

[Do] S. Donaldson, Anti self-dual Yang-Mills connections over complex algebraic surfaces and stable vector bundles, Proc. London Math. Soc. (3), 50 (1985), 1-26

[Dol] A. Dold, Lectures on algebraic topology, Springer Verlag, Grundlehren 200, Berlin (1972)

[Fr] P. Francia, The bicanonical map for surfaces of general type, in "Problems in the theory of surfaces and their classification", Cortona, Italy, Oct. 1988, Symposia Math. 32, Academic Press

[G] A. Grothendieck, Sur quelques points d'algebre homologique, Tohoku Math. J., (1957), 119-221

[GH] P. Griffiths, J. Harris, Residues and zero-cycles on algebraic varieties, Ann. of Math., 108 (1978), 461-505

[Ke] J.H. Keum, Some new surfaces of general type with p_g=0, preprint, 1988

[Ko] S. Kobayashi, Differential geometry of complex vector bundles, Publication of the Mathematical Society of Japan 15, Iwanami Shoten Publishers and Princeton University Press, 1987

[Kod] K. Kodaira, Pluricanonical systems on algebraic surfaces of general type, J. Math. Soc. Japan, 20 (1968), 170-192

[Kot] D. Kotschick, Stable and unstable bundles on algebraic surfaces, in "Problems in the theory of surfaces and their classification", Cortona, Italy, Oct. 1988, Symposia Math. 32, Academic Press

[Mi1] Y. Miyaoka, On numerically Campedelli surfaces, in "Complex Analysis and Algebraic Geometry", Iwanami Shoten, Tokyo (1977), 112-118

[Mi2] Y. Miyaoka, On the Kodaira dimension of minimal threefolds, Math. Ann., 281 (1988), 325-332

[Moi] B. Moishezon, in "Algebraic surfaces", edited by I.R. Shafarevic, Trudy Math. Inst. Steklov, 75 (1965); english transl. in Amer. Math. Soc. Transl. (2), 63 (1967)

[Mu] D. Mumford, Abelian varieties, Tata Institute of fundamental research, Bombay, Oxford University Press, 1974

[Pet] C.A.M. Peters, On certain examples of surfaces with $p_g = 0$ due to Burniat, Nagoya Math. J., 66 (1977), 109-119

[Pet2] C.A.M. Peters, On two types of surfaces of general type with vanishing geometric genus, Inv. Math. 32 (1976)

[Rei] I. Reider, Vector bundles of rank two and linear systems on algebraic surfaces, Ann. of Math., 127 (1988), 309-316

[Sch1] R.E.L. Schwarzenberger, Vector bundles on algebraic surfaces, Proc. London Math., 11 (1961), 601-622

[Sch2] R.E.L. Schwarzenberger, Vector bundles on the projective plane, Proc. London Math., 11 (1961), 623-640

[Se] J.P. Serre, Cours d'arithmetique, Presses Univ. de France, Paris (1970)

[We] L. Weng, A remark on bicanonical maps of surfaces of general type, preprint of the Max Planck Institut fur Mathematik, Bonn (1990)

[Xi] G. Xiao, Surfaces fibrees en courbes de genre deux, Springer L.N.M. 1137, 1985

VERS UNE STRATIFICATION DE L'ESPACE DES MODULES DES VARIETES ABELIENNES PRINCIPALEMENT POLARISEES

par Olivier Debarre

A la conférence sur les Fonctions Thêta organisée par l'American Mathematical Society à Bowdoin en 1987, A. Beauville termina son exposé ([B 1]) sur les variétés de Prym par le tableau suivant –que je me suis permis de traduire et de modifier légèrement :

	Jacobiennes	Pryms
dim Sing Θ	$g-4$	$g-6$
Réductibilité de	$\Theta\cap\Theta_a$	$\Theta\cap\Theta_a\cap\Theta_b$
La variété de Kummer a des	droites trisécantes	plans quadrisécants
Courbes de classe m *fois* la classe minimale	$m=1$	$m=2$

en posant la question : est-il possible de *définir* géométriquement une stratification de l'espace des modules des variétés abéliennes principalement polarisées complexes de dimension g qui compléterait ce tableau ?

Sans répondre complètement à cette question, nous essayons dans cet article de mettre en évidence des relations entre les propriétés suivantes d'une variété abélienne principalement polarisée (X,Θ) de dimension g :

1) Existence d'une "courbe" de m-plans $(m+2)$-sécants à la variété de Kummer $K(X)$ de (X,Θ) (voir Théorème 4.1 pour l'énoncé exact).

2) Existence d'un m-plan $(m+2)$-sécant à $K(X)$.

3) Existence d'une courbe dans X de classe m *fois* la classe minimale.

4) Le lieu singulier de Θ est de dimension au moins $g-2m-2$.

5) La variété (X,Θ) est la variété de Prym-Tjurin associée à une correspondance D symétrique *effective* sans point *fixe* vérifiant $(D-1)(D+m-1)=0$.

Chacune de ces propriétés est vérifiée par les jacobiennes de courbes pour $m=1$ et par les variétés de Prym pour $m=2$. On peut donc espérer les utiliser pour construire la stratification voulue de l'espace des modules des variétés abéliennes principalement polarisées.

Sous des hypothèses restrictives, pour lesquelles nous renvoyons aux énoncés des théorèmes (en particulier, on suppose la plupart du temps la variété abélienne principalement polarisée X suffisamment générale), nous montrons les implications suivantes :

$$\begin{array}{ccccc} 1) & \Rightarrow & 2) & \Rightarrow & 4) \\ \Downarrow & & & & \Uparrow \\ 3) & & \Leftarrow & & 5) \end{array}$$

L'implication 5)⇒3) est due à Welters ([W 1]). La démonstration de 1)⇒3) (Théorème 4.1) est une extension de la démonstration de Gunning du cas $m=1$ ([G 1]). Celle de l'implication 2)⇒4) (Corollaire 3.5) emprunte des idées de [B-D] et celle de 5)⇒4) une idée de A. Bertram ([Be]).

Rappelons que dans le cas $m=1$, la propriété 3) caractérise les jacobiennes parmi les variétés abéliennes principalement polarisées indécomposables ("critère de Matsusaka" [M]). C'est aussi pratiquement le cas de la propriété 1) ("critère de Gunning" [G 1]), mais pas de la propriété 4) ([D,1], [Do]). Welters a conjecturé que la propriété 2) caractérise les jacobiennes ("conjecture de la trisécante" [W 2], [D 2]).

En ce qui concerne le cas $m=2$, on sait que les variétés de Prym ne sont pas les seules variétés abéliennes principalement polarisées qui satisfont 2) ([B-D] Remarque 1) page 617) ou 4) ([D 1]). En décrivant complètement dans [W 1] les variétés abéliennes principalement polarisées satisfaisant 3), Welters montre en particulier que cette propriété ne caractérise pas non plus les variétés de Prym.

En nous basant sur cette étude de Welters et suivant les idées de [G 1], nous proposons pour terminer une caractérisation des variétés de Prym de type 1) (Théorème 5.2).

Pour conclure, je voudrais proposer quelques questions proches des problèmes abordés dans cet article (les deux premières sont dues à Welters [W 1]) : quel est le plus petit entier m tel que toute variété abélienne principalement polarisée de dimension g contienne une courbe de classe m fois la classe minimale ? Pour tout entier $m \geq 1$, quel est le genre maximal des courbes de classe m fois la classe minimale sur une variété abélienne principalement polarisée de dimension g ? Quelles sont les variétés de Prym-Tjurin dont la variété de Kummer admet des m-plans $(m+2)$-sécants ?

On se place dans tout cet article sur le corps des nombres complexes.

1. Jacobiennes de courbes

Soit C une courbe projective lisse de genre g et soit JC sa jacobienne. Pour tous points p, p_1, p_2 et p_3 de C, on a une inclusion schématique :

$$\Theta \cap \Theta_{p_2-p_1} \subset \Theta_{p-p_1} \cup \Theta_{p_2-p_3} ,$$

où Θ est un diviseur thêta sur JC et où, pour tout élément x de JC, Θ_x désigne le translaté de Θ par x. Cette propriété admet l'interprétation géométrique suivante. On peut supposer le diviseur Θ symétrique. Soit $\psi : JC \longrightarrow \mathbb{P}^{2^g-1}$ le morphisme associé au système linéaire $|2\Theta|$. L'image de ψ est la *variété de Kummer* de JC. Soit p_0 un point de C et soit Γ l'image de C dans JC par l'application $p \longmapsto \mathcal{O}_C(p-p_0)$. On a :

(1.1) Pour tous points x_1, x_2 et x_3 de Γ et tout élément ζ de $\frac{1}{2}(\Gamma - x_1 - x_2 - x_3)$, les points $\psi(\zeta+x_1)$, $\psi(\zeta+x_2)$ et $\psi(\zeta+x_3)$ sont alignés.

La variété de Kummer de JC admet donc une famille de dimension 4 de droites trisécantes. Une extension moins connue de ce résultat, due à Gunning ([G 2], [G 3]) est qu'il existe, pour tout $m \geq 2$, une famille de dimension $2m+2$ de m-plans $(m+2)$-sécants.

(1.2) Plus exactement, pour tous points $x_1, \ldots, x_{m+2}, y_1, \ldots, y_m$ de Γ, et $2\zeta = \Sigma y_j - \Sigma x_j$, les points $\psi(\zeta + x_1), \ldots, \psi(\zeta + x_{m+2})$ sont sur un m-plan.

Rappelons pour terminer que le lieu singulier de Θ est partout de codimension inférieure ou égale à 4 dans JC et qu'on a :

(1.3) $\forall x, x' \in \Gamma \qquad \operatorname{Sing}\Theta \subset \Theta_{x-x'}$,

de sorte que $\qquad \dim(\Theta_{x-x'} \cap \operatorname{Sing}\Theta) \geq g-4$.

2. Variétés de Prym

Soit $\widetilde{C} \longrightarrow C$ un revêtement étale double de courbes projectives lisses. On note σ l'involution correspondante de $\widetilde{C}$ et σ^* l'endomorphisme induit sur $J\widetilde{C}$. La variété de Prym associée est la sous-variété abélienne P de $J\tilde{C}$ image de l'endomorphisme $(1-\sigma^*)$. Elle est munie d'une polarisation principale naturelle ([Mu]). Pour tous points p et q de $\tilde{C}$, on notera $[p,q]$ l'élément $p+q-\sigma p-\sigma q=(1-\sigma^*)(p-\sigma q)$ de P.

On rappelle que si C est suffisamment générale ([B-D] Proposition 1) et si p, p_1, p_2, et p_3 sont 4 points de $\tilde{C}$, on a une inclusion schématique :

(2.1) $$\Theta \cap \Theta_{[p,p_1]} \cap \Theta_{[p,p_2]} \subset \Theta_{[p,p_3]} \cup \Theta_{[p_1,p_2]}$$

où Θ est un diviseur thêta de P.

Cela entraîne que pour tout élément ζ de P tel que $2\zeta=[p,p_3]+[p_1,p_2]$, les 4 points $\psi(\zeta)$, $\psi(\zeta-[p,p_1])$, $\psi(\zeta-[p,p_2])$ et $\psi(\zeta-[p,p_3])$ de la variété de Kummer de P sont coplanaires. Il y a donc une famille de dimension 4 de plans quadrisécants.

Il est utile d'exprimer cette propriété sous une forme légérement différente. Le choix d'un point p_0 de $\tilde{C}$ permet de définir un morphisme $p\longmapsto[p,p_0]$ qui envoie $\tilde{C}$ sur une courbe Γ dans P. Pour tous points p_1, p_2, p_3 et p de $\tilde{C}$, on choisit un point ζ de $J\tilde{C}$ tel que :

$$2\zeta = ([p,p_0]-[p_1,p_0]-[p_2,p_0]-[p_3,p_0]).$$

Un petit calcul montre que les 4 points coplanaires précédents sont :

$$\psi(\zeta+[p_1,p_0]),\ \psi(\zeta+[p_2,p_0]),\ \psi(\zeta+[p_3,p_0]) \text{ et}$$
$$\psi(\zeta+[p_1,p_0]+[p_2,p_0]+[p_3,p_0]-[p_0,p_0]).$$

(2.2) Il existe donc une courbe $\Gamma\subset P$ et un point x_0 de Γ tels que, pour tous points x_1, x_2 et x_3 de Γ et tout élément ζ de $\frac{1}{2}(\Gamma-x_1-x_2-x_3)$, les 4 points $\psi(\zeta+x_1)$, $\psi(\zeta+x_2)$, $\psi(\zeta+x_3)$ et $\psi(\zeta+x_1+x_2+x_3-x_0)$ sont coplanaires.

Remarquons que la classe de Γ est deux fois la classe minimale et que $x_0-\Gamma=\Gamma$. On utilisera plus loin la propriété (2.2) pour donner une caractérisation des variétés de Prym.

Comme dans le cas des jacobiennes, on peut généraliser cette propriété des variétés de Prym en montrant qu'il existe une famille de dimension $(2m+2)$ de $2m$-plans $(2m+2)$-sécants, pour tout entier m. Il semble que tous les $(2m+1)$-plans $(2m+3)$-sécants soient dégénérés (i.e. contiennent un $2m$-plan $(2m+2)$-sécant).

Rappelons pour terminer que le lieu singulier de Θ contient une sous-variété notée $\mathrm{Sing}_{st}\Theta$ qui est partout de codimension ≤ 6 dans P ([Be], [D 1]). Elle vérifie de plus :

(2.3) $\quad \forall x, x' \in \Gamma \qquad \mathrm{Sing}_{st}\Theta \subset \Theta_{x-x'}$

de sorte que $\quad \dim(\Theta_{x-x'} \cap \mathrm{Sing}\,\Theta) \geq g-6$.

3. Plans sécants et singularités du diviseur thêta

Soit (X,Θ) une variété abélienne principalement polarisée indécomposable de dimension g , où Θ est un diviseur thêta symétrique, et soit $\{\theta_1,\dots,\theta_{2^g}\}$ une base de l'espace des sections de $\mathcal{O}_X(2\Theta)$. L'image du morphisme associé $\psi : X \longrightarrow \mathbb{P}^{2^g-1}$ est la variété de Kummer K(X) de X .

Soit m un entier strictement inférieur à g . On suppose qu'il existe des points $x_1,\dots,x_{m+2}$ de X non d'ordre 2 tels que leurs images par ψ soient distinctes et situées sur un espace linéaire de dimension m . En choisissant m minimal, on peut toujours supposer que ces images sont en position générale sur ce m-plan. Il existe alors des constantes $\lambda_1,\dots,\lambda_{m+2}$ non nulles telles que :

(3.1) $\quad \forall n \in \{1,\dots,2^g\} \qquad \displaystyle\sum_{i=1}^{m+2} \lambda_i\, \theta_n(x_i) = 0$.

La formule d'addition de Riemann permet de transformer cette égalité en :

$$\forall x \in X \qquad \sum_{i=1}^{m+2} \lambda_i\, \theta(x+x_i)\,\theta(x-x_i) = 0 ,$$

où θ désigne un générateur de $H^0(X,\mathcal{O}_X(\Theta))$. On en déduit l'inclusion schématique :

(3.2) $\quad \Theta_{x_1} \cap \Theta_{x_2} \cap \dots \cap \Theta_{x_{m+1}} \subset \Theta_{x_{m+2}} \cup \Theta_{-x_{m+2}}$.

Choisissons une base $D_1,\dots,D_g$ de champs de vecteurs sur X et, pour tout $j=1,\dots,m+1$, un générateur θ_{x_j} de $H^0(\mathcal{O}_X(\Theta_{x_j}))$. On définit le *lieu jacobien* $\mathrm{Jac}(\Theta_{x_1} \cap \dots \cap \Theta_{x_k})$ comme le schéma des zéros sur $\Theta_{x_1} \cap \ldots \cap \Theta_{x_k}$ des mineurs maximaux de la matrice jacobienne $(D_j\theta_{x_i})_{1\le i\le k, 1\le j\le g}$. Ensemblistement, c'est le lieu où $\Theta_{x_1} \cap \ldots \cap \Theta_{x_k}$ n'est pas lisse de codimension k.

Supposons l'inclusion (3.2) vérifiée. On distingue deux cas :

i) Si $Z = \Theta_{x_1} \cap \ldots \cap \Theta_{x_{m+1}}$ n'est pas de codimension $(m+1)$, on a alors :

$$\dim \mathrm{Jac}(\Theta_{x_1} \cap \ldots \cap \Theta_{x_{m+1}}) \geq g-m .$$

ii) Si par contre Z est de codimension $(m+1)$, le même argument que celui employé dans [B-D] Proposition 2, basé sur la résolution de Koszul de l'idéal de cette intersection, montre que Z n'est pas contenu dans $\Theta_{x_{m+2}}$ ni dans $\Theta_{-x_{m+2}}$. Ceci prouve que $Z \cap \Theta_{x_{m+2}} \cap \Theta_{-x_{m+2}}$ est contenu dans $\mathrm{Jac}(\Theta_{x_1} \cap \ldots \cap \Theta_{x_{m+1}})$, donc que :

$$\dim \mathrm{Jac}(\Theta_{x_1} \cap \ldots \cap \Theta_{x_{m+1}}) \geq g-m-2 . \tag{3.3}$$

L'inclusion (3.2) entraîne donc toujours l'inégalité (3.3). Il apparaît que cette inégalité force l'existence de points singuliers sur le diviseur Θ :

Théorème 3.4. *Soit* X *une variété abélienne et soient* $x_1, \ldots, x_{m+1}$ *des points de* X *tels que, pour tous* $j \neq k$, $(x_j - x_k)$ *engendre* X. *Alors :*

$$\dim \mathrm{Sing}\,\Theta \geq \dim \mathrm{Jac}(\Theta_{x_1} \cap \ldots \cap \Theta_{x_{m+1}}) - m .$$

Corollaire 3.5. *Sous les hypothèses du théorème précédent, on suppose de plus que les images des points* $x_1, \ldots, x_{m+1}$ *et d'un autre point* x_{m+2} *de* X *sur la variété de Kummer sont distinctes et situées sur un* m*-plan. Alors :*

$$\dim \mathrm{Sing}\,\Theta \geq g - 2m - 2 .$$

Remarques 3.6. 1) Il ressort de la démonstration du théorème qu'on a en fait des résultats un peu plus précis. Dans le théorème et le corollaire, il existe un indice j tel que, respectivement :

$$\dim(\mathrm{Jac}(\Theta_{x_1} \cap \ldots \cap \Theta_{x_{m+1}}) \cap \mathrm{Sing}\,\Theta_{x_j}) \geq \dim \mathrm{Jac}(\Theta_{x_1} \cap \ldots \cap \Theta_{x_{m+1}}) - m$$

et :

$$\dim(\Theta_{x_1} \cap \ldots \cap \Theta_{x_{m+1}} \cap \Theta_{x_{m+2}} \cap \Theta_{-x_{m+2}} \cap \mathrm{Sing}\,\Theta_{x_j}) \geq g-2m-2 .$$

2) L'hypothèse que les points $(x_j - x_k)$ engendrent X est essentielle : si (X,Θ) est une variété abélienne principalement polarisée qui contient une courbe elliptique E telle que $\Theta.E = 2$, l'image de E dans la variété de Kummer de X est une conique, alors que Θ est lisse en général ([B-D] Remarque 1) page 617) !

Avant de démontrer le théorème, énonçons un lemme élémentaire qui nous servira à deux reprises. Sa démonstration est laissée au lecteur.

Lemme 3.7. *Soient Y un schéma réduit et $L_1,\dots,L_r$ des faisceaux inversibles sur Y . On suppose que pour chaque $j=1,\dots,r$, il existe n sections $s_1^j,\dots,s_n^j$ de L_j ($n \geq r$) telles que, pour tout y générique dans Y , on ait :*

$$\operatorname{Rang}(s_i^j(y)) = r-1 .$$

Alors il existe, pour $j=1,\dots,r$, des sections λ_j de $L_1 \otimes \dots \otimes \hat{L}_j \otimes \dots \otimes L_r$ non toutes identiquement nulles telles que :

$$\forall i \qquad \forall y \in Y \qquad \sum_{j=1}^{r} \lambda_j(y)\, s_i^j(y) = 0 .$$

Démonstration du théorème 3.4.

On procède par récurrence sur m , le résultat étant évident lorsque m est nul. Soit Z une composante de dimension maximale de $\operatorname{Jac}(\Theta_{x_1} \cap \dots \cap \Theta_{x_{m+1}})_{red}$. Supposons tout d'abord que la variété :

$$B = \bigcup_{j=1}^{m+1} \operatorname{Jac}(\Theta_{x_1} \cap \dots \cap \hat{\Theta}_{x_j} \cap \dots \cap \Theta_{x_{m+1}}) \cap Z ,$$

soit de codimension au moins 2 dans Z . On choisit une courbe générique C dans $Z-B$. Chaque $D_j\theta_{x_j}$ peut être considéré comme une section de $\mathcal{O}_C(\Theta_{x_j})$. Par définition du lieu jacobien, il résulte du lemme 3.7 qu'il existe des sections non toutes nulles λ_j de $\mathcal{O}_C(\Theta_{x_1} + \dots + \hat{\Theta}_{x_j} + \dots + \Theta_{x_{m+1}})$ telles que :

$$\forall z \in C \qquad \sum_j \lambda_j D_i\theta_{x_j}(z) = 0 .$$

Soit $n : N \longrightarrow C$ la normalisation de C. Par construction, les sections $n^*\lambda_j$ ne peuvent être nulles que simultanément. Il en résulte qu'elles ont même diviseur F sur N, et que :

$$\forall j \qquad n^*\mathcal{O}_C(\Theta_{x_1} + \ldots + \hat{\Theta}_{x_j} + \ldots + \Theta_{x_{m+1}}) \simeq \mathcal{O}_N(F) .$$

On déduit alors du théorème du carré que :

$$\forall j,k \qquad n^*\mathcal{O}_C(\Theta_{x_j - x_k} - \Theta) \quad \text{est trivial} .$$

Comme dans [B-D] Proposition 3, cela contredit l'hypothèse que $(x_j - x_k)$ engendre X.

La variété B est donc de codimension au plus 1 dans Z. On peut appliquer l'hypothèse de récurrence, ce qui termine là démonstration du théorème.■

4. Courbes de plans sécants et courbes de classe minimale

On ne sait pas déduire plus que le corollaire 3.5 de l'existence d'*un* m-plan (m+2)-sécant, même dans le cas très étudié d'une droite trisécante ([D 2]).

Cependant, si on suppose qu'il existe une famille à un paramètre de tels plans, les méthodes de [G 1] permettent de construire une courbe dans la variété abélienne de classe de cohomologie m fois la classe minimale $[\Theta]^{g-1}/(g-1)!$.

Notre résultat, qui n'est valable que sous des hypothèses restrictives (que l'on pourrait sans doute améliorer), est le suivant :

Théorème 4.1. *Soit* X *une variété abélienne principalement polarisée dont l'anneau des endomorphismes est* $\mathbb{Z}$. *Soient* $x_1, \ldots, x_{m+2}$ *des points de* X *tels que les* $(x_k - x_1)_{k>1}$ *soient indépendants sur* $\mathbb{Z}$. *On définit :*

$$V = \{ \zeta \in X \mid \psi(\zeta + x_1), \ldots, \psi(\zeta + x_{m+2}) \text{ sont sur un m-plan} \}$$

et on suppose que :

(i) $2V$ *contient une courbe irréductible complète* Γ ,

(ii) *si* $\zeta \in V$ *et* $2\zeta \in \Gamma$, *les* $(m+2)$ *points* $\psi(\zeta + x_j)$ *engendrent un* m*-plan* Π_ζ *tel que le schéma* $\Pi_\zeta . K(X)$ *soit de longueur* $(m+2)$ *et en position générale.*

Alors la classe de Γ *est* m *fois la classe minimale.*

Remarques 4.2. 1) Les points $(-x_j-x_k)$, pour $1 \le j < k \le m+2$, sont toujours dans $2V$. On montre au cours de la démonstration que sous nos hypothèses, ils sont aussi forcément sur Γ et que Γ est *lisse* en ces points. Si on suppose seulement X indécomposable et les $(-x_j-x_k)$ distincts deux à deux pour $j<k$, mais qu'on suppose a priori que Γ contient tous les $(-x_j-x_k)$ et satisfait au reste des hypothèses, on arrive à la conclusion suivante, dans l'esprit de Theorem 2 de [G 1], qui traite le cas $m=1$:

• soit X admet une multiplication complexe F non nulle telle que $F(x_j-x_k)=0$ pour tous j et k,

• soit Γ est de classe m fois la classe minimale.

2) L'exemple des jacobiennes montre que l'hypothèse que les m-plans $(m+2)$-sécants sont non-dégénérés est nécessaire (dans (1.2), faire varier le point y_1 sur la courbe Γ : le m-plan $(m+2)$-sécant contient un $(m-1)$-plan $(m+1)$-sécant lorsque y_1 est égal à l'un des points x_j).

3) Il résulte de la remarque 3.6.1) et du fait que $(-x_j-x_k) \in \Gamma$ pour tous $j \neq k$ que :

$$\exists x_0 \in \Gamma \quad \forall x \in \Gamma \quad \dim(\Theta_{x-x_0} \cap \operatorname{Sing}\Theta) \ge g-2m-2 .$$

Il faut comparer ce résultat aux propriétés (1.3) des jacobiennes et (2.3) des variétés de Prym.

Démonstration du théorème 4.1.

Analysons tout d'abord l'espace tangent à $2V$ en un point $(-x_j-x_k)$, pour $j \neq k$. Prenons par exemple $j=1$, $k=2$ et $2\zeta = -x_1-x_2$. On a alors $\zeta+x_1 = -(\zeta+x_2)$, de sorte que $\zeta \in V$. En différentiant la relation $\psi(\zeta+x_1)\wedge\ldots\wedge\psi(\zeta+x_{m+2})=0$, on vérifie que :

$$\begin{aligned} T_{-x_1-x_2}(2V) &\simeq \{D \in T_{\zeta+x_1}X \mid D\psi(\zeta+x_1)\wedge\psi(\zeta+x_2)\wedge\ldots\wedge\psi(\zeta+x_{m+2})=0\} \\ &\simeq \Pi_\zeta . T_{\zeta+x_1}K(X) . \end{aligned}$$

Mais $\Pi_\zeta.K(X)$ contient déjà les $(m+1)$ points distincts $\psi(\zeta+x_1)=\psi(\zeta+x_2)$ et $\psi(\zeta+x_j)$ pour $j>2$. Nos hypothèses forcent donc :

$$\dim T_{-x_1-x_2}(2V) \le 1 \qquad \text{lorsque} \qquad (-x_1-x_2) \in \Gamma .$$

En particulier, la courbe Γ est lisse en les points $(-x_j-x_k)$ par lesquels elle passe.

Du lemme 3.7, on déduit que pour $j=1,\ldots,m+2$, il existe des sections λ_j de $\mathcal{O}_{\frac{1}{2}\Gamma}((2\Theta)_{-x_1}+\ldots+(2\hat{\Theta})_{-x_j}+\ldots+(2\Theta)_{-x_{m+2}})$ telles que, avec les notations de (3.1) :

$$\forall \zeta \in \tfrac{1}{2}\Gamma \quad \forall n \in (\mathbb{Z}/2)^g \qquad \sum_{j=1}^{m+2} \lambda_j(\zeta)\, \theta_n(\zeta + x_j) = 0 .$$

Le diviseur de λ_j est stable par translation par tout point d'ordre 2 de X. On peut donc écrire $\mathrm{Div}(\lambda_j) = 2_X^* D_j$, où 2_X est la multiplication par 2 sur X et où D_j est un diviseur sur Γ. La section λ_j est nulle si et seulement si les $(m+1)$ points $\psi(\zeta+x_p)$, $p \neq j$; sont sur un $(m-1)$-plan. Par hypothèse, cela ne peut se produire que si deux d'entre eux sont confondus, c'est-à-dire lorsque :

$$\begin{array}{ll} \exists\, p, q \neq j & \zeta + x_p = -(\zeta + x_q) \\ \text{soit} & 2\zeta = -x_p - x_q . \end{array}$$

Rappelons que Γ est lisse en un tel point. Un calcul local montre que si λ_j s'annule à l'ordre au moins deux en un tel point, les $(m-1)$ points $\psi(\zeta+x_r)$ pour $r \neq j, p, q$ et la droite $\psi_*(T_{-x_p-x_q}\Gamma)$ sont sur un $(m-1)$-plan, ce qui est interdit par nos hypothèses. Le diviseur D_j est donc contenu dans le lieu lisse de Γ et vaut :

$$\sum_{\substack{p,q\neq j;\, p<q \\ -x_p-x_q \in \Gamma}} (-x_p - x_q) .$$

Remarquons tout de suite que le fait que $\mathcal{O}_\Gamma(D_j-D_k) \simeq \mathcal{O}_\Gamma(\Theta_{x_j-x_k} - \Theta)$ est de degré 0 entraîne que l'entier :

$$\mathrm{Card}\,\{\, p \neq j \mid (-x_j-x_p) \in \Gamma \,\}$$

est indépendant de j. On le note M.

A la courbe Γ est associé un endomorphisme $\alpha_{\Gamma,\Theta}$ défini par :

$$\forall x \in X \qquad \alpha_{\Gamma,\Theta}(x) = \text{somme dans } X \text{ des points du diviseur } \mathcal{O}_\Gamma(\Theta_x - \Theta) .$$

Il s'ensuit que :

$$(4.3)\quad \alpha_{\Gamma,\Theta}(x_j-x_k) = \sum_{\substack{p,q\neq j;\, p<q \\ -x_p-x_q\in\Gamma}} (-x_p-x_q) \;-\; \sum_{\substack{p,q\neq k;\, p<q \\ -x_p-x_q\in\Gamma}} (-x_p-x_q)$$

$$= -\sum_{\substack{p\neq j,k \\ -x_k-x_p\in\Gamma}} (x_k+x_p) \;+\; \sum_{\substack{q\neq j,k \\ -x_j-x_p\in\Gamma}} (x_j+x_q)$$

L'endomorphisme $\alpha_{\Gamma,\Theta}$ est par hypothèse la multiplication par un entier N . On a donc :

$$N(x_j-x_k) = M(x_j-x_k) + \sum_{\substack{p\neq j,k \\ -x_j-x_p\in\Gamma}} x_p \;-\; \sum_{\substack{q\neq j,k \\ -x_k-x_q\in\Gamma}} x_q \qquad \text{si } -x_j-x_k \notin \Gamma$$

$$= (M-1)(x_j-x_k) + \sum_{\substack{p\neq j,k \\ -x_j-x_p\in\Gamma}} x_p \;-\; \sum_{\substack{q\neq j,k \\ -x_k-x_q\in\Gamma}} x_q \qquad \text{si } -x_j-x_k \in \Gamma .$$

L'hypothèse que les combinaisons linéaires non triviales des (x_p-x_q) sont non nulles entraîne alors que :

- soit $N=M$ et aucun (x_j-x_k) n'est sur Γ , ce qui est absurde car M serait alors nul, donc aussi $\alpha_{\Gamma,\Theta}$,
- soit $N=M-1$ et tous les (x_j-x_k) sont sur Γ . On a alors $M=m+1$ et $\alpha_{\Gamma,\Theta}=m_X$. Ceci est équivalent au fait que la classe de Γ est m fois la classe minimale et le théorème est démontré.■

5. Courbes de plans quadrisécants et variétés de Prym

Les variétés de Prym fournissent des exemples de variétés abéliennes principalement polarisées dont la variété de Kummer admet des 2-plans quadrisécants (2.2). Un raffinement du théorème 4.1. dans le cas $m=2$ va nous permettre d'utiliser cette propriété pour donner une caractérisation des variétés de Prym.

Théorème 5.1. *Soit* X *une variété abélienne principalement polarisée indécomposable et soient* x_1 , x_2 , x_3 *et* x_4 *des points de* X *tels que chacun des points* x_1-x_2 , x_1-x_3 *et* x_1-x_4 *engendre* X . *On suppose de plus que les points* $(-x_i-x_j)$ *sont non nuls et distincts 2 à 2 pour* $i<j$. *On définit :*

$$V = \{ \zeta \in X \mid \psi(\zeta+x_1),\ldots,\psi(\zeta+x_4) \text{ sont sur un 2-plan } \}$$

et on suppose que :

(i) $2V$ *contient une courbe irréductible complète* Γ ,

(ii) *si* $\zeta \in V$ *et* $2\zeta \in \Gamma$, *les 4 points* $\psi(\zeta+x_j)$ *engendrent un 2-plan* Π_ζ *tel que le schéma* $\Pi_\zeta.K(X)$ *soit de longueur 4 et en position générale.*

Alors la classe de Γ *est 2 fois la classe minimale.*

■ La formule (4.3) exprimant les images des points (x_j-x_k) par l'endomorphisme $\alpha_{\Gamma,\Theta}$ est toujours valable. Passons en revue les différentes valeurs possibles de M :

i) $M=0$. Tous les $\alpha_{\Gamma,\Theta}(x_j-x_k)$ sont nuls, ce qui est incompatible avec le fait que (x_1-x_2) engendre X .

ii) $M=1$. Quitte à renuméroter x_2 , x_3 et x_4 , on peut supposer que seuls $(-x_1-x_2)$ et $(-x_3-x_4)$ sont sur Γ . La formule (4.3) donne alors $\alpha_{\Gamma,\Theta}(x_1-x_2)=0$, ce qui est de nouveau incompatible avec nos hypotèses.

iii) $M=2$. On peut supposer $(-x_1-x_2)$ et $(-x_1-x_3)$ sur Γ . La seule situation compatible avec la définition de M est alors : $(-x_1-x_2)$ et $(-x_2-x_3) \notin \Gamma$ et $(-x_4-x_2)$ et $(-x_4-x_3) \in \Gamma$. La formule (4.3) donne alors $\alpha_{\Gamma,\Theta}(x_1-x_4)=2(x_1-x_4)$. Comme (x_1-x_4) engendre X , ceci entraîne $\alpha_{\Gamma,\Theta}=2_X$. La formule (4.3) donne alors $2(x_1-x_2) = \alpha_{\Gamma,\Theta}(x_1-x_2)= x_1-x_2+x_3-x_4$, soit $-x_2-x_3=-x_1-x_4$, ce qui contredit nos hypothèses.

iv) On a donc $M=3$, de sorte que $\alpha_{\Gamma,\Theta}(x_j-x_k)=2(x_j-x_k)$ pour tous j et k , ce qui termine la démonstration du théorème.■

Welters a fait dans [W 1] une étude exhaustive des courbes dans une variété abélienne principalement polarisée dont la classe est 2 fois la classe minimale. Nous nous appuyons sur ses résultats pour énoncer la caractérisation suivante des variétés de Prym, analogue à la caractérisation des jacobiennes qu'il donne dans [W 3] Corollary 1.8.

Théorème 5.2. *Soit* X *une variété abélienne principalement polarisée indécomposable et soit* Γ *une courbe irréductible complète contenue dans* X *, qui engendre* X *. On suppose qu'il existe des points* x_0 *,* x_2 *et* x_3 *sur* Γ *, avec* x_0 *,* $2x_2$ *,* x_2+x_3 *et* $2x_3$ *distincts deux à deux tels que, pour tout point* x_1 *général sur* Γ *, les propriétés suivantes soient vérifiées :*

(i) Pour tout $\zeta \in \frac{1}{2}(\Gamma - x_1 - x_2 - x_3)$ *, les points* $\psi(\zeta+x_1),\dots,\psi(\zeta+x_4)$ *engendrent un 2-plan* Π_ζ *tel que le schéma* $\Pi_\zeta.K(X)$ *soit de longueur 4 et en position générale.*

(ii) Tout point x *sur* Γ *vérifiant* $2x=x_0$ *est singulier sur* Γ *.*

Alors Γ *est une courbe stable de genre arithmétique* $2\dim X+1$ *, elle admet un translaté symétrique* $\Gamma_0 = \Gamma - \frac{1}{2}x_0$ *, le revêtement* $\Gamma_0 \longrightarrow \Gamma_0/\pm 1$ *est admissible au sens de Beauville ([B 2]) et sa variété de Prym est isomorphe à* X *.*

■ On reprend les notations précédentes. Nos hypothèses entraînent que pour x_1 assez général sur Γ , les hypothèses du théorème 5.1 sont satisfaites avec $x_4 = x_1+x_2+x_3-x_0$. On fixe un tel x_1 et on note $\Gamma' = \Gamma - x_1 - x_2 - x_3$. Par hypothèse, Γ' engendre X et les points $(-x_2-x_3)$, $(-x_1-x_3)$ et $(-x_1-x_2)$ sont sur Γ' . Il ressort de la démonstration du théorème 5.1. qu'on a $M=3$. En particulier :

$$- x_1 - (x_1+x_2+x_3-x_0) \in \Gamma' ,$$

de sorte que $x_0 - x_1 \in \Gamma$ et $x_0 - \Gamma = \Gamma$. Tout translaté $\Gamma_0 = \Gamma - \frac{1}{2}x_0$ de Γ est donc symétrique. On est dans le cas II^+ de Welters (Theorem 3.1.4) de [W 1]) et c'est là qu'intervient l'hypothèse (ii). Elle assure que tout point d'ordre 2 sur Γ_0 est singulier, donc ([W 1], (3.1), (3.12), (3.20)) que Γ_0 est stable de genre arithmétique $2\dim X+1$ et qu'on est dans la situation de Prym-Beauville.■

6. Variétés de Prym-Tjurin.

Ces variétés, introduites par Tjurin dans [T] et étudiées par la suite par Bloch et Murre, Puts et Kanev, sont définies de la façon suivante. Soit C une courbe lisse et soit $D \in \mathrm{Div}(C\times C)$ une correspondance *symétrique, effective et sans point fixe,* telle que l'endomorphisme associé i de JC satisfasse l'égalité :

$$(i-1)(i+m-1) = 0 ,$$

où m est un entier supérieur ou égal à 2. La variété de Prym-Tjurin associée au couple (C,D) est la sous-variété abélienne $P = \mathrm{Im}(1-i)$ de JC. Les variétés de Prym usuelles correspondent au cas $m=2$.

Kanev montre dans [K] les généralisations suivantes de propriétés classiques des variétés de Prym :

- Il existe un diviseur symétrique Θ_C induisant la polarisation principale naturelle sur JC tel que $\Theta_C.P = m\Theta$, où Θ est un diviseur sur P qui induit une *polarisation principale* sur P.

- Soit κ la thêtacaractéristique sur C associée à Θ_C. On a, pour tout point x de P :

$$h^0(C,\kappa+x) = 0 \quad \Leftrightarrow \quad x \notin \Theta$$

(6.1) $$h^0(C,\kappa+x) \geq m \quad \Leftrightarrow \quad x \in \Theta$$

(6.2) $$h^0(C,\kappa+x) \geq m+1 \quad \Rightarrow \quad x \in \mathrm{Sing}\,\Theta .$$

Ces résultats entraînent :

Proposition 6.3. *Soit* (P,Θ) *la variété de Prym-Tjurin associée à une correspondance symétrique, effective et sans point fixe sur une courbe lisse* C *telle que l'endomorphisme* i *de* JC *associé vérifie* $(i-1)(i+m-1)=0$. *Alors :*

$$\dim \mathrm{Sing}\,\Theta \geq \dim P - 2m - 2 .$$

■ Nous utilisons la même version améliorée de Corollary 3 de [F-H-L] que A. Bertram dans [Be] : notons g le genre de C et, pour tout entier naturel r, $W^r = \{ L \in \mathrm{Pic}^{g-1}C \mid h^0(C,L) > r \}$; pour toute sous-variété Σ de W^r, on a alors :

$$\dim(\Sigma \cap W^{r+1}) \geq \dim \Sigma - 2r - 3 .$$

On applique ce résultat à $\Sigma = \kappa + \Theta$, qui est contenu dans W^{m-1} (6.1). Par (6.2), l'intersection $(\Sigma \cap W^m)$ est contenue dans $\mathrm{Sing}\,\Theta$, de sorte que :

$$\begin{aligned} \dim \mathrm{Sing}\,\Theta &\geq \dim(\Sigma \cap W^m) \\ &\geq \dim \Sigma - 2m - 1 \\ &= \dim P - 2m - 2 . \blacksquare \end{aligned}$$

D'autre part, le choix d'un point x_0 de C permet de définir un morphisme $x \mapsto (1-i)(x-x_0)$ de C sur une courbe Γ dans P. Welters prouve dans [W1] que *la classe de* Γ *est* m *fois la classe minimale.*

Comme mentionné dans l'introduction, il serait donc très intéressant de pouvoir compléter ce tableau en étudiant les variétés de Prym-Tjurin dont les variétés de Kummer admettent des m-plans $(m+2)$-sécants.

REFERENCES

[B 1] A. BEAUVILLE : Prym Varieties : A Survey, in "Theta functions, Bowdoin 1987", Proc. of Symposia in Pure Mathematics 49, Part 1 (1989), 607-520.

[B 2] A. BEAUVILLE : Prym varieties and the Schottky problem. Invent. Math. 41 (1977), 149-196.

[B-D] A. BEAUVILLE, O. DEBARRE : Sur le problème de Schottky pour les variétés de Prym. Ann. della Sc. Norm. Sup. di Pisa, Serie IV, 14 (1987), 613-623.

[Be] A. BERTRAM : An existence theorem for Prym special divisors. Invent. Math. 90 (1987), 669-671.

[D 1] O. DEBARRE : Sur les variétés abéliennes dont le diviseur thêta est singulier en codimension 3. Duke Math. J. 56 (1988), 221-273.

[D 2] O. DEBARRE : The trisecant conjecture for Pryms, in "Theta functions, Bowdoin 1987", Proc. of Symposia in Pure Mathematics 49, Part 1 (1989), 621-626.

[Do] R. DONAGI : The tetragonal construction. Bull. Amer. Math. Soc., vol. 4 (1981), 181-185.

[F-H-L] W. FULTON, J. HARRIS, R. LAZARSFELD : Excess linear series on an algebraic curve. Proc. Am. Math. Soc. 92 (1984), 320-322.

[G 1] R.C. GUNNING : Some curves in abelian varieties. Invent. Math. 66 (1982), 377-389.

[G 2] R.C. GUNNING : Analytic identities for theta functions, in "Theta functions, Bowdoin 1987", Proc. of Symposia in Pure Mathematics 49, Part 1 (1989), 503-515.

[G 3] R.C. GUNNING : On generalized theta functions. Amer. J. Math. 103 (1981), 411-435.

[K] V. KANEV : Principal polarizations of Prym-Tjurin varieties. Comp. Math. 64 (1987), 243-270.

[M] T. MATSUSAKA : On a characterization of a Jacobian variety. Mem. Coll. of Sci. Kyoto, Ser. A, 32 (1959), 1-19.

[Mu] D. MUMFORD : Prym varieties I. Contributions to analysis. Academic Press, New-York (1974).

[T] A.N. TJURIN : Five lectures on three dimensional varieties. Russ. Math. Surveys 27 (1972), 1-53.

[W 1] G. WELTERS : Curves of twice the minimal class on principally polarized varieties. Proc. Kon. Ned. Akad. van Wetenschappen, Indagationes Math. 49 (1987), 87-109.

[W 2] G. WELTERS : A criterion for Jacobi varieties. Ann. of Math. 120 (1984), 497-504.

[W 3] G. WELTERS : On flexes of the Kummer variety. Proc. Kon. Ned. Akad. van Wetenschappen, Indagationes Math. 45 (1983), 501-520.

Singular hermitian metrics on positive line bundles

Jean-Pierre Demailly

Université de Grenoble I,
Institut Fourier, BP 74,
Laboratoire associé au C.N.R.S. n° 188,
F-38402 Saint-Martin d'Hères

Abstract. — *The notion of a singular hermitian metric on a holomorphic line bundle is introduced as a tool for the study of various algebraic questions. One of the main interests of such metrics is the corresponding L^2 vanishing theorem for $\overline{\partial}$ cohomology, which gives a useful criterion for the existence of sections. In this context, numerically effective line bundles and line bundles with maximum Kodaira dimension are characterized by means of positivity properties of the curvature in the sense of currents. The coefficients of isolated logarithmic poles of a plurisubharmonic singular metric are shown to have a simple interpretation in terms of the constant ε of Seshadri's ampleness criterion. Finally, we use singular metrics and approximations of the curvature current to prove a new asymptotic estimate for the dimension of cohomology groups with values in high multiples $\mathcal{O}(kL)$ of a line bundle L with maximum Kodaira dimension.*

1. Introduction

Our purpose is to show that several important concepts of algebraic geometry have a nice interpretation in differential geometric terms, once we admit hermitian metrics with singularities, and especially plurisubharmonic weights with *logarithmic poles.*

A *singular (hermitian) metric* on a line bundle L is simply a hermitian metric which is given in any trivialization by a weight function $e^{-\varphi}$ such that φ is locally integrable. We then have a well-defined *curvature current* $c(L) = \frac{i}{\pi}\partial\overline{\partial}\varphi$ and the case when $c(L) \geq 0$ as a current is especially interesting. One of the main reasons for this is the basic L^2 existence theorem of Hörmander-Andreotti-Vesentini for solutions of $\overline{\partial}$ equations with plurisubharmonic weights. With relatively few efforts, the L^2 theory gives strong vanishing theorems (of Kawamata-Viehweg type) and existence results for sections of the adjoint line bundle $K_X + L$; here K_X denotes the canonical line bundle of the base manifold X, and we use an additive notation for the group $\mathrm{Pic}(X) = H^1(X, \mathcal{O}^\star)$. These techniques can be used in combination with the Calabi-Yau theorem to obtain explicit numerical criteria for very ample line bundles; we refer to [De 90] for results in this direction.

On a projective algebraic manifold X, the real vector space generated by the Neron-Severi group $H^2(X,\mathbb{Z}) \cap H^{1,1}(X)$ contains two canonical closed convex cones $\Gamma_+ \supset \Gamma_a$, which are generated by cohomology classes of effective or ample divisors, respectively. For a line bundle L on X, we show that L has a singular metric with positive curvature current if and only if $c_1(L) \in \Gamma_+$. We also give similar differential geometric descriptions of the line bundles for which $c_1(L)$ belongs to Γ_a (numerically effective line bundles), or to the open cone Γ_+° (line bundles with $\kappa(L) = n$), or else to $\Gamma_+^\circ \cap \Gamma_a$ (big and nef line bundles).

The well-known Seshadri ampleness criterion asserts that a line bundle L on a projective manifold X is ample if and only if there is a constant $\varepsilon > 0$ such that $L \cdot C \geq \varepsilon\, m(C)$ for every curve $C \subset X$, where $m(C)$ is the maximum of the multiplicity of the singular points of C. We show that the optimal constant $\varepsilon(L)$ is precisely equal to the supremum of coefficients γ for which a plurisubharmonic weight on L may have an isolated logarithmic pole of slope γ at any point. This result is then refined by introducing "local" Seshadri constants $\varepsilon(L,x)$ which measure ampleness along curves passing though a fixed point x.

Finally we use approximation techniques for singular metrics, combined with the general holomorphic Morse inequalities of [De 85], to obtain an asymptotic upper bound for the dimensions of cohomology groups $H^q(X,kL)$ when L is a line bundle of maximum Kodaira dimension $\kappa(L) = n$ and k tends to $+\infty$. If some multiple of L is written as $mL \simeq \mathcal{O}(A+D)$ where A (resp. D) is an ample (resp. effective) divisor, the upper bound is expressed in a simple way in terms of the first Chern class $c_1(L)$, the multiplicities of the singular points of D and the curvature of the tangent bundle TX.

Several results of the present article have been worked out while the author was visiting the Tata Institute in Bombay in August 1989. The author wishes to thank this institution for its hospitality, and especially Prof. M.S. Narasimhan, R.R Simha and S. Subramanian for stimulating discussions.

2. Notion of singular hermitian metrics

Let L be a holomorphic line bundle over a complex manifold X. We are mostly interested in the case of a compact manifold, but this restriction is irrelevant in the present section.

DEFINITION 2.1. — *A singular (hermitian) metric on L is a metric which is given in any trivialization $\theta : L_{\restriction\Omega} \xrightarrow{\simeq} \Omega \times \mathbb{C}$ by*

$$\|\xi\| = |\theta(\xi)|\, e^{-\varphi(x)}, \qquad x \in \Omega,\ \xi \in L_x$$

where $\varphi \in L^1_{\rm loc}(\Omega)$ is an arbitrary function, called the weight of the metric with respect to the trivialization θ.

If $\theta' : L_{\restriction\Omega'} \longrightarrow \Omega' \times \mathbb{C}$ is another trivialization, φ' the associated weight and $g \in \mathcal{O}^\star(\Omega \cap \Omega')$ the transition function, then $\theta'(\xi) = g(x)\theta(\xi)$ for $\xi \in L_x$, and so

$\varphi' = \varphi + \log|g|$ on $\Omega \cap \Omega'$. The curvature form of L is then given by the closed (1,1)-current $c(L) = \frac{i}{\pi}\partial\overline{\partial}\varphi$ on Ω, if we compute formally $c(L) = \frac{i}{2\pi}D^2$ as in the smooth case; $c(L)$ is, of course, a global current on X which is independent of the choice of trivializations. Our assumption $\varphi \in L^1_{\mathrm{loc}}(\Omega)$ guarantees that $c(L)$ exists in the sense of distribution theory. Then the De Rham cohomology class of $c(L)$ is the image of the first Chern class $c_1(L) \in H^2(X, \mathbb{Z})$ in $H^2_{DR}(X, \mathbb{R})$ (De Rham cohomology can be computed either by means of smooth differential forms or by means of currents). Before going further, we discuss two basic examples.

Example 2.2. — Let $D = \sum \alpha_j D_j$ be a divisor with coefficients $\alpha_j \in \mathbb{Z}$ and let $L = \mathcal{O}(D)$ be the associated invertible sheaf of meromorphic functions f such that $\mathrm{div}(f) + D \geq 0$; the corresponding line bundle can be equipped with the singular metric defined by $||f|| = |f|$. If g_j is a generator of the ideal of D_j on an open set $\Omega \subset X$ then $\theta(f) = f\prod g_j^{\alpha_j}$ defines a trivialization of $\mathcal{O}(D)$ over Ω, thus our singular metric is associated to the weight $\varphi = \sum \alpha_j \log|g_j|$. By the Lelong-Poincaré equation, we find

$$(2.3) \qquad c(\mathcal{O}(D)) = \frac{i}{\pi}\partial\overline{\partial}\varphi = [D],$$

where $[D] = \sum \alpha_j [D_j]$ denotes the current of integration over D (cf. [Le 57] and [Le 69]).

Example 2.4. — Assume that $\sigma_1, \ldots, \sigma_N$ are non zero holomorphic sections of L. Then we can define a natural (possibly singular) hermitian metric on $L^\star$ by

$$||\xi^\star||^2 = \sum_{1\leq j\leq n} |\xi^\star.\sigma_j(x)|^2 \quad \text{for } \xi^\star \in L^\star_x.$$

The dual metric on L is given by

$$||\xi||^2 = \frac{|\theta(\xi)|^2}{|\theta(\sigma_1(x))|^2 + \cdots + |\theta(\sigma_N(x))|^2}$$

with respect to any trivialization θ. The associated weight function is thus given by $\varphi(x) = \log\big(\sum_{1\leq j\leq N} |\theta(\sigma_j(x))|^2\big)^{1/2}$. In this case φ is a plurisubharmonic function, so $c(L)$ is a (closed) positive current. □

It is worth observing that the weight functions φ have logarithmic poles in both examples. In the second case, the set of poles is the base locus $\bigcap \sigma_j^{-1}(0)$ of the linear system generated by the sections $\sigma_1, \ldots, \sigma_N$. In the sequel, we always suppose that the curvature current $c(L)$ is positive, i.e. that the weight functions φ are plurisubharmonic.

DEFINITION 2.5. — *A singular metric on L with positive curvature current $c(L) \geq 0$ is said to have a logarithmic pole of coefficient γ at a point $x \in X$ if the Lelong number*

$$\nu(\varphi, x) = \liminf_{z\to x} \frac{\varphi(z)}{\log|z - x|}$$

is non zero and if $\nu(\varphi, x) = \gamma$.

For the basic properties of Lelong numbers, we refer to [Le 69], [Siu 74] and [De 87]. It is well known that $\nu(\varphi, x)$ is always equal to the Lelong number of the associated current $T = \frac{i}{\pi}\partial\overline{\partial}\varphi$, defined by $\nu(T,x) = \lim_{r\to 0_+} \nu(T,x,r)$ with

$$\nu(T,x,r) = \frac{1}{(2\pi r^2)^{n-1}} \int_{B(x,r)} T(z) \wedge (i\partial\overline{\partial}|z|^2)^{n-1}. \tag{2.6}$$

Finally, for every $c > 0$, we consider the sublevel sets

$$E_c(T) = \{x \in X\,;\, \nu(T,x) \geq c\}. \tag{2.7}$$

By a theorem of [Siu 74], $E_c(T)$ is a (closed) analytic subset of X. If $T = \frac{i}{\pi}\partial\overline{\partial}\varphi$ on an open set $\Omega \subset X$, we denote accordingly $E_c(\varphi) = E_c(T) \cap \Omega$. The following simple lemma is very useful in this context.

LEMMA 2.8. — *If φ is a plurisubharmonic function on X, then $e^{-2\varphi}$ is integrable in a neighborhood of x as soon as $\nu(\varphi, x) < 1$ and non integrable as soon as $\nu(\varphi, x) \geq n$.*

Proof. — If $\nu(\varphi, x) = \gamma$, the usual convexity properties of plurisubharmonic functions show that

$$\varphi(z) \leq \gamma \log|z-x| + O(1) \quad \text{near } x, \tag{2.9}$$

thus $e^{-2\varphi(z)} \geq C|z-x|^{-2\gamma}$ is non integrable as soon as $\gamma \geq n$. For a proof that $e^{-2\varphi}$ is integrable when $\nu(\varphi, x) < 1$, we refer to Skoda [Sk 72]. Both bounds are best possible as the examples $\varphi(z) = (n-\varepsilon)\log|z|$ and $\varphi(z) = \log|z_1|$ in $\mathbb{C}^n$ easily show; in the first case $\nu(\varphi, 0) = n - \varepsilon$ but $e^{-2\varphi}$ is integrable at 0; in the second case $\nu(\varphi, 0) = 1$ but $e^{-2\varphi}$ is non integrable at 0. □

3. L^2 vanishing theorem and criterion for the existence of sections

One of the main reasons for which singular metrics are especially interesting is the powerful existence theorem of Hörmander for solutions of equations $\overline{\partial}u = v$.

THEOREM 3.1. — *Suppose that X is a Stein or compact projective manifold equipped with a Kähler metric ω. Let L be a holomorphic line bundle equipped with a singular metric associated to plurisubharmonic weight functions φ such that $c(L) \geq \varepsilon\omega$ for some $\varepsilon > 0$. For every $q \geq 1$ and every (n,q) form v with values in L such that $\overline{\partial}v = 0$ and $\int_X |v|^2 e^{-2\varphi} dV_\omega < +\infty$, there is a $(n, q-1)$-form u with values in L such that $\overline{\partial}u = v$ and*

$$\int_X |u|^2 e^{-2\varphi} dV_\omega \leq \frac{1}{2\pi q\varepsilon} \int_X |v|^2 e^{-2\varphi} dV_\omega.$$

Here dV_ω stands for the Kähler volume element $\omega^n/n!$ and $|u|^2 e^{-2\varphi}$ denotes somewhat abusively the pointwise norm of $u(z)$ at each point $z \in X$, although φ is only defined on an open set in X. The constant 2π comes from the fact that we have included 2π in the definition of $c(L)$.

Proof. — The result is standard when X is Stein and L is the trivial bundle (see [AV 65] and [Hö 66]); the proof can then be reduced to the case of a smooth metric, because any plurisubharmonic function is the limit of a decreasing sequence of smooth plurisubharmonic functions. In general, there exists a hypersurface $H \subset X$ such that $X \setminus H$ is Stein and L is trivial over $X \setminus H$. We then solve the equation $\overline{\partial} u = v$ over $X \setminus H$ and observe that the solution extends to X thanks to the L^2 estimate (cf. [De 82], lemma 6.9). □

From this general theorem, we can easily derive an abstract vanishing theorem for the adjoint line bundle $K_X + L$ and a criterion for the existence of sections of $K_X + L$.

COROLLARY 3.2. — *Let L be a line bundle with a singular hermitian metric. Assume that $c(L) \geq \varepsilon\omega$ and that the metric (i.e. the weight $e^{-2\varphi}$) is integrable near all but finitely many points of X . Then $H^q(X, K_X + L) = 0$ for $q \geq 1$.*

Proof. — Let $x_1, \ldots, x_m$ be the points where the metric is not integrable and let v be a smooth (n,q)–form with values in L such that $\overline{\partial} v = 0$. Let u_j be a smooth solution of $\overline{\partial} u_j = v$ in a neighborhood of V_j of x_j and ψ_j a cut-off function with support in V_j such that $\psi_j(x_j) = 1$. Then $v' = v - \sum \overline{\partial}(\psi_j u_j)$ satisfies $\overline{\partial} v' = 0$ and $\int_X |v'|^2 e^{-2\varphi} d\sigma < +\infty$, because $e^{-2\varphi}$ is integrable except at the x_j's and v' vanishes in a neighborhood of x_j . By theorem 3.1, there is a L^2 solution u' of $\overline{\partial} u' = v'$, hence $u = u' + \sum \psi_j u_j$ is a solution of $\overline{\partial} u = v$. It is well-known that the existence of a L^2_{loc} solution implies the existence of a smooth one, whence the corollary. □

COROLLARY 3.3. — *Let L be a line bundle with a singular metric such that $c(L) \geq \varepsilon\omega$, $\varepsilon > 0$. If the weight function φ is such that $\nu(\varphi, x) \geq n + s$ at some point $x \in X$ which is an isolated point of $E_1(\varphi)$, then $H^0(X, K_X + L)$ generates all s-jets at x.*

Proof. — The proof is a straightforward adaptation of the Hörmander-Bombieri-Skoda technique ([Bo 70], [Sk 75]). We have $e^{-2\varphi(z)} \geq C|z - x|^{-2(n+s)}$ near x by (2.9), and in particular $e^{-2\varphi}$ is non integrable near x. Since x is supposed to be isolated in $E_1(\varphi)$, we infer from lemma 2.8 that $e^{-2\varphi}$ is locally integrable on a small punctured neighborhood $V \setminus \{x\}$. Let $P(z)$ be an arbitrary polynomial of degree $\leq s$ in given analytic coordinates $(z_1, \ldots, z_n)$ on V. Fix a smooth cut-off function χ with compact support in V such that $\chi = 1$ near x and a non vanishing local section $h \in H^0(V, K_X + L)$. Then $v = P\overline{\partial}\chi \otimes h$ can be viewed as a $\overline{\partial}$-closed $(n,1)$-form on X with values in L, such that $\int_X |v|^2 e^{-2\varphi} dV_\omega < +\infty$; indeed v is smooth, has compact support in V and vanishes near x. The solution u of $\overline{\partial} u = v$ is smooth and we have $|u(z)| = o(|z - x|^s)$ near x thanks to the L^2 estimate. Therefore

$$f = \chi P h - u \in H^0(X, K_X + L)$$

has the prescribed s-jet Ph at x. □

4. Numerical cones associated to positive line bundles

We suppose here that X is a projective algebraic manifold and denote $n = \dim X$. It is well known that an integral cohomology class in $H^2(X,\mathbb{Z})$ is the first Chern class of a holomorphic (or algebraic) line bundle if and only if this class is of type $(1,1)$. Hence the so-called *Neron-Severi* group

$$NS(X) = H^2(X,\mathbb{Z}) \cap H^{1,1}(X) \subset H^2(X,\mathbb{R})$$

is the set of cohomology classes of algebraic line bundles (or of integral divisors).

DEFINITION 4.1. — *A holomorphic line bundle L over X is said to be numerically effective, nef for short, if $L \cdot C = \int_C c_1(L) \geq 0$ for every curve $C \subset X$, and in this case L is said to be big if $L^n = \int_X c_1(L)^n > 0$.*

If L is nef, it is well known that $L^p \cdot Y = \int_Y c_1(L)^p \geq 0$ for any p-dimensional subvariety $Y \subset X$ (see e.g. [Ha 70]). The Nakai-Moishezon ampleness criterion then shows that $L + A$ is ample as soon as A is ample. In fact, if A is ample, it is easy to see that L is nef if and only if $kL + A$ is ample for all integers $k \geq 1$. We are going to describe a simple dictionary relating these concepts to similar concepts in the context of differential geometry.

Let $NS_{\mathbb{R}}(X)$ be the real vector space $NS(X) \otimes \mathbb{R} \subset H^2(X,\mathbb{R})$ and let $\Gamma_+ \subset NS_{\mathbb{R}}(X)$, resp. $\Gamma_a \subset \Gamma_+$, be the closed convex cone generated by cohomology classes of effective (resp. ample) divisors D; denote by Γ_+° (resp. Γ_a°) the interior of Γ_+ (resp. Γ_a). We will call Γ_+ and Γ_a° respectively the *effective cone* and the *ample cone* of X. Finally, recall that the *Kodaira dimension* $\kappa(L)$ is the supremum of the rank of the canonical maps

$$\Phi_m : X \setminus Z_m \longrightarrow P(V_m^\star), \qquad x \longmapsto H_x = \{\sigma \in V_m \,;\, \sigma(x) = 0\}, \qquad m \geq 1$$

with $V_m = H^0(X, mL)$ and $Z_m = \bigcap_{\sigma \in V_m} \sigma^{-1}(0) =$ base locus of V_m. If $V_m = \{0\}$ for all $m \geq 1$, we set $\kappa(L) = -\infty$. Then we have $h^0(X, mL) \leq O(m^{\kappa(L)})$ for $m \geq 1$, and $\kappa(L)$ is the smallest constant for which this estimate holds.

PROPOSITION 4.2. — *If L is a holomorphic line bundle on X, ω a Kähler metric and $\varepsilon > 0$, we have the following equivalent properties:*

(a) $c_1(L) \in \Gamma_+ \Longleftrightarrow L$ has a singular metric with $c(L) \geq 0$;

(b) $c_1(L) \in \Gamma_+^\circ \Longleftrightarrow \exists\varepsilon$, L has a singular metric with $c(L) \geq \varepsilon\,\omega \Longleftrightarrow \kappa(L) = n$;

(c) $c_1(L) \in \Gamma_a \Longleftrightarrow \forall\varepsilon$, L has a smooth metric with $c(L) \geq -\varepsilon\,\omega \Longleftrightarrow L$ *is nef*;

(d) $c_1(L) \in \Gamma_a^\circ \Longleftrightarrow \exists\varepsilon$, L has a smooth metric with $c(L) \geq \varepsilon\,\omega \Longleftrightarrow L$ *is ample.*

Proof. — It is well known that L is ample if and only if L has a smooth metric with positive definite curvature, and this gives the last equivalence in (4.2d).

(4.2a) Suppose that $c_1(L) \in \Gamma_+$. By definition $c_1(L)$ is a limit of cohomology classes of effective real divisors $D_k = \sum \lambda_{j,k} D_{j,k}$. Then $[D_k]$ is a sequence of closed positive currents, with a uniform bound of the mass $\int_X [D_k] \wedge \omega^{n-1}$, for this integral converges to $\int_X c_1(L) \wedge \omega^{n-1}$. By weak compactness, there is a subsequence $[D_{k_\nu}]$ converging to some closed positive current T of bidegree $(1,1)$, such that the cohomology class of T is equal to $c_1(L)$. Therefore, if L is equipped with an arbitrary smooth metric, we find $T = c(L) + \frac{i}{\pi}\partial\overline{\partial}\psi$ for some function $\psi \in L^1(X)$, and T is the curvature current of the singular metric obtained by multiplication of the original smooth metric by $e^{-\psi}$. Conversely, if L has a singular metric with $c(L) \geq 0$, we fix a point $x_0 \in X$ such that the associated weight satisfies $\nu(\varphi_L, x_0) = 0$. Let ψ_0 be a smooth function on $X \setminus \{x_0\}$ which is equal to $n \log |z - x_0|$ (in some coordinates) near x_0, and let A be a fixed ample line bundle, equipped with a smooth metric of positive curvature. For m_0 large enough, we have $m_0 c(A) + \frac{i}{\pi}\partial\overline{\partial}\psi_0 \geq \omega$ and the tensor product metric on $kL + m_0 A$ multiplied by $e^{-2\psi_0}$ is associated to a weight $\varphi_k = k\varphi_L + m_0\varphi_A + \psi_0$ such that

$$\frac{i}{\pi}\partial\overline{\partial}\varphi_k = k\, c(L) + m_0 c(A) + \frac{i}{\pi}\partial\overline{\partial}\psi_0 \geq k\, c(L) + \omega \geq \omega \qquad \forall k \geq 1.$$

Moreover $\nu(\varphi_k, x_0) = n$, whereas $\nu(\varphi_k, x) = k\,\nu(\varphi_L, x) < 1$ for $x \neq x_0$ near x_0. We infer from corollary 3.3 that $kL + m_0 A + K_X$ has non zero sections for all $k \geq 1$. Let D_k be the divisor of any such section and $\{D_k\}$ its cohomology class. Then

$$c_1(L) = \frac{1}{k}\Big(\{D_k\} - m_0 c_1(A) - c_1(K_X)\Big) = \lim_{k \to +\infty} \frac{1}{k}\{D_k\}$$

and therefore $c_1(L) \in \Gamma_+$.

(4.2b) Without loss of generality, we may suppose that the cohomology class $\{\omega\}$ is integral, i.e. $\{\omega\} \in NS(X)$. The first equivalence in (4.2b) is then an immediate consequence of (4.2a), since a class $\{\alpha\}$ is in the interior Γ_+° if and only if $\{\alpha - \varepsilon\,\omega\} \in \Gamma_+$ for ε small enough ($\{\omega\}$ is obviously an interior point). If $c(L) \geq \varepsilon\,\omega$, we can construct as above a singular metric φ_k on $kL - K_X$ such that $\frac{i}{\pi}\partial\overline{\partial}\varphi_k \geq \omega$ for $k \geq k_0$, $\nu(\varphi_k, x_0) = n+1$ and $\nu(\varphi_k, x) < 1$ for $x \neq x_0$ near x_0. Then corollary 3.3 shows that kL has sections with arbitrary 1-jets at x_0, hence $\kappa(L) = n$. Conversely, if $\kappa(L) = n$, then $h^0(X, kL) \geq ck^n$ for $k \geq k_0$ and $c > 0$. Let A be a smooth ample divisor. The exact cohomology sequence

$$0 \longrightarrow H^0(X, kL - A) \longrightarrow H^0(X, kL) \longrightarrow H^0(A, kL_{\restriction A})$$

where $h^0(A, kL_{\restriction A}) = O(k^{n-1})$ shows that $kL - A$ has non zero sections for k large. If D is the divisor of such a section, then $kL \simeq \mathcal{O}(A + D)$. If we select a smooth metric on A such that $c(A) \geq \varepsilon_0\omega$ and the singular metric on $\mathcal{O}(D)$ described in example 2.2, then $c(L) = \frac{1}{k}(c(A) + [D]) \geq (\varepsilon_0/k)\,\omega$, as desired.

(4.2c) If $c_1(L) \in \Gamma_a$, then L is nef, because the condition that a cohomology class of type $(1,1)$ has nonnegative integrals over curves is preserved through convex combinations and limits. Conversely, when L is nef, $kL + A$ is ample as soon as A is ample; thus $c_1(kL + A)$ is the cohomology class of an ample rational divisor D_k and

$c_1(L) = \lim_{k\to+\infty} k^{-1}\{D_k\} \in \Gamma_a$. Moreover, arbitrary choices of smooth metrics with positive curvature on A and $kL + A$ yield a smooth metric on L such that

$$c(L) = \frac{1}{k}\big(c(kL + A) - c(A)\big) \geq -\frac{1}{k}c(A);$$

in this way the negative part can be made smaller than $\varepsilon\,\omega$ for any $\varepsilon > 0$ if we take k large enough. Finally, if $c(L) \geq -\varepsilon\,\omega$ for every $\varepsilon > 0$, then $L\cdot C \geq -\varepsilon\int_C \omega$ and we conclude that $L \cdot C \geq 0$, thus L is nef.

(4.2d) Only the first equivalence remains to be checked: this is an immediate consequence of (4.2c) and the fact that $\{\omega\} \in \Gamma_a^\circ$. □

COROLLARY 4.3. — *If L is nef, then $\kappa(L) = n$ if and only if $L^n > 0$. Moreover, the following properties are equivalent:*

(a) *L is nef and big;*

(b) *$c_1(L) \in \Gamma_a \cap \Gamma_+^\circ$;*

(c) *for every $\delta > 0$, L has a singular metric such that $c(L) \geq \varepsilon\,\omega$ for some $\varepsilon > 0$ and such that $\max_{x\in X} \nu(\varphi, x) \leq \delta$.*

The metrics obtained in (c) can be chosen to be smooth on the complement of a fixed divisor D, with logarithmic poles along D.

Proof. — The first statement is well-known and obtained as follows: if L is nef, the Hilbert polynomial of $\chi(X, kL)$ has leading coefficient $L^n/n! \geq 0$; by the Kodaira-Nakano vanishing theorem, we have an exact sequence

$$H^{j-1}(A, (kL + A)_{\restriction A}) \longrightarrow H^j(X, kL) \longrightarrow H^j(X, kL + A) = 0, \quad j \geq 1$$

whenever A is a smooth divisor chosen sufficiently ample so that $A - K_X$ is ample; thus $h^j(X, kL) = O(k^{n-1})$ for $j \geq 1$ and

$$h^0(X, kL) = (L^n/n!)\,k^n + O(k^{n-1}),$$

therefore L is big if and only if $\kappa(L) = n$. The equivalence of (a) and (b) follows immediately from this and from (4.2b), (4.2c).

We also observe that $H^0(X, kL - A)$ is the kernel of the restriction morphism $H^0(X, kL) \to H^0(A, kL_{\restriction A})$ in which the target has dimension $O(k^{n-1})$. If $L^n > 0$, we infer $H^0(X, kL - A) \neq 0$ for k large, so there is an effective divisor D such that $kL \simeq \mathcal{O}(A + D)$. Now, $pL + A$ is ample for every $p \geq 0$, so $pL + A$ has a smooth metric with $c(pL + A) \geq \varepsilon_p\omega$ and the isomorphism $(k + p)L \simeq pL + A + D$ gives a metric on L such that

$$c(L) = (k + p)^{-1}c(pL + A) + (k + p)^{-1}[D] \geq (k + p)^{-1}\varepsilon_p\,\omega.$$

Observe that the singular part $(p + k)^{-1}[D]$ can be chosen as small as desired by taking p large, so $\max_X \nu(\varphi, x) \leq (k+p)^{-1} \max_D \nu(D, x)$ can be made arbitrarily small. Hence (a) implies (c).

Finally, if property (c) holds, the regularization theorem of [De 91] applied to $T = c(L)$ shows that L has smooth metrics such that the regularized curvature form T_ε has arbitrary small negative part. Hence L is nef by (4.2c) and $\kappa(L) = n$ by (4.2b). Therefore (c) implies (a). □

5. The Kawamata-Viehweg vanishing theorem

To illustrate the strength of theorem 3.1, we give below a very simple derivation of the Kawamata-Viehweg vanishing theorem [Ka 82], [Vi 82]. Only the case of maximum Kodaira dimension will be treated here (the general case can be easily deduced by a slicing argument and an induction on $\dim X$, cf. [De 89]).

DEFINITION 5.1. — *We say that a divisor $D = \sum \alpha_j D_j$ with rational coefficients $\alpha_j \in \mathbb{Q}$ is integrable at a point $x_0 \in X$ if the function $\prod |g_j|^{-2\alpha_j}$ associated to local generators g_j of the ideal of D_j at x_0 is integrable on a neighborhood of x_0 .*

Observe that $\prod |g_j|^{-2\alpha_j} = e^{-2\varphi}$ where φ is the weight function of the natural singular metric on $\mathcal{O}(D)$ described in example 2.2. When D has normal crossings, the g_j's can be taken to be coordinates at x_0; thus D is integrable if and only if $\alpha_j < 1$ for all j . When D is effective and has arbitrary singularities, lemma 2.8 shows that a sufficient condition for the integrability of D at x_0 is that the multiplicity (or Lelong number)

$$\nu(D, x_0) = \sum \alpha_j \, \nu(D_j, x_0)$$

be < 1. If neither D has normal crossings nor $\nu(D, x_0) < 1$, the integrability condition can be checked by means of a sequence of blowing-ups which lift D into a divisor with normal crossings (this is always possible by [Hi 64]). Taking into account the jacobian divisor J of the blow-up morphism π, we get at the end a divisor $D' = \pi^\star D - J$ with normal crossings which is integrable if and only if D is integrable. A consequence of this is that integrability is an open condition : if E is an arbitrary effective divisor and if D is integrable at x_0, then $D + p^{-1}E$ is again integrable at x_0 for p large enough. With these definitions, we have:

THEOREM 5.2 (Kawamata-Viehweg). — *Let L be a line bundle over a projective manifold X with $\kappa(L) = n$. Assume that some positive multiple mL can be written $mL = \mathcal{O}(F + D)$ where F is a nef line bundle and D an effective divisor such that $m^{-1}D$ is integrable on $X \setminus \{$finite set$\}$. Then*

$$H^q(X, K_X + L) = 0 \quad \text{for} \quad q \geq 1.$$

Proof. — By the proof of (4.2b), there is an ample divisor A and an effective divisor E such that $kL \simeq \mathcal{O}(A + E)$. Then $(pm + k)L \simeq \mathcal{O}(pF + A + pD + E)$ where $pF + A$ is ample and

$$(pm + k)^{-1}(pD + E) \leq m^{-1}D + (pm + k)^{-1}E$$

is integrable on $X \setminus \{\text{finite set}\}$ for $p \geq 1$ large enough (integrability is an open condition). If we select a smooth metric on $pF + A$ with positive curvature $\omega_p = c(pF + A)$ and take the singular metric on $\mathcal{O}(pD + E)$ described in example 2.2, we find a singular metric on L such that

$$c(L) = (pk + m)^{-1}\omega_p + (pk + m)^{-1}(p[D] + [E]),$$

and the associated weight $e^{-2\varphi}$ is locally integrable on $X \setminus \{\text{finite set}\}$. Hence we can apply corollary 3.2 to conclude that $H^q(X, K_X + L) = 0$ for $q \geq 1$. □

6. Seshadri constants of nef line bundles

Let L be a nef line bundle over a projective algebraic manifold X. To every point $x \in X$, we attach the number

$$\varepsilon(L, x) = \inf_{C \ni x} \frac{L \cdot C}{\nu(C, x)} \tag{6.1}$$

where the infimum is taken over all irreducible curves C passing through x and $\nu(C, x)$ is the multiplicity of C at x. The infimum

$$\varepsilon(L) = \inf_{x \in X} \varepsilon(L, x) = \inf_{C} \frac{L \cdot C}{\nu(C)} \quad \text{where} \quad \nu(C) = \max_{x \in C} \nu(C, x) \tag{6.1'}$$

will be called the *Seshadri constant* of L. It is well known that L is ample if and only if $\varepsilon(L) > 0$ (Seshadri's criterion [Ha 70]). For two nef line bundles L_1, L_2 we have $\varepsilon(L_1 + L_2) \geq \varepsilon(L_1) + \varepsilon(L_2)$, in particular this shows again that $L_1 + L_2$ is ample if L_1 or L_2 is ample.

If L is a nef line bundle, we are especially interested in singular metrics with isolated logarithmic poles: we say that a logarithmic pole x of the weight φ is *isolated* if φ is finite and continuous on $V \setminus \{x\}$ for some neighborhood V of x and we define

$$\gamma(L, x) = \sup \left\{ \begin{array}{l} \gamma \in \mathbb{R}_+ \text{ such that } L \text{ has a singular metric with } ic(L) \geq 0 \\ \text{and with an isolated log pole of coefficient } \gamma \text{ at } x \end{array} \right\}; \tag{6.2}$$

if there are no such metrics, we set $\gamma(L, x) = 0$.

The numbers $\varepsilon(L, x)$ and $\gamma(L, x)$ will be seen to carry a lot of useful information about the global sections of L and its multiples kL. To make this precise, we first introduce some further definitions. Let $s(L, x)$ be the largest integer $s \in \mathbb{N}$ such that the global sections in $H^0(X, L)$ generate all s-jets $J^s_x L = \mathcal{O}_x(L)/\mathcal{M}^{s+1}_x \mathcal{O}_x(L)$. If L_x is not generated, i.e. if all sections of L vanish at x, we set $s(L, x) = -\infty$. We also introduce the limit value

$$\sigma(L, x) = \limsup_{k \to +\infty} \frac{1}{k} s(kL, x) = \sup_{k \in \mathbb{N}^*} \frac{1}{k} s(kL, x) \tag{6.3}$$

if $s(kL,x) \neq -\infty$ for some k, and $\sigma(L,x) = 0$ otherwise. The limsup is actually equal to the sup thanks to the superadditivity property

$$s(L_1 + L_2, x) \geq s(L_1, x) + s(L_2, x).$$

The limsup is in fact a limit as soon as kL spans at x for $k \geq k_0$, e.g. when L is ample.

THEOREM 6.4. — *Let L be a nef line bundle over X. For every point $x \in X$ we have*

$$\varepsilon(L,x) \geq \gamma(L,x) \geq \sigma(L,x).$$

If L is ample, the equality holds for every $x \in X$. If L is nef and big, the equality holds outside any divisor D prescribed by corollary 4.3.

Proof. — Fix a point $x \in X$ and a coordinate system $(z_1, \ldots, z_n)$ centered at x. If $s = s(kL,x)$, then $H^0(X,kL)$ generates all s-jets at x and we can find holomorphic sections $f_1, \ldots, f_N$ whose s-jets are all monomials z^α, $|\alpha| = s$. We define a global singular metric on L by

$$|\xi| = \Big(\sum_{1 \leq j \leq N} |f_j(z) \cdot \xi^{-k}|^2\Big)^{-1/2k}, \quad \xi \in L_z \tag{6.5}$$

associated to the weight function $\varphi(z) = \frac{1}{2k}\log\sum\big|\theta(f_j(z))\big|^2$ in any trivialization $L_{\restriction\Omega} \simeq \Omega \times \mathbb{C}$. Then φ has an isolated logarithmic pole of coefficient s/k at x, thus

$$\gamma(L,x) \geq \frac{1}{k} s(kL,x)$$

and in the limit we get $\gamma(L,x) \geq \sigma(L,x)$.

Now, suppose that L has a singular metric with an isolated log pole of coefficient $\geq \gamma$ at x. Set $c(L) = \frac{i}{\pi}\partial\overline{\partial}\varphi$ on a neighborhood Ω of x and let C be an irreducible curve passing through x. Then all weight functions associated to the metric of L must be locally integrable along C (since φ has an isolated pole at x). We infer

$$L \cdot C = \int_C c(L) \geq \int_{C\cap\Omega} \frac{i}{\pi}\partial\overline{\partial}\varphi \geq \gamma\,\nu(C,x)$$

because the last integral is larger than the Lelong number of the current $[C]$ with respect to the weight φ (cf. [De 87]) and we may apply the comparison theorem with the ordinary Lelong number associated to the weight $\log|z - x|$. Therefore

$$\varepsilon(L,x) = \inf \frac{L \cdot C}{\nu(C,x)} \geq \sup \gamma = \gamma(L,x).$$

Finally, we show that $\sigma(L,x) \geq \varepsilon(L,x)$ when L is ample. This is done essentially by same arguments as in the proof of Seshadri's criterion, as explained in [Ha 70]. Consider the blow-up $\pi : \widetilde{X} \to X$ at point x, the exceptional divisor $E = \pi^{-1}(x)$ and the line bundles $F_{p,q} = \mathcal{O}(p\pi^\star L - q\,E)$ over $\widetilde{X}$, where $p, q > 0$. Recall that

$\mathcal{O}(-E)_{\restriction E}$ is the canonical line bundle $\mathcal{O}_E(1)$ over $E \simeq \mathbb{P}^{n-1}$, in particular we have $E^n = \mathcal{O}_E(-1)^{n-1} = (-1)^{n-1}$. For any irreducible curve $\widetilde{C} \subset \widetilde{X}$, either $\widetilde{C} \subset E$ and

$$F_{p,q} \cdot \widetilde{C} = \mathcal{O}(-q\,E) \cdot \widetilde{C} = q\,\mathcal{O}_E(1) \cdot \widetilde{C} = q \deg \widetilde{C}$$

or $\pi(\widetilde{C}) = C$ is a curve and

$$F_{p,q} \cdot C = p\,L \cdot C - q\,\nu(C,x) \geq \big(p - q/\varepsilon(L,x)\big) L \cdot C.$$

Thus $F_{p,q}$ is nef provided that $p \geq q/\varepsilon(L,x)$. Since $F_{p,q}$ is ample when p/q is large, a simple interpolation argument shows that $F_{p,q}$ is ample for $p > q/\varepsilon(L,x)$. In that case, the Kodaira-Serre vanishing theorem gives

$$H^1(\widetilde{X}, k\,F_{p,q}) = H^1\big(\widetilde{X}, \mathcal{O}(kp\,\pi^\star L - kq\,E)\big) = 0$$

for k large. Hence we get a surjective map

$$H^0(\widetilde{X}, kp\,\pi^\star L) \longrightarrow H^0\Big(\widetilde{X}, \mathcal{O}(kp\,\pi^\star L) \otimes \big(\mathcal{O}/\mathcal{O}(-kq\,E)\big)\Big) \simeq J_x^{kq-1}(kp\,L),$$

that is, $H^0(X, kp\,L)$ generates all $(kq-1)$ jets at x. Therefore $p > q/\varepsilon(L,x)$ implies $s(kp\,L, x) \geq kq - 1$ for k large, so $\sigma(L,x) \geq q/p$. At the limit we get $\sigma(L,x) \geq \varepsilon(L,x)$.

Assume now that L is nef and big and that $\varepsilon(L,x) > 0$. By the proof of lemma 4.3, there exist an integer $k_0 \geq 1$ and effective divisors A, D such that $k_0 L \simeq A + D$ where A is ample. Then $a\,\pi^\star A - E$ is ample for a large. Hence there are integers $a, b > 0$ such that $a\,\pi^\star A - b\,E - K_{\widetilde{X}}$ is ample. When $F_{p,q}$ is nef, the sum with any positive multiple $k\,F_{p,q}$ is still ample and the Akizuki-Nakano vanishing theorem gives

$$H^1(\widetilde{X}, k\,F_{p,q} + a\,\pi^\star A - b\,E) = H^1\big(\widetilde{X}, (kp + k_0 a)\,\pi^\star L - a\,\pi^\star D - (kq + b)E\big) = 0$$

when we substitute $A = k_0 L - D$. As above, this implies that we have a surjective map

$$H^0\big(X, (kp + k_0 a)\,L - a\,D\big) \longrightarrow J_x^{kq+b-1}\big((kp + k_0 a)\,L - a\,D\big)$$

when $p \geq q/\varepsilon(L,x)$. Since $\mathcal{O}(-aD) \subset \mathcal{O}$, we infer $s\big((kp + k_0 a)L, x\big) \geq kq + b - 1$ at every point $x \in X \setminus D$ and at the limit $\sigma(L,x) \geq \varepsilon(L,x)$. $\square$

Remark 6.6. — Suppose that L is ample. The same arguments show that if $\pi : \widetilde{X} \to X$ is the blow-up at two points x, y and if $E_x + E_y$ is the exceptional divisor, then $F_{p,q} = p\,\pi^\star L - q\,E_x - E_y$ is ample for $p > q/\varepsilon(L,x) + 1/\varepsilon(L,y)$. In that case, $H^0(X, kp\,L)$ generates $J_x^{kq-1}(kp\,L) \oplus J_y^{k-1}(kp\,L)$ for k large. Take $p > q/\varepsilon(L,x) + 1/\varepsilon(L)$ and let y run over $X \setminus \{x\}$. For k large, we obtain sections $f_j \in H^0(X, kp\,L)$ whose jets at x are all monomials z^α, $|\alpha| = kq - 1$, and with no other common zeros. Moreover, formula (6.5) produces a metric on L which is smooth and has positive definite curvature on $X \setminus \{x\}$, and which has a log pole of coefficient $(kq-1)/kp$ at x. Therefore the supremum $\gamma(L,x) = \sup\{\gamma\}$ is always achieved by metrics that are smooth and have positive definite curvature on $X \setminus \{x\}$.

Remark 6.7. — If Y is a p-dimensional algebraic subset of X passing through x, then

$$L^p \cdot Y \geq \varepsilon(L,x)^p \nu(Y,x)$$

where $L^p \cdot Y = \int_Y c_1(L)^p$ and $\nu(Y,x)$ is the multiplicity of Y at x (equal by Thie's theorem [Th 67] to the Lelong number of the integration current $[Y]$). If L is ample, we can take a metric on L which is smooth on $X \setminus \{x\}$ and defined on a neighborhood Ω of x by a weight function φ with a log pole of coefficient γ at x. By the comparison theorem for Lelong numbers, we get

$$L^p \cdot Y \geq \int_{Y\cap\Omega} \left(\frac{i}{\pi}\partial\overline{\partial}\varphi\right)^p \geq \gamma^p \nu(Y,x)$$

and γ can be chosen arbitrarily close to $\varepsilon(L,x)$. If L is nef, we apply the inequality to $k\,L + M$ with M ample and take the limit as $k \to +\infty$. □

The numbers $\varepsilon(L,x)$ and Seshadri's constant $\varepsilon(L) = \inf \varepsilon(L,x)$ are especially interesting because they provide effective results concerning the existence of sections of $K_X + L$. The following proposition illustrates this observation.

PROPOSITION 6.8. — *Let L be a big nef line bundle over X.*

(a) *If $\varepsilon(L,x) > n + s$, then $H^0(X, K_X + L)$ generates all s-jets at x.*

(b) *If $\varepsilon(L) > 2n$, then $K_X + L$ is very ample.*

Proof. — By the proof of theorem 6.4, the line bundle $\pi^\star L - q\,E$ is nef for $q \leq \varepsilon(L,x)$. Moreover, its n-th self intersection is equal to $L^n + (-q)^n E^n = L^n - q^n$ and as $L^n \geq \varepsilon(L,x)^n$ by remark 3.5, we see that $\pi^\star L - q\,E$ is big for $q < \varepsilon(L,x)$. The Kawamata-Viehweg vanishing theorem 5.2 then gives

$$H^1(\widetilde{X}, K_{\widetilde{X}} + \pi^\star L - q\,E) = H^1(\widetilde{X}, \pi^\star K_X + \pi^\star L - (q-n+1)E) = 0,$$

since $K_{\widetilde{X}} = \pi^\star K_X + (n-1)E$. Thus we get a surjective map

$$\begin{array}{ccc} H^0(\widetilde{X}, \pi^\star K_X + \pi^\star L) & \longrightarrow\!\!\!\!\!\rightarrow & H^0(\widetilde{X}, \pi^\star\mathcal{O}(K_X + L) \otimes \mathcal{O}/\mathcal{O}(-(q-n+1)E)) \\ \| & & \| \\ H^0(X, K_X + L) & \longrightarrow\!\!\!\!\!\rightarrow & J_x^{q-n}(K_X + L) \end{array}$$

provided that $\varepsilon(L,x) > q$. The first statement is proved. To show that $K_X + L$ is very ample, we blow up at two points x, y. The line bundle $\pi^\star L - n\,E_x - n\,E_y$ is ample for $1/\varepsilon(L,x) + 1/\varepsilon(L,y) < 1/n$, a sufficient condition for this is $\varepsilon(L) > 2n$. Then we see that

$$H^0(X, K_X + L) \longrightarrow (K_X + L)_x \oplus (K_X + L)_y$$

is also surjective. □

These results are related to a conjecture of Fujita [Fu 88], asserting that if L is an ample line bundle, then $K_X + mL$ is spanned for $m \geq n+1$ and very ample for

$m \geq n+2$. The answer is positive for surfaces, thanks to I. Reider's numerical criterion (a deep extension of Bombieri's work [Bo 73] on pluricanonical embeddings of surfaces of general type). Our paper [De 90] describes a new method which gives partial results in the higher dimensional case. By proposition 6.8 above, we know that $K_X + mL$ generates s-jets for $m > (n+s)/\varepsilon(L)$ and is very ample for $m > 2n/\varepsilon(L)$. It is easy to see for example that $\varepsilon(L) \geq 1$ for any flag manifold, in which the ample cone has a very simple structure. In general, unfortunately, it seems to be a rather hard problem to compute the Seshadri constant $\varepsilon(L)$, even in the case of surfaces. An answer to the following question would be urgently needed.

QUESTION 6.9. — *Given a projective algebraic manifold X, is there always a universal lower bound for $\varepsilon(L)$ when L runs over all ample line bundles of X ? In this case, is it possible to compute explicitly such a lower bound in terms of geometric invariants of X ?*

7. Asymptotic estimates of cohomology groups

Let X be a compact Kähler manifold, E a holomorphic vector bundle of rank r and L a line bundle over X. If L is equipped with a smooth metric of curvature form $c(L)$, we define the q-index set of L to be the open subset

$$(7.1) \qquad X(q,L) = \left\{ x \in X \ ; \ c(L)_x \text{ has } \begin{matrix} q & \text{negative eigenvalues} \\ n-q & \text{positive eigenvalues} \end{matrix} \right\}, \qquad 0 \leq q \leq n.$$

It is shown in [De 85] that the cohomology groups $H^q(X, E \otimes \mathcal{O}(kL))$ satisfy the asymptotic "Morse inequalities"

$$(7.2) \qquad h^q(X, E \otimes \mathcal{O}(kL)) \leq r\frac{k^n}{n!} \int_{X(q,L)} (-1)^q (c(L))^n + o(k^n) \qquad \text{as } k \to +\infty.$$

One difficulty in the application of this result is that the curvature integral is in general quite uneasy to compute, since it is neither a topological nor a holomorphic invariant. However, when $c_1(L) \in \Gamma_+$, the results of [De 91] allow us to measure the distance of L to the ample cone Γ_a. In that case, a use of singular metrics combined with the approximation theorem of [De 91] produces smooth metrics on L for which an explicit bound of the negative part of the curvature is known. It follows that (7.2) gives an explicit upper bound of the cohomology groups of $E \otimes \mathcal{O}(kL)$ in terms of a polynomial in the first Chern class $c_1(L)$ (related techniques have already been used in [Su] in a slightly different context). To state the precise result, we need the notion of nefness for a real $(1,1)$ cohomology class which does not necessarily belong to $NS_{\mathbb{R}}(X)$: we say that $\{u\} \in H^{1,1}(X)$ is nef if $\{u\}$ belongs to the closed convex cone generated by classes of Kähler forms (the so-called *Kähler cone* of $H^{1,1}(X)$).

THEOREM 7.3. — *Suppose that there is a nef cohomology class $\{u\}$ in $H^{1,1}(X)$ such that $c_1(\mathcal{O}_{TX}(1)) + \pi^\star\{u\}$ is nef over the hyperplane bundle $P(T^\star X)$. Suppose* *moreover that L is equipped with a singular metric such that $T = c(L) \geq 0$. For $p = 1, 2, \ldots, n, n+1$ set*

$$b_p = \inf\{c > 0\,;\ \operatorname{codim} E_c(T) \geq p\},$$

with $b_{n+1} = \max_{x \in X} \nu(T, x)$. Then for any holomorphic vector bundle E of rank r over X we have

$$h^q(X, E \otimes \mathcal{O}(kL)) \leq A_q r\, k^n + o(k^n)$$

where A_q is the cup product

$$A_q = \frac{1}{q!\,(n-q)!}\big(b_{n-q+1}\{u\}\big)^q \cdot \big(c_1(L) + b_{n-q+1}\{u\}\big)^{n-q}$$

in $H^{2n}(X, \mathbb{R})$, identified to a positive number.

Remark 7.4. — When X is projective algebraic and $\kappa(L) = n$, the proof of (4.2b) shows that $mL \simeq \mathcal{O}(A+D)$ with A ample and D effective, for some $m \geq 1$. Then we can choose a singular metric on L such that $T = c(L) = \omega + m^{-1}[D]$, where $\omega = m^{-1}c(A)$ is a Kähler metric. As $\nu(T, x) = m^{-1}\nu(D, x)$ at each point, the constants b_j of theorem 7.3 are obtained by counting the multiplicities of the singular points of D ; for example, if D only has isolated singularities, then $b_1 = 0$, $b_2 = \ldots = b_n = 1/m$. Observe moreover that the nefness assumption on $\mathcal{O}_{TX}(1)$ is satisfied with $\{u\} = c_1(G)$ if G is a nef line bundle such that $\mathcal{O}(TX) \otimes \mathcal{O}(G)$ is nef, e.g. if $\mathcal{O}(S^m TX) \otimes \mathcal{O}(mG)$ is spanned by sections for some $m \geq 1$.

Proof of theorem 7.3. — By definition, we have $0 = b_1 \leq b_2 \leq \ldots \leq b_n \leq b_{n+1}$, and for $c \in \,]b_p, b_{p+1}]$, $E_c(T)$ has codimension $\geq p$ with some component(s) of codimension p exactly. Let ω be a fixed Kähler metric on X. By adding $\varepsilon\omega$ to u if necessary, we may assume that $u \geq 0$ and that $\mathcal{O}_{TX}(1)$ has a smooth hermitian metric such that $c\big(\mathcal{O}_{TX}(1)\big) + \pi^\star u \geq 0$.

Under this assumption, the main approximation theorem of [De 91] shows that the metric of L can be approximated by a sequence of smooth metrics such that the associated curvature forms T_j satisfy the uniform lower bound

$$T_j \geq -\lambda_j(x)\, u(x) - \varepsilon_j\, \omega(x) \tag{7.5}$$

where $\lim\!\downarrow_{j \to +\infty} \varepsilon_j = 0$ and $(\lambda_j)_{j>0}$ is a decreasing sequence of continuous functions on X such that $\lim_{j \to +\infty} \lambda_j(x) = \nu(T, x)$ at each point.

Estimate (7.2) cannot be used directly with $T = c(L)$ because wedge products of currents do not make sense in general. Therefore, we replace $c(L)$ by its approximations T_j and try to find an upper bound for the limit.

LEMMA 7.6. — *Let $U_j = X(q, T_j)$ be the q-index set associated to T_j and let c be a positive number. On the open set $\Omega_{c,j} = \{x \in X\,;\ \lambda_j(x) < c\}$ we have*

$$(-1)^q \mathbb{1}_{U_j} T_j^n \leq \frac{n!}{q!\,(n-q)!}\big(c\,u + \varepsilon_j\,\omega\big)^q \wedge \big(T_j + c\,u + \varepsilon_j\,\omega\big)^{n-q}.$$

Proof. — Write $v = c\,u + \varepsilon_j\,\omega > 0$ and let $\alpha_{1,j} \le \ldots \le \alpha_{n,j}$ be the eigenvalues of T_j with respect to v at each point. Then $T_j^n = \alpha_{1,j}\ldots\alpha_{n,j}\,v^n$ and

$$v^q \wedge (T_j + v)^{n-q} = \frac{q!\,(n-q)!}{n!} \sum_{1\le i_1<\ldots<i_{n-q}\le n} (1+\alpha_{i_1,j})\ldots(1+\alpha_{i_{n-q},j})\,v^n.$$

On $\Omega_{c,j}$ we get $T_j \ge -v$ by inequality (7.5), thus $\alpha_{i,j} \ge -1$; moreover, we have $\alpha_1 \le \ldots \le \alpha_q < 0$ and $0 < \alpha_{q+1} \le \ldots \le \alpha_n$ on U_j. On $\Omega_{c,j}$ we thus find

$$0 \le (-1)^q \mathbb{1}_{U_j}\alpha_{1,j}\ldots\alpha_{n,j} \le \mathbb{1}_{U_j}\alpha_{q+1,j}\ldots\alpha_{n,j} \le (1+\alpha_{q+1,j})\ldots(1+\alpha_{n,j}),$$

therefore $(-1)^q \mathbb{1}_{U_j} T_j^n \le \big(n!/q!\,(n-q)!\big)\,v^q \wedge (T_j+v)^{n-q}$. □

End of the proof of theorem 7.3. — Set $\Lambda = \max_X \lambda_1(x)$. By lemma 7.6 applied with an arbitrary $c > \Lambda$ we have

$$(-1)^q \mathbb{1}_{U_j} T_j^n \le \frac{n!}{q!(n-q)!}(\Lambda\,u + \varepsilon_1\omega)^q \wedge (T_j + \Lambda\,u + \varepsilon_1\omega)^{n-q} \quad \text{on } X.$$

Then estimate (7.2) and lemma 7.6 again imply

$$\begin{aligned} h^q(X, E\otimes\mathcal{O}(kL)) &\le r\frac{k^n}{n!}\int_X (-1)^q \mathbb{1}_{U_j} T_j^n + o(k^n) \\ &\le \frac{r\,k^n}{q!\,(n-q)!}\Big(\int_{\Omega_{c,j}} (c\,u+\varepsilon_j\,\omega)^q \wedge (T_j + c\,u + \varepsilon_j\,\omega)^{n-q} \\ &\quad + \int_{X\setminus\Omega_{c,j}} (\Lambda\,u+\varepsilon_1\omega)^q \wedge (T_j+\Lambda\,u+\varepsilon_1\omega)^{n-q}\Big) + o(k^n). \end{aligned} \tag{7.7}$$

Since $\lambda_j(x)$ decreases to $\nu(T,x)$ as $j \to +\infty$, the set $X \setminus \Omega_{c,j}$ decreases to $E_c(T)$. Now, $T_j + \Lambda\,u + \varepsilon_1\omega$ is a closed positive $(1,1)$-form belonging to a fixed cohomology class, so the mass of any wedge power $(T_j + \Lambda\,u + \varepsilon_1\omega)^p$ with respect to ω is constant. By weak compactness, there is a subsequence (j_ν) such that $(T_{j_\nu} + \Lambda\,u + \varepsilon_1\omega)^p$ converges weakly to a closed positive current Θ_p of bidegree (p,p), for each $p = 1,\ldots,n$. For $c > b_{p+1}$, we have $\operatorname{codim} E_c(T) \ge p+1$, hence $\mathbb{1}_{E_c(T)}\Theta_p = 0$. It follows that the integral over $X\setminus\Omega_{c,j}$ in (7.7) converges to 0 when $c > b_{n-q+1}$. For the same reason the integral over $\Omega_{c,j}$ converges to the same limit as its value over X: observe that $(T_j + c\,u + \varepsilon_j\,\omega)^{n-q}$ can be expressed in terms of powers of u, ω and of the positive forms $(T_j + \Lambda\,u + \varepsilon_1\omega)^p$ with $p \le n-q$; thus the limit is a linear combination with smooth coefficients of the currents Θ_p, which carry no mass on $E_c(T)$. In the limit, we obtain

$$h^q(X, E\otimes\mathcal{O}(kL)) \le \frac{r\,k^n}{q!\,(n-q)!}(c\{u\})^q \cdot \big(c_1(L) + c\{u\}\big)^{n-q} + o(k^n),$$

and since this is true for every $c > b_{n-q+1}$, theorem 7.3 follows. □

References

[AV 65] A. ANDREOTTI and E. VESENTINI. — *Carleman estimates for the Laplace-Beltrami equation in complex manifolds*, Publ. Math. I.H.E.S., **25** (1965), 81–130.

[Bo 70] E. BOMBIERI. — *Algebraic values of meromorphic maps*, Invent. Math., **10** (1970), 267–287 and *Addendum*, Invent. Math. **11** (1970), 163–166.

[Bo 73] E. BOMBIERI. — *Canonical models of surfaces of general type*, Publ. Math. IHES, **42** (1973), 171–219.

[De 82] J.-P. DEMAILLY. — *Estimations L^2 pour l'opérateur $\overline{\partial}$ d'un fibré vectoriel holomorphe semi-positif au dessus d'une variété kählérienne complète*, Ann. Sci. Ec. Norm. Sup., **15** (1982), 457–511.

[De 85] J.-P. DEMAILLY. — *Champs magnétiques et inégalités de Morse pour la d''-cohomologie*, Ann. Inst. Fourier (Grenoble), **35** (1985), 189–229.

[De 87] J.-P. DEMAILLY. — *Nombres de Lelong généralisés, théorèmes d'intégralité et d'analyticité*, Acta Math., **159** (1987), 153–169.

[De 89] J.-P. DEMAILLY. — *Transcendental proof of a generalized Kawamata-Viehweg vanishing theorem*, C. R. Acad. Sci. Paris Sér. I Math., 309 (1989), 123–126 and: Proceedings of the Conference "Geometrical and algebraical aspects in several complex variables" held at Cetraro, Univ. della Calabria, June 1989.

[De 90] J.-P. DEMAILLY. — *A numerical criterion for very ample line bundles*, Preprint n° 153, Institut Fourier, Univ. Grenoble I, December 1990, submitted to J. Differential Geom.

[De 91] J.-P. DEMAILLY. — *Regularization of currents and intersection theory*, Preprint Institut Fourier, Univ. Grenoble I, January 1991.

[Fu 88] T. FUJITA. — *Problem list*, Conference held at the Taniguchi Foundation, Katata, Japan, August 1988.

[Ha 70] R. HARTSHORNE. — *Ample subvarieties of algebraic varieties*, Lecture Notes in Math. n° 156, Springer-Verlag, Berlin, 1970.

[Hi 64] H. HIRONAKA. — *Resolution of singularities of an algebraic variety over a field of characteristic zero*, Ann. of Math., **79** (1964), 109–326.

[Hö 66] L. HÖRMANDER. — *An introduction to Complex Analysis in several variables*, 1966, 2nd edition, North-Holland Math. Libr., vol.7, Amsterdam, London, 1973.

[Ka 82] Y. KAWAMATA. — *A generalization of Kodaira-Ramanujam's vanishing theorem*, Math. Ann., **261** (1982), 43–46.

[Le 57] P. LELONG. — *Intégration sur un ensemble analytique complexe*, Bull. Soc. Math. France, **85** (1957), 239–262.

[Le 69] P. LELONG. — *Plurisubharmonic functions and positive differential forms*, Gordon and Breach, New-York, and Dunod, Paris, 1969.

[Re 88] I. REIDER. — *Vector bundles of rank 2 and linear systems on algebraic surfaces*, Ann. of Math., **127** (1988), 309–316.

[Siu 74] Y.T. SIU. — *Analyticity of sets associated to Lelong numbers and the extension of closed positive currents*, Invent. Math., **27** (1974), 53–156.

[Sk 72] H. SKODA. — *Sous-ensembles analytiques d'ordre fini ou infini dans $\mathbb{C}^n$*, Bull. Soc. Math. France, **100** (1972), 353–408.

[Sk 75] H. SKODA. — *Estimations L^2 pour l'opérateur $\overline{\partial}$ et applications arithmétiques*, Séminaire P. Lelong (Analyse), année 1975/76, Lecture Notes in Math. n° 538, Springer-Verlag, Berlin (1977), 314–323.

[Su 87] K-I. SUGIYAMA. — *A geometry of Kähler cones*, preprint University of Tokyo in Hongo, September 1987.

[Th 67] P. THIE. — *The Lelong number of a point of a complex analytic set*, Math. Annalen, **172** (1967), 269–312.

[Vi 82] E. VIEHWEG. — *Vanishing theorems*, J. Reine Angew. Math., **335** (1982), 1–8.

(August 17, 1990, revised on December 21, 1990)

On adjoint bundles of ample vector bundles

TAKAO FUJITA

In this paper we prove the following

MAIN THEOREM. *Let $\mathcal{E}$ be an ample vector bundle on a compact complex manifold M such that $r = \text{rank}(\mathcal{E}) \geq n = \dim M$. Let $A = \det \mathcal{E}$ and let K be the canonical bundle of M. Then $K + A$ is ample except the following cases:*

(1) $(M, \mathcal{E}) \cong (\mathbf{P}^n, \bigoplus^{n+1} \mathcal{O}(1))$.
(2) $(M, \mathcal{E}) \cong (\mathbf{P}^n, \bigoplus^{n} \mathcal{O}(1))$.
(3) *There is a vector bundle $\mathcal{F}$ over a curve C such that $M \cong \mathbf{P}(\mathcal{F})$ and $\mathcal{E}_F \cong \bigoplus^{n} \mathcal{O}(1)$ for any fiber $F(\cong \mathbf{P}^{n-1})$ of $M \to C$.*
(4) $(M, \mathcal{E}) \cong (\mathbf{P}^n, \mathcal{O}(2) \oplus (\bigoplus^{n-1} \mathcal{O}(1)))$.
(5) *$M \cong \mathbf{P}^n$ and $\mathcal{E}$ is the tangent bundle of it.*
(6) *$M \cong \mathbf{Q}^n$, a smooth hyperquadric in $\mathbf{P}^{n+1}$, and $\mathcal{E} \cong \bigoplus^{n} \mathcal{O}(1)$.*

This gives an affirmitive answer to a conjecture of Mukai [**Mu**], and improves upon a result in [**YZ**], from which we borrow the idea of the proof. Our argument depends heavily on the methods in [**CS**] too. I would like to express my hearty thanks to these authors, who kindly sent me their interesting preprints. I should also remark that many techniques are originated by Mori in his famous theory on extremal rays. We use Kawamata's theory in [**KMM**] too.

When I was in Bayreuth, I found that this topic was studied by several authors. Wiśniewski [**W**] considered the case in which $\mathcal{E}$ is spanned. Peternell [**P**] studied the case where $K + A = 0$, and he gave a completely different argument proving the same result in this case. I would like to thank them and M. Schneider and F. Campana for the stimulating discussions during the stay in Bayreuth.

§1. Preliminaries

We recall several results in the theory of extremal rays.

(1.1) **Fibration Theorem.** *Let L be an ample line bundle on M such that $K + L$ is nef, which means, $(K + L)Z \geq 0$ for any curve Z in M. Then there exists a surjective morphism $\Phi : M \to W$ onto a normal variety W together with an ample line bundle H on W such that $K + L = \Phi^* H$ and $\Phi_* \mathcal{O}_M = \mathcal{O}_W$.*

This follows immediately from the Base-Point-Free Theorem of Kawamata [**KMM**]. In fact, it suffices to assume that L is nef and big (instead of the ampleness), M may have certain mild singularities, and similar assertion is true in the relative situation too.

(1.2) **Theorem.** *Suppose that $L \in \text{Pic}(M)$ is nef but $K + L$ is not nef. then there exists an extremal rational curve Z such that $(K + L)Z < 0$ and $-KZ \leq n + 1$.*

This follows from Mori's Cone Theorem [**Mo2**;(1.4)]. Recall that the last assertion is based on the following observation.

(1.3) **Lemma.** *Let Z be a rational curve in M and let p be a smooth point on Z. If $-KZ > n + 1$, then the cycle Z can be deformed to a sum of more than one rational curves, one of which contains p.*

For a proof, see [**Mo1**; Theorem 4]. This is further strengthened as follows.

(1.4) **Ionescu's Lemma.** *Let Z be an extremal curve. Suppose that there exists a subset E of M such that every curve which is numerically proportional to Z is contained in E. Then $-KZ \leq n+1-2\mathrm{codim}E$.*

For a proof, see [**I**;(0.4)].

(1.5) The following criterion is very important.

Mori's Bur-Nut Lemma. *Let p be a simple point on a rational curve Z in M. Suppose that any deformation Y containing p of the cycle Z is irreducible and reduced, and that the restriction of the tangent bundle Θ of M to Y is ample for every such Y. Then $M \cong \mathbf{P}^n$.*

This is not explicitly stated by Mori, but is essentially proved by him in [**Mo1**;§3], as was pointed out by Lazarsfeld [**L**], who used this lemma to prove the following beautiful result.

(1.6) **Theorem**(cf.[**L**;§4]). *Let $f: \mathbf{P}^n \to X$ be a finite surjective morphism onto a complex manifold X. Then $X \cong \mathbf{P}^n$.*

The lemma (1.5) (in fact, an improved version of it) is very important in the proof of the following result too.

(1.7) **Theorem**(cf. [**CS**]). *Let $f: \mathbf{Q}^n \to X$ be a finite surjective morphism from a smooth hyperquadric $\mathbf{Q}^n$ in $\mathbf{P}^{n+1}$ onto a complex manifold X. Then $X \cong \mathbf{P}^n$ unless f is an isomorphism.*

(1.8) We need also the following results.

Theorem (cf. [**KO**]). *Let H be an ample line bundle on a compact complex manifold F with $n = \mathrm{dim}F$. Suppose that the canonical bundle $K(F)$ of F is $-rH$ for some integer r. Then*

1) $r \leq n+1$ and the equality holds if and only if $(F, H) \cong (\mathbf{P}^n, \mathcal{O}(1))$.

2) $r = n$ if and only if $(F, H) \cong (\mathbf{Q}^n, \mathcal{O}(1))$.

(1.9) **Theorem** (cf. [**F2**;(2.5)]). *Let $f: M \to W$ be the contraction morphism of an extremal ray R in M. Suppose that f is birational and that $k = \mathrm{dim}f^{-1}(x) > 0$ for some $x \in W$. Then $(K+kH)R \geq 0$ for any ample line bundle H on M.*

(1.10) *Remark.* Suppose in addition that $(K+kH)R = 0$. Then, for any irreducible component X of $f^{-1}(x)$ with $\mathrm{dim}X = k$, we have $(\tilde{X}, H_{\tilde{X}}) \cong (\mathbf{P}^k, \mathcal{O}(1))$ for the normalization $\tilde{X}$ of X.

Indeed, as in [**F2**;(2.3),(2.4)], we have $H^k(X, -tH) = 0$ for any $t \leq k$. Therefore [**F3**;(2.2)] applies.

(1.11) **Trivial Lemma.** *$AZ \geq r = \mathrm{rank}(\mathcal{E})$ for any rational curve Z in M.*

Indeed, on the normalization $\tilde{Z}(\cong \mathbf{P}^1)$, $\mathcal{E}$ is a direct sum of line bundles of positive degrees, hence $AZ = c_1(\mathcal{E}_{\tilde{Z}}) \geq r$.

§2. Proof of the Main Theorem; the case $r > n$.

(2.1) If $K+A$ is not nef, then, by (1.2), $(K+A)Z < 0$ for some extremal rational curve Z such that $-KZ \leq n+1$. This contradicts the Trivial Lemma when $r > n$. Thus $K+A$ is nef in case $r > n$.

(2.2) Let $\Phi : M \to W$ and H be as in (1.1) such that $\Phi^*H = K+A$. Set $w = \mathrm{dim}W$. We assume that $K+A$ is not ample, or equivalently, Φ is not a finite morphism.

(2.3) Suppose that $w = n$, or equivalently, Φ is birational. Then $\Phi(Z)$ is a point for some extremal rational curve Z in M. For any curve Y which is numerically proportional to Z, we have $\Phi^*H \cdot Y = 0$ and $\Phi(Y)$ is a point. So Y is contained in the exceptional set E of Φ. Hence $-KZ \leq n-1$ by Ionescu's Lemma (1.4). But this contradicts $(K+A)Z = 0$ and the Trivial Lemma (1.11).

Thus $w < n$ and Φ is of fibration type.

(2.4) Let F be a general fiber of Φ. Then $d = \dim F = n - w$. The canonical bundle of F is the restriction of K to F, hence $-KZ \leq d+1$ for some extremal rational curve Z of F. Clearly $(K+A)Z = 0$, so we get $r = d+1 = n+1$ by the Trivial Lemma (1.11). This means that W is a point and $K + A = 0$.

(2.5) In the above case we obtain $(M, \mathcal{E}) \cong (\mathbf{P}^n, \bigoplus^{n+1}\mathcal{O}(1))$. This is Theorem 1 of **[YZ]**. Here we outline the proof for the convenience of the reader.

Let $P = \mathbf{P}(\mathcal{E})$ and set $H = \mathcal{O}_P(1)$. Then the canonical bundle of P is $-rH$. In particular P is a Fano variety. Since $\rho(P) > 1$, there is an extremal ray R such that its contraction morphism $\phi : P \to N$ is different from $\pi : P \to M$.

Suppose that ϕ is birational. Take a point x on N such that $\dim\phi^{-1}(x) = k > 0$. Since $K(P) = -rH$, we have $k \geq r = n+1$ by (1.9). On the other hand, any curve in $X = \phi^{-1}(x)$ is proportional to R and hence is not mapped to a point by π. So $\pi_X : X \to \pi(X) \subset M$ is a finite morphism. Thus we get a contradiction.

Now we see that ϕ is of fibration type. Let F be a general fiber of it. Then $\pi_F : F \to M$ is a finite morphism by the same reasoning as above. On the other hand $K(F) = -rH_F$. Therefore, by Kobayashi-Ochiai's theorem (1.8), we obtain $F \cong \mathbf{P}^n$. Applying Lazarsfeld's theorem (1.6) to π_F, we infer $M \cong \mathbf{P}^n$.

For any line l in M, we have $\mathcal{E}_l \cong \bigoplus^{n+1}\mathcal{O}(1)$. This implies $\mathcal{E} \cong \bigoplus^{n+1}\mathcal{O}(1)$ (see, e.g., **[OSS]**).

§3. The case $r = n$; first step.

(3.1) From now on we assume $r = n$. If $K+A$ is not nef, then $(M, \mathcal{E}) \cong (\mathbf{P}^n, \bigoplus^n\mathcal{O}(1))$. This is Theorem 2 in **[YZ]**, and is proved by similar arguments as in §2.

Thus, we assume that $K + A$ is nef, but not ample.

(3.2) Let $\Phi : M \to W$ and H be as in the fibration theorem. Φ is not birational by the same reasoning as in (2.3).

We further see $w = \dim W \leq 1$. To see this, let F be a general fiber of Φ. Then, as in (2.4), we have $-KZ \leq n+1-w$ for some extremal rational curve Z in F. Since $(K+A)Z = 0$ and $AZ \geq n$, this implies $w \leq 1$.

(3.3) Assume that $w = 1$. $\mathcal{E}_F$ is ample and $r > \dim F$ for any smooth fiber F of Φ. Hence $(F, \mathcal{E}_F) \cong (\mathbf{P}^{n-1}, \bigoplus^n\mathcal{O}(1))$ by the result in §2.

Let $S \subset W$ be the locus of singular fibers of Φ. Then Φ is a $\mathbf{P}^{n-1}$-bundle over $U = W - S$. Since $H^2(U, \mathcal{O}_U^\times) = 0$, Φ is associated to a vector bundle over U, and there is a line bundle H on $\Phi^{-1}(U)$ such that $H_F \cong \mathcal{O}_F(1)$ for any fiber F over U. H can be extended to a line bundle on M, which will be denoted by H by abuse of notation.

Suppose that some fiber X of Φ has two components X_1, X_2. Then $X_1Z < 0$ for some curve Z in X. By Cone Theorem in **[Mo2]**, we may assume that Z is an extremal rational curve. Then $-KZ \leq n-1$ by Ionescu's Lemma (1.4). But this contradicts the Trivial Lemma (1.11), so every fiber is irreducible as a set.

In fact, every fiber X is reduced too. Indeed, if $X = mX_0$ for some $m > 1$ in $\mathrm{Div}(M)$, we have $\chi(\mathcal{O}_{X_0}[tX_0]) = \chi(\mathcal{O}_{X_0})$ for any integer t since $[X_0]_{X_0}$ is numerically trivial. Hence $\chi(\mathcal{O}_X) = m\chi(\mathcal{O}_{X_0})$, contradicting $\chi(\mathcal{O}_X) = \chi(\mathcal{O}_F) = 1$.

Now we have $A \equiv nH$ modulo fibers of Φ, since every fiber is irreducible and reduced. Hence H is Φ-ample. As in [**F1**;§5], we have $\Delta(X, H_X) \leq \Delta(F, H_F) = 0$ for any fiber X by the upper-semicontinuity theorem, so $(X, H_X) \cong (\mathbf{P}^{n-1}, \mathcal{O}(1))$. Hence $S = \emptyset$.

Thus we are in the case (3) of the Main Theorem.

(3.4) From now on, we study the case $w = 0$ and $K + A = 0$. First we review a few results in [**YZ**].

(3.5) **Fact.** *If the Picard number* $\rho(M) > 1$, *then* $M \cong \mathbf{Q}^2 \cong \mathbf{P}^1 \times \mathbf{P}^1$ *and* $\mathcal{E} \cong \mathcal{O}(1) \oplus \mathcal{O}(1)$.

For a proof, see [**YZ**;§4.4]. We assume $\rho(M) = 1$ from now on.

(3.6) *Observation. It suffices to show* $M \cong \mathbf{P}^n$ *or* $\mathbf{Q}^n$.

Indeed, if $M \cong \mathbf{P}^n$, we have $\mathcal{E}_l \cong \mathcal{O}(2) \oplus (\bigoplus^{n-1}\mathcal{O}(1))$ for every line l. So we are in the case (4) or (5) by the theory of uniform vector bundles. When $M \cong \mathbf{Q}^n$, we similarly infer $\mathcal{E} \cong \bigoplus^n \mathcal{O}(1)$. See [**YZ**] for details.

(3.7) Let $P = \mathbf{P}(\mathcal{E})$ and set $H = \mathcal{O}_P(1)$ as in (2.5). Then $\dim P = 2n - 1$, the canonical bundle $K(P)$ of P is $-nH$ and P is a Fano variety. As before, let $\phi : P \to N$ be a contraction morphism of an extremal ray which is different from $\pi : P \to M$.

(3.8) When ϕ is birational, take a point x on N with $\dim \phi^{-1}(x) = k > 0$ and let X be a component of $\phi^{-1}(x)$ with $\dim X = k$. Using (1.9) we infer $k \geq n$ as in (2.5). The equality must hold since π_X is a finite morphism. Hence the normalization $\tilde{X}$ of X is $\mathbf{P}^n$ by (1.10). Applying Lazarsfeld's Theorem (1.6), we infer $M \cong \mathbf{P}^n$. This is in fact the case (4) of the Main Theorem.

(3.9) From now on we assume that ϕ is of fibration type. Let F be its general fiber. Then $\pi_F : F \to M$ is a finite morphism. On the other hand, since $K(F) = K(P)_F = -nH_F$, we have $\dim F \geq n-1$ by Kobayashi-Ochiai's theorem (1.8). Therefore $\dim F = n$ or $n - 1$.

(3.10) If $\dim F = n$, we have $F \cong \mathbf{Q}^n$ by (1.8). Hence $M \cong \mathbf{Q}^n$ or $\mathbf{P}^n$ by (1.7). Thus we are done in this case. In fact we are in the case (6) of the Main Theorem.

(3.11) If $\dim F = n - 1$, we have $(F, H_F) \cong (\mathbf{P}^{n-1}, \mathcal{O}(1))$ by (1.8). Using J. Kollàr's vanishing theorem, [**YZ**] proves that ϕ is equidimensional. In the next section we will show that we are in the case (5) of the Main Theorem.

§4. The remaining case in which $r = n$.

(4.1) Let things be as in (3.11). Applying [**F2**;(2.12)], we infer that ϕ is a $\mathbf{P}^{n-1}$-bundle over N and N is smooth. $\mathcal{F} = \phi_*\mathcal{O}_P(H)$ is a locally free sheaf on N and (P, H) is the scroll associated to $\mathcal{F}$. Moreover, since $K(P) = -nH$, the canonical bundle $K(N)$ of N is $-\det\mathcal{F}$. If we can show $N \cong \mathbf{P}^n$, it is easy to see that $\mathcal{F}$ is the tangent bundle, and hence $M \cong \mathbf{P}^n$. Thus the problem is completely symmetric with respect to the exchange of π and ϕ.

(4.2) Let L_α (resp. L_β) be a line in a fiber of π (resp. ϕ), and set $B = \det\mathcal{F}$. We may assume $\pi^*A \cdot L_\beta \geq \phi^*B \cdot L_\alpha > 0$ by the above symmetry. In the sequel we often omit to write π^* and ϕ^*.

(4.3) Take an extremal rational curve Z in N such that $BZ \leq BZ'$ for any other rational curve Z' in N. Then $BZ = n$ or $n+1$ by (1.3) and (1.11). We treat these cases separately.

Case (a): $BZ = n + 1$.

(a.1) Let $\tilde{Z}$ be the normalization of Z and let X be the fiber product of P and $\tilde{Z}$ over N. Since $X \cong \mathbf{P}(\mathcal{F}_{\tilde{Z}})$ and $\mathcal{F}_{\tilde{Z}} \cong \mathcal{O}(2) \oplus (\bigoplus^{n-1}\mathcal{O}(1))$, X is isomorphic to the blowing-up of $\mathbf{P}^n_\eta$ along a linear subspace of codimension two. Let H_η and H_β be the pull-backs to X of $\mathcal{O}(1)$ of $\mathbf{P}^n_\eta$ and $\tilde{Z} \cong \mathbf{P}^1$. Then the pull-back H_X of H is $H_\eta + H_\beta$, and the exceptional divisor E of $X \to \mathbf{P}^n_\eta$ is the unique member of the linear system $|H_\eta - H_\beta|$.

(a.2) Set $\pi_X^* A = aH_\eta + bH_\beta$ for some integers a, b. Then $b \geq 0$ since $A_X = \pi_X^* A$ is nef. Moreover $a > 0$ since the restriction of π to a fiber of $X \to \tilde{Z}$ is finite. Therefore A_X is nef and big, and hence $\pi_X : X \to M$ is surjective.

If $b = 0$, then the Stein factorization of π_X yields a finite morphism $\mathbf{P}^n_\eta \to M$, so $M \cong \mathbf{P}^n$ by (1.6). Hence we may assume that $b > 0$. In this case A_X is ample and π_X is a finite morphism.

(a.3) Let R be the ramification locus of π_X and set $S = \pi_X(R \cup E)$. Take a rational curve Y in M not in S such that $AY \leq AY'$ for any other such curve Y'. Choose a smooth point p on Y not in S. Then $AY \leq n+1$ by (1.3), while $AY \geq n$ by (1.11). Any deformation of the cycle Y in M containing p is irreducible and reduced by the minimality of the value AY. If the restriction of the tangent bundle $\Theta(M)$ of M to every such deformation is ample, Bur-Nut Lemma (1.5) applies. Therefore, replacing Y if necessary, we may assume that $\Theta(M)_Y$ is not ample.

(a.4) Take a curve C in X such that $\pi(C) = Y$. Then $C \not\subset R \cup E$, since $Y \not\subset S$. Hence the natural homomorphism $\Theta(X) \to \pi_X^*\Theta(M)$ is generically surjective on C. Furthermore, as in [L], $\Theta(M)_C$ is ample if so is $\Theta(X)_C$. But then $\Theta(M)_Y$ would be ample, contradicting the assumption in (a.3). Therefore $\Theta(X)_C$ is not ample.

Let Θ' be the relative tangent bundle of $\phi_X : X \to \tilde{Z}(\cong \mathbf{P}^1_\beta)$. Then we have the following exact sequences

$$0 \longrightarrow \Theta' \longrightarrow \Theta(X) \longrightarrow \phi_X^*\Theta(\tilde{Z}) \longrightarrow 0$$

$$0 \longrightarrow \mathcal{O}_X \longrightarrow \phi_X^*\mathcal{F}^\vee \otimes H_X \longrightarrow \Theta' \longrightarrow 0.$$

If $\phi(C)$ is not a point, then $H_\beta C > 0$ and $(H_\eta - H_\beta)C = EC \geq 0$. If $EC = 0$, then $C \cap E = \emptyset$ and $\Theta(X)_C = \Theta(\mathbf{P}^n_\eta)_C$ would be ample. So $EC > 0$. Then $\mathcal{F}_X^\vee \otimes H_X \cong [H_\eta - H_\beta] \oplus (\bigoplus^{n-1} H_\eta)$ is ample on C. Hence Θ'_C and $\Theta(X)_C$ are ample. This is ruled out by assumption, and so $\phi(C)$ is a point.

Thus C is contained in a fiber F of ϕ. Let d be the degree of this curve in $F \cong \mathbf{P}^{n-1}$.

(a.5) Let δ be the mapping degree of the morphism $\tilde{C} \to \tilde{Y}$ between normalizations. Since $\tilde{Y} \cong \mathbf{P}^1$, $\mathcal{E}_{\tilde{Y}}$ is the direct sum of $\mathcal{O}(1)$'s and $\mathcal{O}(2)$, and hence $\mathcal{E}_{\tilde{C}}$ is the direct sum of line bundles of degrees $\geq \delta$. So $d = HC = \deg(H_C) \geq \delta$ since H is a quotient bundle of $\pi^*\mathcal{E}$.

We have $\delta AY = AC = dAL_\beta$ for any line L_β in F, so $AY \geq AL_\beta$. On the other hand, $AY \leq A \cdot \pi(L_\beta) \leq AL_\beta$ by the minimality of AY. Therefore we must have equalities, in particular $d = \delta$, and $L_\beta \to \pi(L_\beta)$ is birational. Moreover $a = AL_\beta = AY$, which is n or $n+1$.

(a.6) Take a point on the center of the blowing-up $X \to \mathbf{P}^n_\eta$ and let V be the fiber over this point. Then $V \cong \mathbf{P}^1$ and $V \subset E$. As a 1-cycle in P, we can set $V \equiv xL_\alpha + yL_\beta$ ($\equiv$ denotes the numerical equivalence), since $\rho(P) = \rho(M) + 1 = 2$.

We have

$$x + y = 1 \tag{i}$$

since $HV = (H_\eta + H_\beta)V = 1$. Moreover

$$n \leq AV = yAL_\beta = ay \tag{ii}$$

since π_X is finite and $\mathcal{E}_V$ is ample. Similarly we have

$$n \leq BV = xBL_\alpha, \tag{iii}$$

since $H_\beta V = 1$ and $\mathcal{F}_V$ is ample. Therefore $x > 0$, hence $y < 1$ by (i), which implies $a > n$ by (ii). Now, by (a.5), we get $a = n+1$. Then $AV = n$ by (ii) since $y < 1$. Hence $y = n/(n+1)$ and $x = 1/(n+1)$ by (i), so $BL_\alpha \geq n(n+1)$ by (iii). This contradicts the assumption $AL_\beta \geq BL_\alpha$ in (4.2). Thus we are done in this case (a).

Case (b): $BZ = n$.

(b.1) We proceed as in the case (a) and employ similar notation. First we have $\mathcal{F}_{\tilde{Z}} \cong \bigoplus^n \mathcal{O}(1)$ and hence $\tilde{X} \cong \mathbf{P}_\xi^{n-1} \times \tilde{Z}$. On X we have $H_X = H_\xi + H_\beta$.

(b.2) Set $\pi_X^* A = aH_\xi + bH_\beta$. Restricting to a fiber of ϕ we infer $a > 0$. Moreover $b \geq 0$ since A_X is nef.

Assume that $b = 0$. Then $\dim(\pi(X)) = n - 1$, so $D_z = \pi(\phi^{-1}(z)) = \pi(X)$ for every point z on Z. Let V be a fiber of $X \to \mathbf{P}_\xi^{n-1}$. Then V is contained in a fiber G of π. Let μ be the degree of this curve V in $G \cong \mathbf{P}^{n-1}$. Then $BZ = BV = \mu BL_\alpha$. By the minimality of the value BZ, we infer $\mu = 1$ and $L_\alpha \to \phi(L_\alpha)$ is birational. Hence we may replace Z by $\phi(L_\alpha)$ if necessary. In particular, $D_z = \pi(\phi^{-1}(z))$ does not depend on the point z on $\phi(L_\alpha)$.

The above is true for any line in G, so D_z is the same for every $z \in \phi(G)$. Therefore, setting $D = \pi^{-1}(\pi(X))$, we get $\phi^{-1}(\phi(D)) = D$. This implies $DL_\beta = 0$. On the other hand $DL_\alpha = 0$ since $D = \pi^{-1}(\pi(D))$. But this is impossible since the group of numerical equivalence classes of 1-cycles of P is generated by L_α and L_β.

Thus we conclude $b > 0$.

(b.3) π_X is a finite morphism by (b.2). Let R be the ramification locus of it and set $S = \pi_X(R)$. As in (a.3), we take a curve Y not contained in S such that the number AY attains the minimum among such curves. We may further assume that $\Theta(M)_Y$ is not ample.

(b.4) If π_X is étale, we have $\chi(\mathcal{O}_X) = \deg\pi_X \cdot \chi(\mathcal{O}_M)$ by the Riemann-Roch Theorem. Hence π_X is an isomorphism since $\chi(\mathcal{O}_X) = \chi(\mathcal{O}_M) = 1$. This contradicts $\rho(M) = 1$. Hence $S \neq \emptyset$. Moreover S is an ample divisor on M. Therefore $Y \cap S \neq \emptyset$, so $\#(\pi_X^{-1}(y)) < \deg(\pi_X)$ for some point y on Y.

(b.5) Let C be a curve in X such that $\pi(C) = Y$. Then $\Theta(X)_C$ is not ample, hence C is either a fiber of $X \to \mathbf{P}_\xi^{n-1}$ or a curve contained in a fiber F of $X \to \tilde{Z}$.

Suppose that every irreducible component C of $\pi_X^{-1}(Y)$ is of the former type. Then, since $HC = (H_\xi + H_\beta)C = 1$, as in (a.5), we infer that $C \to \pi(C)$ is birational, so the number of such curves is $\deg(\pi_X)$, hence $\#(\pi_X^{-1}(y)) \geq \deg(\pi_X)$ for every y on Y. This contradicts (b.4).

Thus we may assume that $C \subset F$, replacing C if necessary.

(b.6) Let d be the degree of C in $F \cong \mathbf{P}^{n-1}$, and let δ be the mapping degree of $C \to Y$. Then we get $d = \delta$, $a = AL_\beta = AY = n$ or $= n+1$ as in (a.5).

Let V be a fiber of $X \to \mathbf{P}_\xi^{n-1}$ and set $V \equiv xL_\alpha + yL_\beta$ as 1-cycle in P. Then we get a contradiction as in (a.6). Thus we complete the proof.

Remark. As a matter of fact, we should be in the case (5) of the Main Theorem. Therefore the above case (b) does not occur actually.

Comments. We conjecture that our Main Theorem is true even if M is allowed to have log-terminal singularities. Note that K is a $\mathbf{Q}$-bundle in this case and the ampleness of $K + A$ is well-defined. Of course, in the exceptional case (6), M may be singular.

There will be some results in case $r < n$ too. There are various exceptional cases, but we should be able to classify and describe them.

It will be interesting to study the case in which $K + A$ is not nef. In **[YZ]**, the cases $r \geq n - 1$ are studied.

It is conjectured that $K + tH$ is spanned by global sections if H is an ample line bundle and if $t > n$. However, $K + A$ is not always spanned even if $r > n$. Indeed, for any $r > 0$, we have an ample vector bundle of rank r with $c_1 = 1$ on an elliptic curve.

Probably we will have analogous results in positive characteristic cases too.

References

[CS] K. Cho and E. Sato, *Smooth projective varieties dominated by G/P*, Preprint, Kyushu Univ., 1989.

[F1] T.Fujita, *On the structure of polarized varieties with Δ-genera zero*, J. Fac. Sci. Univ. Tokyo Sect. IA Math. **22** (1975), 103–115.

[F2] ———, *On polarized manifolds whose adjoint bundles are not semipositive*, in "Algebraic Geometry Sendai 1985," Advanced Studies in Pure Math. **10**, Kinokuniya, 1987, pp. 167–178.

[F3] ———, *Remarks on quasi-polarized varieties*, Nagoya Math. J. **115** (1989), 105–123.

[H] R. Hartshorne, *Ample vector bundles*, Publ. Math. IHES **29** (1966), 63–94.

[I] P. Ionescu, *Generalized adjunction and applications*, Math. Proc. Camb. Phil. Soc. **99** (1986), 457–472.

[KMM] Y. Kawamata, K. Matsuda and K. Matsuki, *Introduction to the minimal model problem*, in "Algebraic Geometry Sendai 1985," Advanced Studies in Pure Math. **10**, 1987, pp. 283–360.

[KO] S. Kobayashi and T. Ochiai, *Characterizations of complex projective spaces and hyperquadrics*, J. Math. Kyoto Univ. **13** (1973), 31–47.

[L] R. Lazarsfeld, *Some applications of the theory of positive vector bundles*, in "Complete Intersections," Lecture Notes in Math. **1092**, Springer, 1984, pp. 29–61.

[Mo1] S. Mori, *Projective manifolds with ample tangent bundles*, Ann. of Math. **110** (1979), 593–606.

[Mo2] ———, *Threefolds whose canonical bundle is not numerically effective*, Ann. of Math. **116** (1982), 133–176.

[Mu] S.Mukai, *Problems on characterization of the complex projective space*, in Proceedings of 23rd Taniguchi Symposium at Katata, 1988.

[OSS] C. Okonek, M. Schneider and H. Spindler, "Vector Bundles on Complex Projective Spaces," Progress in Math. **3**, Birkhäuser, 1980.

[P] T. Peternell, *Ample vector bundles on Fano manifolds*, Preprint, 1990.

[W] J. A. Wiśniewski, *Length of extremal rays and generalized adjunction*, Math. Z. **200** (1989), 409–427.

[YZ] Y. G. Ye and Q. Zhang, *On ample vector bundles whose adjunction bundles are not numerically effective*, Preprint, Duke Univ., 1989, to appear in Duke Math. J..

Note. E. Sato communicated to me that the main result in [CS] is obtained independently by the paper below:

[PS] K. H. Paranjape and V. Srinivas, *Self maps of homogeneous spaces*, Invent. math. **98** (1989), 425–444.

Department of Mathematics
Tokyo Institute of Technology
Oh-okayama, Meguro, Tokyo
152 Japan

Moderate Degenerations of Algebraic Surfaces

YUJIRO KAWAMATA

In this paper we shall study degenerations of complex surfaces over a disk. We shall define *permissible degenerations* of surfaces in §1, which have slightly general singularities than semistable degenerations. The reason to introduce this new concept is in Theorem 1.3: for an arbitrary algebraic degeneration of non-ruled surfaces $f : X \to \Delta$, after a base change, there exists a bimeromorphically equivalent degeneration $g : Y \to \Delta$ which is permissible and minimal in the sense that the canonical divisor is relatively nef (cf. [KMM]). Thus, permissible degenerations are two dimensional counterpart of semistable degenerations of curves in some sense. We note that g is not necessarily projective although the fibers of g are proved to be projective. A *moderate degeneration* is defined to be a permissible degeneration whose fibers have only isolated singularities.

We shall study local properties of permissible degenerations in §2. It will be proved that the cohomology groups with coefficients in $\mathbf{Q}$ and their Hodge structures behave similarly as in the case of semistable degenerations. In §3, we shall review a strange relationship by Wahl [W] between elliptic surfaces and the singularities of the central fibers of permissible degenerations. By using this, in §4 we shall classify central fibers of minimal moderate degenerations whose general fibers have the Kodaira dimension 0 or 1. In §5, we shall prove that the indices of singularities are bounded for all the permissible degenerations of a given deformation type of surfaces with $\kappa = 1$.

In Appendix, we shall prove that the degenerate locus of a contraction morphism is covered by rational curves. This problem was discussed with Professors F. Campana and T. Peternell after the conference. The author would like to thank them and Professors M. Schneider and K. Hulek for the hospitality.

§1. Minimal models.

Let d and r be positive integers and the $a_j (1 \leqq j \leqq d)$ integers. Supppose that a group $G = \mathbf{Z}/(r)$ acts on a vector space $D = \mathbf{C}^d$ with coordinates $(t_1, ..., t_d)$ by

$$g^*(t_1, ..., t_d) = (\zeta_r^{a_1}.t_1, ..., \zeta_r^{a_d}.t_d)$$

for generators g and ζ_r of G and μ_r, the group of r-th roots of unity, respectively. Then we say that G acts on D with weights $1/r(a_1, ..., a_d)$, and the singularity of the quotient space D/G at the image of $(0, ..., 0)$ is called a quotient singularity of type $1/r(a_1, ..., a_d)$. If a holomorphic function f on D is invariant under the action of G, then f descends to a holomorphic function on D/G, which will again be denoted by f by abuse of notation.

As a generalization of the notion of normal crossing divisors on threefolds, we shall consider three types of pairs consisting of threefolds V_i with isolated singularities and holomorphic functions f_i on them for $i = 1, 2, 3$ in the following definition, which will be justified by Theorem 1.3.

Typeset by $\mathcal{A}_{\mathcal{M}}\mathcal{S}$-TEX

Definition 1.1. A proper surjective morphism $f : X \longrightarrow \Delta$ from a 3-dimensional normal complex space X to a disk $\Delta = \{t \in C; |t| < 1\}$ is called a *permissible* (resp. *moderate*) *degeneration* of surfaces, if the fibers $f^{-1}(t)$ for $t \in \Delta^* = \Delta - \{0\}$ are connected smooth surfaces, $f^{-1}(0)$ is reduced, and if each singular point x of $f^{-1}(0)$ has an open neighborhood U in X such that the pair $(U, f^*(t))$ is biholomorphic to an open neighborhood of a singular point of one of the $(V_i\ ,\ f_i)$ (resp. $(V_1\ ,\ f_1)$) defined below.

(1.1.1) Let r, a and n be positive integers such that $(r, a) = 1$, and let a group $G = \mathbf{Z}/(r)$ act on $D_4 = \mathbf{C}^4$ with weights $1/r(a, -a, 1, 0)$. Let

$$\tilde{V}_1 = \tilde{V}_1(r; n) = \{t \in D_4; t_1t_2 + t_3^r + t_4^n = 0\},$$

$V_1 = V_1(r, a; n)$ the image of $\tilde{V}_1$ in D_4/G, and let $f_1 = t_4$.

(1.1.2) Let r and a be positive integers, and let G act on $D_3 = \mathbf{C}^3$ with weights $1/r(a, -a, 1)$. Let $\tilde{V}_2 = D_3$, $V_2 = V_2(r, a) = \tilde{V}_2/G$, and $f_2 = t_1t_2$.

(1.1.3) Let $V_3 = D_3$ and $f_3 = t_1t_2t_3$.

The singularities of the V_i defined above are all terminal singularities. By abuse of notation, we sometimes write V_i instead of the pair $(V_i\ ,\ f_i)$ in case the function f_i is obvious.

Lemma 1.2. ([Ka1, p.156]) *The surface*

$$T_1(r,\ a\ ;\ n)\ =\ \{\ t\ \in\ V_1(r,\ a\ ;\ n)\ ;\ f_1(t)\ =\ 0\ \}$$

has a cyclic quotient singularity of type $1/r^2(a, r - a)$.

Proof. The isomorphism is given by

$$t_1 = u^r, t_2 = -v^r, t_3 = uv,$$

where (u, v) are coordinates on $\mathbf{C}^2$ with an action of $\mathbf{Z}/(r^2)$ with weights $1/r^2(a, r - a)$. Q.E.D.

The following theorem will be obtained by the stable analytic $\mathbf{Q}$-factorialization from a projective minimal model of a given degeneration of surfaces.

Theorem 1.3. *Let $f : X \longrightarrow \Delta$ be a surjective projective morphism from a complex manifold X of dimension 3 to a disk Δ such that the fibers $f^{-1}(t)$ for $t \in \Delta^*$ are connected smooth surfaces which are not birationally equivalent to ruled surfaces. Then there exist a positive integer m and a permissible degeneration of surfaces $g : Y \longrightarrow \Delta$ which satisfy the following conditions.*

(1) g is bimeromorphically equivalent to the morphism $f_m : X_m \longrightarrow \Delta$ obtained from f by the base change $\tau : \Delta \longrightarrow \Delta$ defined by $\tau^*(t) = t^m$.

(2) The fibers $g^{-1}(t)$ for $t \in \Delta^*$ are smooth projective surfaces, the central fiber $S = g^{-1}(0) = div(g^*(t))$ is a reduced projective surface, and all the irreducible components of S are normal $\mathbf{Q}$-Cartier divisors.

(3) g is minimal in the sense that the canonical divisor K_Y is g-nef, i.e., $(K_Y.C) \geqq 0$ for any curve C on Y such that g(C) is a point.

Proof. By the semistable reduction theorem [KKMS] and by [Ka1, 10.1' and 10.4], there exist a positive integer m' and a projective surjective morphism $g' : Y' \to \Delta$ from a complex space Y' with only terminal singularities which satisfy conditions (1), (2) and (3) of the theorem.

Let $y' \in S' = \mathrm{div}(g'^*(t))$ be a singular point of S', and r the index of the singularity (Y', y') (cf. [KMM]). Let U' be a small open neighborhood of y' in Y' which has a canonical cover $\pi : \tilde{U}' \to U'$, i.e., π is a finite surjective morphism of degree r which is etale over $U' - \{y'\}$ and $\tilde{U}'$ has at most an isolated cDV singularity.

Let S'_1 , ..., S'_d be irreducible components of $S' \cap U'$ through y'. Since the S'_j are **Q**-Cartier, their intersections are purely 1-dimensional. Hence by [Ka1, 10.4 and 9.6], d is at most 3. The pull-backs $\tilde{S}'_j = \pi^{-1}(S')_j$ are also **Q**-Cartier, and hence by [Ka1, 5.1], they are Cartier divisors.

If $d = 3$, then by [Ka1, 10.4 and 9.6], the S'_j are nonsingular, and hence by [Ka1, 10.8], we have $r = 1$. Since U' has a smooth Cartier divisor, U' itself is nonsingular. Thus $(U', g'^*(t))$ has a singularity of type (1.1.3).

If $d = 2$, then by [Ka1, 10.4, 10.8 and 9.6], the $\tilde{S}'_j$ are irreducible and nonsingular. Hence $\tilde{U}'$ is nonsingular and we have type (1.1.2).

Let us consider the case $d = 1$. By [Ka1, 10.9], we know already that $S' \cap U'$ has at most rational double points or cyclic quotient singularities. In order to simplify the singularity of $(U', g'^*(t))$, we shall take some multiple $m = km'$ of m'. Let $g'_k : Y'_k \to \Delta$ be the morphism obtained from g' by the base change $\sigma : \Delta \to \Delta$ defined by $\sigma^*(t) = t^k$. Then $g : Y \to \Delta$ will be obtained by a partial resolution of singularities $\mu : Y \to Y'_k$. We shall prove that μ is crepant, i.e., the canonical divisor K_Y is relatively numerically trivial for μ, so that the condition (3) is preserved. We shall also check that the fiber $S = \mathrm{div}(g^*(t))$ is still reduced along new irreducible components which are exceptional with respect to μ. We shall note that μ is not projective, but prove that it is projective when restricted on S.

The rational double points of S' can be resolved by the simultaneous resolution of g'_k by Brieskorn-Tyurina ([B], [T]) for suitably divisible k: for a small open neighborhood U'_k in Y'_k of the point y'_k over y', there exists a projective surjective morphism $\mu : U \to U'_k$ from a nonsingular complex space U which induces a minimal resolution on ${g'_k}^{-1}(0) \cap U'_k$ and an isomorphism from $U - \mu^{-1}(y'_k)$ onto $U'_k - \{y'_k\}$. μ is small in the sense that it has only a 1-dimensional degenerate locus. Hence it is automatically crepant.

But after the base change by σ, Y'_k becomes singular along the curves which lie over the double curves of S'. We shall resolve these new singularities later. Before that, we treat the case $d = 1$ and $r > 1$.

Let r and a be positive integers such that $(r, a) = 1$, and $h(x, y)$ a germ of a holomorphic function near $(0, 0) \in \mathbf{C}^2$. Let $G = \mathbf{Z}/(r)$ act on $D_4 = \mathbf{C}^4$ with weights $1/r(a, -a, 1, 0)$ as in (1.1.1), and let $\tilde{V}' = \tilde{V}'(r; h)$ be a germ of a hypersurface in D_4 near (0, 0, 0, 0) defined by an equation

$$t_1 t_2 + h(t_3^r, t_4) = 0.$$

Let $V' = V'(r,a;h)$ be the image of $\tilde{V}'$ in D_4/G. Thus $V_1(r,a;n) = V'(r,a;x+y^n)$ as a germ of a singularity.

Lemma 1.4. *With the notation as above, if $d=1$ and $r>1$, then the pair $(U', g'^*(t))$ is biholomorphic near y' to $(V'(r,a;h), t_4)$ for some r, a and h such that $h(x,y)$ is reduced and $h(x,0) \neq 0$.*

Proof. We have already shown that U' is biholomorphic to some $V'(r,a;h)$ in the proof of [Kal, 10.9] except in the case where $r=2$. But this case can be treated in a similar manner. In fact, according to [Mo], there are three special types of terminal singularities of index 2, and the equation of their canonical covers in $\mathbf{C}^4$ induce on any invariant nonsingular hypersurface an equation of the form t_1^2+(higher terms of degree $\geqq 4$). Hence U' does not have these kinds of singularities. Now, the equation for $S' \cap U'$ is written as t_4+ (higher terms). After a suitable coordinate change, it is transformed to t_4 . Since the singularity of U' is isolated, h is reduced. The condition $h(x,0) \neq 0$ corresponds to the irreducibility of $S' \cap U'$. Q.E.D.

We shall generalize the process of simultaneous resolution to the case $r>1$. If we take the neighborhood U' small enough, then there exist positive integers k and b such that we have a decomposition

$$h_k(x,y) = h(x,y^k) = \text{unit.} \prod_{j=1}^{b} (x + h_k^{(j)}(y)).$$

Let $\tilde{V}'_k = \tilde{V}'(r;h_k)$ and $V'_k = \tilde{V}'_k/G$. Let $\tilde{\nu}_1 : \tilde{W}_1 \to \tilde{W}_0 = \tilde{V}'_k$ be the blow-up along the ideal $(t_1, t_3^r + h_k^{(1)}(t_4))$. Then the action of G lifts to $\tilde{W}_1$, and one can check that the quotient $W_1 = \tilde{W}_1/G$ with the pull-back of the function t_4 on $W_0 = \tilde{W}_0/G$ has two singular points of types $V'(r,a;\prod_{j=2}^{b}(x+h_k^{(j)}(y)))$ and $V_1(r,a;n)$ for $n = \mathrm{ord}(h_k^{(1)})$. The induced morphism $\nu_1 : W_1 \to W_0 = V'_k$ is projective, and the degenerate locus of ν_1 consists of one nonsingular rational curve. If $b>2$, we blow up W_1 again. Repeating this process, we obtain a projective morphism $\mu : V_1 = W_{b-1} \to W_0$ whose degenerate locus consists of $b-1$ nonsingular rational curves and such that V_1 with the pull-back of t_4 has b singular points of types $V_1(r,a;n_j)$ for $n_j = \mathrm{ord}(h_k^{(j)})(j=1,...,b)$.

By replacing U'_k by V_1 as above, we construct a desired partial resolution of Y'_k near y'_k. Although the morphism thus obtained is not projective globally, it is projective when restricted to the fibers over $t=0$, because there exists a relatively ample Cartier divisor supported on the degenerate locus.

Now, let us resolve the 1-dimensional singular locus of Y'_k . First, let us consider the base change of the pair of type (1.1.2). The singularity is of type $V'(r,a;y^k)$. The blow-up along the ideal (t_1,t_4) of the canonical cover $\tilde{V}'$ induces a morphism $\nu_1 : W_1 \to W_0 = U'_k$. The degenerate locus of ν_1 is a singular rules surface which is mapped to the singular locus of W_0 by ν_1 , and the reduced part of an arbitrary positive dimensional fiber of ν_1 is isomorphic to $\mathbf{P}^1$.W_1 has 1-dimensional singular locus of type $V'(r,a;y^{k-1})$ and one isolated singularity of type $V_2(r,a)$. Repeating this process $k-1$ times, we obtain a partial resolution $\mu : V_2 = W_{k-1} \to W_0$. There are only type (1.1.2) singularities on V_2 , and there is a μ -ample Cartier divisor supported on the degenerate locus of μ. Since μ coincides with the minimal resolution of rational double points of type A_{k-1} along the general points of the singular locus of U'_k , we can patch μ globally. Since there is no divisor on V_2 mapped to a point on U'_k by μ, we check that μ is crepant and the new fiber of g over $t = 0$ is reduced.

Finally, we resolve the singularity of Y'_k over the singularity of type (1.1.3) of Y'. This is a singularity of a hypersurface

$$V'_k = \{t \in \mathbf{C}^4; t_1t_2t_3 + t_4^k = 0\}$$

with a holomorphic function t_4. Though it is not hard to construct a resolution explicitly (starting with the bliow-up at the ideal (t_1, t_4)), we use a general result on minimal models of toric varieties by Reid here. Let $S'_k = \{t \in V'_k; t_4 = 0\}$. Then $T = V'_k - S'_k \subset V'_k$ is a torus embedding. Hence V'_k has only rational singularities. Since it is also Gorenstein, it has only canonical singularities. Then by [R, 0.2] ($A = X$ there), there exists a projective crepant toric morphism $\mu : V_3 \longrightarrow V'_k$ from a toric veriety V_3 with only $\mathbf{Q}$-factorial terminal singularities. Since V'_k is Gorenstein, V_3 is also Gorenstein, and hence smooth by [MS] or [D]. Let $S = \mu^* S'_k$. Then $\mu^*(K_{V'_k} + S'_k) = K_{V_3} + S$. Hence S is reduced. We note that the toric morphism μ has a relatively ample divisor whose support is contained in the degenerate locus of μ. Combining the above arguments together, we obtain a desired partial resolution $\mu : Y \longrightarrow Y'_k$. Q.E.D. of Theorem 1.3.

For example, if $f' : X' \longrightarrow \Delta$ is a "flower pot" type semistable degeneration of Enriques surfaces (cf. [P, p.85]), then by contracting all the irreducible components of the central fiber to points except the "pot", we obtain a moderate degeneration with index $r = 2$. The central fiber S in this case has only quotient singularities $x_1, ..., x_n$ of type $1/4(1,1)$, and there is a generically two to one morphism $\pi : \tilde{S} \longrightarrow S$ from a K3 surface $\tilde{S}$ with ordinary double points $\tilde{x}_1, ..., \tilde{x}_n$ such that $\pi^{-1}(x_j) = \tilde{x}_j$ and π is etale outside the $\tilde{x}_j$.

§2. Topology and Hodge structures.

We shall see that the topology of permissible (resp. moderate) degenerations are very similar to that of semistable degenerations (resp. smooth families). First, we calculate the Milnor fibers.

Theorem 2.1. *In the notation of (1.1), let* $M_i = \{t \in V_i, f_i(t) = 1\}$. *Then* M_1 *(resp.* M_2*) is homotopy equivalent to the quotient space* $\bar{\Delta}/\sim$ *(resp. 1-shere* S^1*), where* $\bar{\Delta} = \{t \in \mathbf{C}; |t| \leqq 1\}$ *and* $t \sim t'$ *for* $t \neq t'$ *if and only if* $|t| = 1$ *and* $(t/t')^r = 1$. *In particular,* $H^p(M_1, \mathbf{Q}) = 0$ *for* $p > 0$.

Proof. Let $\tilde{N}_1 = \{t \in \tilde{V}_1; t_4 = 1, t_1 + \bar{t}_2 = 0, |t_1| \leqq 1\}$ and $N_1 = \tilde{N}_1/G \subset M_1$. Then $\tilde{N}_1$ is homeomorphic to the union of r discs of dimension 2 glued along the boundaries, and the group G acts freely on $\tilde{N}_1$ permuting the r sheets of discs in a cyclic way so that $N_1 \simeq \bar{\Delta}/\sim$. We shall construct a G-equivariant deformation retraction from $\tilde{M}_1 = \{t \in \tilde{V}_1; t_4 = 1\}$ to $\tilde{N}_1$. $\tilde{M}_1$ is an r-fold covering of the (t_1, t_2)-plane $B \simeq \mathbf{C}^2$ branched along the curve C defined by $t_1t_2+1=0$. So it is enough to construct a deformation retraction of the pair (B,C) to (B_0,C_0), where $B_0=\{t\in B; t_1+\bar{t}_2=0, |t_1|\leqq 1\}$ and $C_0=B_0\cap C$. First, fixing the value of t_1t_2, we move (t_1,t_2) until the equality $|t_1|=|t_2|$ holds. Next, preserving this eqality and fixing those points (t_1,t_2) such that $t_1t_2\in[-1,0]$, we move (t_1,t_2) so that $\mathrm{Re}(t_1t_2)\leqq 0$, then $t_1t_2\in\mathbf{R}_{<0}$, and finally $t_1t_2\in[-1,0]$ holds. Thus, N_1 is a deformation retract of M_1, and we are done in this case. As for M_2, we take $\tilde{N}_2=\{t\in\tilde{V}_2; t_3=0, t_1=\bar{t}_2, |t_1|=1\}$. Q.E.D.

The above theorem says that there are no vanishing cycles over $\mathbf{Q}$ for moderate degenerations. In particular, the Betti numbers of fibers are constant. Note that this is not the case for simpler singularities of families such as ordinary double points.

Corollary 2.2. *Let S be a normal compact complex surface whose singular points $P_\alpha(\alpha=1,...,\lambda)$ are all quotient singularities of types $1/r_\alpha^2(a_\alpha,r_\alpha-a_\alpha)$ for some positive integers r_α and a_α such that $(r_\alpha,a_\alpha)=1$. Then the self-intersection number (K_S^2) of the canonical divisor is an integer, and the Noether formula holds:*

$$\chi(S,\mathcal{O}_S)=1/12((K_S^2)+e(S)),$$

where $e(S)$ is the Euler number.

Proof. Let $\sigma: M\longrightarrow S$ be the minimal resolution, and let $\sigma^*K_S=K_M+\sum_\alpha\Xi_\alpha$, where the Ξ_α are $\mathbf{Q}$-divisors lying over P_α. Since S has only rational singularities, $\chi(S,\mathcal{O}_S)=\chi(M,\mathcal{O}_M)$. By the usual Noether formula for M, we see that it is enough to prove the following equality

$$(\Xi_\alpha^2)+e(\mathrm{Supp}(\Xi_\alpha))=1.$$

Thus the question is local, and we may assume that S is the central fiber of a moderate degeneration $f: X\longrightarrow\Delta$. We have $\chi(S,\mathcal{O}_S)=\chi(X_t,\mathcal{O}_{X_t})$, $(K_S^2)=(K_{X_t}^2)$ and $e(S)=e(X_t)$ for a fiber $X_t=f^{-1}(t)$ with $t\neq 0$, and the corollary follows from the usual Noether formula. Q.E.D.

Let $f: X\longrightarrow\Delta$ be a moderate degeneration of surfaces, and let S be the central fiber. S is said to be in class $\mathcal{C}$ if it is bimeromorphic to a compact Kähler manifold. In our case, S is in class $\mathcal{C}$ if and only if the first Betti number $b_1(S)$ is even. For, $\sigma: M\longrightarrow S$ being the minimal resolution, we have $b_1(S)=b_1(M)$, since the exceptional locus of σ consists of trees of nonsingular rational curves, and it is known that M is Kählerian if and only if $b_1(M)$ is even ([MK], [Mi1] and [Si]).

We recall some results in [St]. We define the sheaf $\tilde{\Omega}_S^p$ on S as the double dual of the sheaf of Kähler differential p-forms for $p\geqq 0$. Then $\tilde{\Omega}_S^2$ coincides with the dualizing sheaf ω_S, and we have the duality $\mathbf{R}\mathrm{Hom}_{\mathcal{O}_S}(\tilde{\Omega}_S^p,\omega_S)\simeq\tilde{\Omega}_S^{2-p}$. The complex $\tilde{\Omega}_S$ gives a resolution of the constant sheaf $\mathbf{C}_S$, and the cohomology groups with coefficients in $\mathbf{Q}$ satisfy the Poincaré duality. Moreover, if S is in class $\mathcal{C}$, then the Hodge spectral sequence

$$E_1^{p,q}=H^q(S,\tilde{\Omega}_S^p)\Rightarrow H^{p+q}(S,\mathbf{C})$$

degenerates at E_1, and there is the Hodge symmetry between $H^q(S,\tilde{\Omega}^p_S)$ and $H^p(S,\tilde{\Omega}^q_S)$.

Similarly, let $\tilde{\Omega}^p_{X/\Delta}$ be the double dual of the sheaf of relative Kähler p-forms with respect to f.

Theorem 2.3. *Let $f : X \to \Delta$ be a moderate degeneration of surfaces. Then the following hold.*

(1) $\tilde{\Omega}^p_{X/\Delta} \otimes_{O_X} O_S \simeq \tilde{\Omega}^p_S$ for $p \geqq 0$.

(2) $\tilde{\Omega}^2_{X/\Delta}$ coincides with the relative dualizing sheaf $\omega_{X/\Delta}$, and $\mathbf{R}Hom_{O_X}(\tilde{\Omega}^p_{X/\Delta},\omega_{X/\Delta}) \simeq \tilde{\Omega}^{2-p}_{X/\Delta}$.

(3) The complex $\tilde{\Omega}^{\cdot}_{X/\Delta}$ gives a resolution of the inverse image sheaf $f^{\cdot}O_\Delta$.

(4) If $b_1(S)$ is even, then the sheaves $R^qf_*\tilde{\Omega}^p_{X/\Delta}$ are locally free, and the Hodge spectral sequence

$$E_1^{p,q} = R^qf_*\tilde{\Omega}^p_{X/\Delta} \Rightarrow R^{p+q}f_*\mathbf{C}_X \otimes O_\Delta$$

degenerates at E_1, giving a variation of Hodge structures over Δ.

Proof. By Theorem 2.1, $R^pf_*\mathbf{C}_X$ is a constant sheaf for any p. Hence (4) follows from (1), (3) and [St] by the upper semi-continuity. Since the assertions of (1), (2) and (3) are local, we may assume that $X = V_1(r,a;n)$ in the following.

(1): Since the depth of $\tilde{\Omega}^p_{X/\Delta}$ is at least 2 for $0 \leqq p \leqq 2$, the sheaf $\tilde{\Omega}^p_{X/\Delta}\otimes O_S$ is torsion free. Hence it is enough to prove that the restriction homomorphisms $\varepsilon^p : \tilde{\Omega}^p_X \to \tilde{\Omega}^p_S$ are surjective. Let the group $H = \mathbf{Z}/(r^2)$ act on $\mathbf{C}^2$ with weights $1/r^2(a, r-a)$ so that $S \simeq \mathbf{C}^2/H$, and let $\gamma : \mathbf{C}^2 \to S$ be the quotient morphism given by $\gamma^*(t_1,t_2,t_3) = (u^r, -v^r, uv)$ for coordinates (u,v) on $\mathbf{C}^2$. Then $\tilde{\Omega}^p_S$ can be identified with the sheaf of H-invariant p-forms on $\mathbf{C}^2$.

Let $\omega_1 = u^{i-1}v^j du$ be an H-invariant 1-form for some integers i,j with $i > 0$, $j \geqq 0$. Then $ia + j(r-a) \equiv 0 \pmod{r^2}$, hence $i \equiv j \pmod r$. Let $i = j + kr$. Then $ak + j \equiv 0 \pmod r$. If $k > 0$, then $r\omega_1 = \varepsilon_1(t_1^{k-1}t_3^j dt_1)$. If $k = 0$, then $i \geqq r$ and $r\omega_1 = \varepsilon_1(-t_2t_3^{i-r}dt_1)$. If $k < 0$, then $r\omega_1 = \varepsilon_1(r(-t_2)^{-k}t_3^{i-1}dt_3 - (-t_2)^{-k-1}t_3^i dt_2)$. By symmetry, H-invariant 1-forms of type $u^iv^{j-1}dv$ are also in the image of ε_1 . Since these 1-forms generate $\tilde{\Omega}^1_S$, ε_1 is surjective.

Let $\omega_2 = u^{i-1}v^{j-1}du \wedge dv$ be an H-invariant 2-form for some integers $i,j > 0$. Similarly, we let $i = j + kr$. If $k > 0$, then $r\omega_2 = \varepsilon_2(t_1^{k-1}t_3^{j-1}dt_1 \wedge dt_3)$. If $k = 0$, then $i \geqq r$ and $r\omega_2 = \varepsilon_2(-t_2t_3^{i-r-1}dt_1 \wedge dt_3)$. By symmetry, the case $k < 0$ is also OK, and ε_2 is surjective.

(2): $\tilde{\Omega}^2_{X/\Delta}$ and $\omega_{X/\Delta}$ coincide outside the singular points of X, and hence over X. Since the depth of $\tilde{\Omega}^p_{X/\Delta}$ is 3 for $0 \leqq p \leqq 2$ by (1), we have also the second assertion.

(3): It is clear that $\tilde{\Omega}^{\cdot}_{X/\Delta}$ is a complex and $\mathcal{H}^0(\tilde{\Omega}^{\cdot}_{X/\Delta}) \simeq f^{\cdot}O_\Delta$. Let ω be a section of $\tilde{\Omega}^p_{X/\Delta}$ such that $d\omega = 0$ in $\tilde{\Omega}^{p+1}_{X/\Delta}$ for some $p > 0$. By (2) and the exactness of $\tilde{\Omega}^{\cdot}_S$ at degree p, we can write $\omega = d\theta + f^*(t)\omega'$ for some sections θ and ω' of $\tilde{\Omega}^{p-1}_{X/\Delta}$ and $\tilde{\Omega}^p_{X/\Delta}$, respectively. Since $d\omega' = 0$ in $\tilde{\Omega}^{p+1}_{X/\Delta}$, we have $\mathcal{H}^p(\tilde{\Omega}^{\cdot}_{X/\Delta}) = f^*(t)\mathcal{H}^p(\tilde{\Omega}^{\cdot}_{X/\Delta})$, and we are done by Nakayama's lemma. Q.E.D.

We note that V_2 in (1.1.2) is a V-manifold and $\{t \in V_2; f_2 = 0\}$ is a divisor with V-normal crossings in the sense of [St, 1.16].

Lemma 2.4. *Let $f : X \to \Delta$ be a moderate degeneration of surfaces, $\sigma : M \to S$ the minimal resolution of the central fiber $S = f^{-1}(0)$, and $\tau : M \to N$ the contraction morphism to a relatively minimal model. Assume that $b_1(S)$ is even. Then $h^p(\mathcal{O}_{X_t}) = h^p(\mathcal{O}_N)$ for any fiber $X_t = f^{-1}(t)(t \in \Delta)$ and for all p. Moreover, if f is not smooth and if K_X is f-nef, then there exist positive integers m_1 and m_2 such that inequalities of m-genara $P_m(X_t) > P_m(N)$ for $t \in \Delta^*$ hold for positive integers m with $m_1|m$ and $m_2 < m$.*

Proof. Since the singularities of S are rational, we have $h^p(\mathcal{O}_{X_t}) = h^p(\mathcal{O}_S) = h^p(\mathcal{O}_M) = h^p(\mathcal{O}_N)$. By [N1, 6.3], there exist m_1 and m_2 such that the linear system $|mK_X|$ is free and $h^0(mK_{X_t}) = h^0(mK_S)$ for any positive integer m with $m_1|m, m_2 < m$ and for $t \in \Delta^*$. Since S has singularities other than rational double points, not all m-canonical forms on S can be lifted to M, i.e., we have $h^0(mK_S) > h^0(mK_M) = h^0(mK_N)$. Q.E.D.

§3. Singular elliptic surfaces.

There is a close relationship between the surface singularities of type $1/r^2(a, r-a)$ for $(r,a) = 1$ and singular fibers of elliptic surfaces. J. Kollár informed the author that the following lemma was already found by Wahl [W, 2.8.2]. The proofs of Lemma 3.1 and Corollary 3.2 are easy.

Lemma 3.1. *Let N be a smooth complex surface, and L a compact rational curve on N whose singular locus consists of one node and such that $L^2 = 0$. Let n be a positive integer and*

$$M_n \xrightarrow{\tau_n} M_{n-1} \to \dots \to M_1 \xrightarrow{\tau_1} M_0 = N$$

a sequence of quadratic transformations such that

(i) the center of τ_1 is the singular point of L, and

(ii) the center of τ_j for $j > 1$ is one of the two singular points of the support of the total transform L_{j-1} of L on M_{j-1} which also lie on the exceptional curve E_{j-1} of τ_{j-1}.

Let $\sigma_n : M_n \to S_n$ be a proper bimeromorphic morphism obtained by contracting all the irreducible components of L_n except E_n to a normal singular point $x_n \in S_n$. Then the germ (S_n, x_n) is a quotient singularity of type $1/r^2(a, r-a)$ for some positive integers r and a such that $r \geqq 2$, $(r,a) = 1$ and $0 < a < r$. Conversely, an arbitrary quotient singularity of this type is obtained in this way.

Corollary 3.2. *With the notation as in Lemma 3.1, assume that the germ (S_n, x_n) has a quotient singularity of type $1/r^2(a, r-a)$. Let b be an integer such that $ab \equiv 1 \pmod r$ and $0 < b < r$. Let $F_j (j = 1, ..., n)$ be the exceptional curves of σ_n, and*

$$\sigma_n^* K_{S_n} = K_{M_n} + \sum_{j=1}^{n} d_j F_j, L_n = m_0 E_n + \sum_{j=1}^{n} m_j F_j.$$

Then $d_1 = (r-b)/r, d_n = b/r, m_0 = r, m_1 = b$ and $m_n = r-b$ hold up to the permutation of F_1 and F_n.

By the above corollary, we observe that the type of the singularity S_n is determined by d_1 or d_n , since $(r, b) = 1$. The following lemma is also easy.

Lemma 3.3. *Let $\rho : N \longrightarrow \Gamma$ be a $\mathbf{P}^1$ – bundle over a smooth curve Γ with two disjoint sections Γ_1 and Γ_2 , and let $L = \rho^{-1}(\gamma_0)$ for some $\gamma_0 \in \Gamma$. Let n be an integer $\geqq 2$, and*

$$M_n \xrightarrow{\tau_n} M_{n-1} \longrightarrow \dots \longrightarrow M_1 \xrightarrow{\tau_1} M_0 = N$$

a sequence of quadratic transformations such that
(i) the center of τ_1 is $L \cap \Gamma_1$,
(ii) the center of τ_j for $j > 1$ is one of the singular points of the support of the total transform L_{j-1} of L on M_{j-1} which also lie on the exceptional curve E_{j-1} of τ_{j-1} .
Let $\sigma_n : M_n \longrightarrow S_n$ be a proper bimeromorphic morphism obtained by contracting all the irreducible components of L_n except E_n and then by identifying the strict transforms Γ'_ε of Γ_ε for $\varepsilon = 1, 2$ by the following equivalence relation: $\gamma_1 \sim \gamma_2$ for $\gamma_\varepsilon \in \Gamma'_\varepsilon$ if and only if $\rho'(\gamma_1) = \rho'(\gamma_2)$, where $\rho' : M_n \longrightarrow \Gamma$ is induced from ρ, so that S_n has depth 2 at every point and has normal crossings along the images of the Γ'_ε except at the point $x_n = \sigma_n(L_n \cap \Gamma'_1) = \sigma_n(L_n \cap \Gamma'_2)$ Then the germ (S_n, x_n) is biholomorphic to the germ of a divisor on a quotient singularity of type $1/r(a, -a, 1)$ defined by $t_1 t_2 = 0$ as in (1.1.2) for some positive integers r and a such that $r \geqq 2$, $(r, a) = 1$ and $0 < a < r$. The converse also holds.

§4. Minimal moderate degenerations.

Theorem 4.1. *Let $f : X \longrightarrow \Delta$ be a moderate degeneration of surfaces. Assume that $b_1(f^{-1}(0))$ is even, f is minimal in the sense that K_X is f-nef and that the Kodaira dimension $\kappa(f^{-1}(t)) = 0$ for $t \in \Delta^*$. Then f is smooth except in the case of a flower pot degeneration, i.e., the fibers $f^{-1}(t)$ for $t \in \Delta^*$ are Enriques surfaces and $f^{-1}(0)$ is a singular rational surface with only quotient singularities of type $1/4(1, 1)$.*

Proof. We use the notation of Lemma 2.4. We assume that f is not smooth. If $p_g(X_t) = 1$ for $t \in \Delta^*$, then $p_g(N) = 1$ and $P_m(N) = 0$ for some $m > 0$ by Lemma 2.4, a contradiction. Next, suppose that the X_t for $t \in \Delta^*$ are hyperelliptic surfaces. Then by Lemma 2.4, N is an elliptic ruled surface. Since σ contracts only rational curves, the strict transform of a general fiber of the ruling $N \longrightarrow E$ gives a nonsingular rational curve whose self-intersection number is zero on S, a contradiction. Finally, assume that the X_t are Enriques surfaces. By Lemma 2.4 again, N is a rational surface. Since $K_S \equiv 0$ and $h^0(2K_S) = 1$, we have $r = 2$. Q.E.D.

Singular fibers of elliptic surfaces are classified by [Kol], and their smooth deformations are studied in [I]. A new phenomenon, a confluence of a multiple fiber and non-multiple singular fibers, occurs for moderate degenerations (cf. Example 4.5).

Theorem 4.2. *Let $f : X \longrightarrow \Delta$ be a minimal moderate degeneration of surfaces. Assume that $b_1(f^{-1}(0))$ is even and $\kappa(f^{-1}(t)) = 1$ for $t \in \Delta^*$. Then there exist a smooth complex surface B, a proper surjective morphism $g : X \longrightarrow B$ whose general fibers are elliptic curves, and a proper smooth morphism $h : B \longrightarrow \Delta$ such that $f = h \circ g$. Let $S = f^{-1}(0)$ be the central fiber, $\sigma : M \longrightarrow S$ the minimal resolution, $g_S : S \longrightarrow \Gamma = g(S)$ and $g_M : M \longrightarrow \Gamma$ the morphisms induced by g, and $\tau : M \longrightarrow N$ the morphism to the relative minimal model $g_N : N \longrightarrow \Gamma$ of g_M . Let L be a scheme theoretic fiber of g_S which passes through some singular points of X. Let $L_M = \sigma^* L$ and $L_N = \tau_* L_M$. Let $\tilde{m}$ be the greatest common divisor of the coefficients of the Weil divisor L on S, and denote $L = \tilde{m} L_{\mathrm{red}}$. Then one of the following holds.*

(1) (type ${}_m I_d(r,a)$; for positive integers m, d, r and a such that $0 < a < r$ and $(r,a) = 1$): $\tilde{m} = mr$, L_{red} is a reduced cycle of d nonsingular rational curves (resp. a reduced rational curve with one node) if $d > 1$ (resp. $d = 1$), and Sing(S)$\cap$ Supp(L) consists of quotient singularities of type $1/r^2(a, r-a)$ at d double points of L_{red} . L_N is a singular fiber of type ${}_m I_d$ of the elliptic surface $g_N : N \longrightarrow \Gamma$ in the sense of [Ko1], and Supp(L_M) is also a cycle of nonsingular rational curves.

(2) (type II(r) ; for r = 2 or 3): $\tilde{m} = r$, L_{red} is a rational curve with one cusp, and Sing(S)$\cap$ Supp(L) consists of a quotient singularity of type $1/r^2(1, r-1)$ at the cusp of L_{red} . L_N is of type II, and Supp(L_M) consists of r nonsingular rational curves intersecting tangentially (resp. pairwise transversally) at a poiont for r = 2 (resp. 3).

(3) (type II(r) ; for r = 4 or 5): $\tilde{m} = rr'$, r' = 2 (resp. 5) if r = 4 (resp. 5), L_{red} is a non-singular rational curve, and Sing(S)$\cap$ Supp(L) consists of two quotient singularities of types $1/r^2(3, r-3)$ and $1/r'^2(1, r'-1)$. L_N is of type II, and Supp(L_M) is a tree of nonsingular rational curves whose dual graph is as follows:

type II(4)

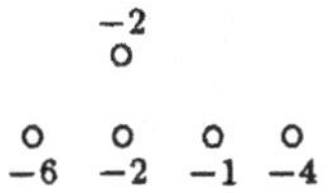

type II(5)

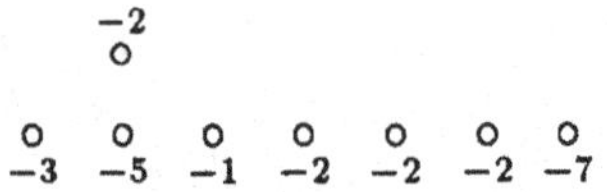

(4) (type III(2)): $\tilde{m}$ = 2, $L_{\mathrm{red}} = F_1 + 2F_2$, where the F_j are nonsingular rational curves intersecting transversally at a point, and Sing(S)$\cap$ Supp(L) consists of two quotient singularities of type $1/4(1,1)$ on nonsingular points of Supp(L) on F_2 . L_N is of type III, and Supp(L_M) is a tree of nonsingular rational curves as follows:

type III(2)

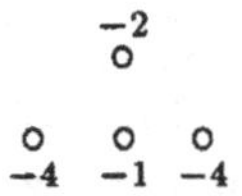

(5) (typ III(3)): $\tilde{m}$ = 9, $L_{\mathrm{red}} = F_1 + F_2$, where the F_j are nonsingular rational curves intersecting transversally at a point, and Sing(S)$\cap$ Supp(L) consists of three quotient singularities of type $1/9(1,2)$ such that one of them is at $F_1 \cap F_2$ and the other two are

at nonsingular points of L_{red} *one on each* F_j . L_N *is of type III, and* $Supp(L_M)$ *is a tree of nonsingular rational curves as follows:*

type III(3)

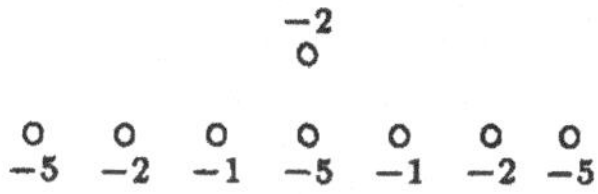

(6) (type IV(2)): $\tilde{m} = 4$, $L_{\text{red}} = F_1 + F_2 + F_3$, *where the* F_j *are nonsingular rational curves intersecting transversally at one point, and* $Sing(S) \cap Supp(L)$ *consists of four quotient singularities of type* $1/4(1,1)$ *such that one of them is at* $F_1 \cap F_2 \cap F_3$ *and the other three are at nonsingular points of* L_{red} *one on each* F_j . L_N *is of type IV, and* $Supp(L_M)$ *is a tree of nonsingular rational curves as follows:*

type IV(2)

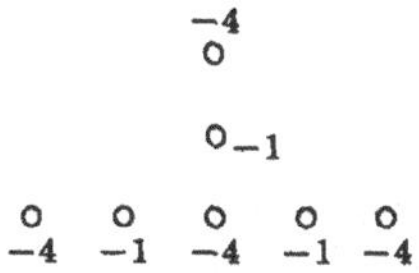

Proof. By [N1, 6.3], the linear system $|mK_X|$ is free for some $m > 0$, and gives the morphism $g : X \longrightarrow B$. Since the fiber $h^{-1}(0)$ is nonsingular, so is B.

The morphism $\tau : M \longrightarrow N$ is a composition of quadratic transformations $\tau_j : M_j \longrightarrow M_{j-1}$ for $j = 1, ..., n$, where $M_0 = N$ and $M_n = M$. Let E_j be the exceptional curves of τ_j , $g_j : M_j \longrightarrow \Gamma$ the induced morphisms, and $L_j = g_j^*(g_S(L))$ the total transforms of L. We have $L_M = L_n$ and $L_N = L_0$.

Let us write $\sigma^* K_S = K_M + \sum d_i \Theta_i$ for exceptional curves Θ_i of σ. Since σ is a minimal resolution and since S has only non-Gorenstein quotient singularities, we have $0 < d_i < 1$ for all i.

We write $\Theta = \Theta^{(n)} = \sum d_i \Theta_i$ and define $\Theta^{(j)} (0 \leqq j < n)$ inductively by $\tau_{j*}\Theta^{(j)} = \Theta^{(j-1)}$. We have $(K_{M_j}.E_j) = -1$ and $((K_{M_j} + \Theta^{(j)}).E_j) = 0$. Since the coefficients of the effective $\mathbf{Q}$-divisor $\Theta^{(j)}$ are less than 1, $\tau_j(E_j)$ is a singular point of $\text{Supp}(\Theta^{(j-1)}) \subset \text{Supp}(L_{j-1})$.

Lemma 4.3. *The fiber* $L_N = L_0$ *of the elliptic surface* $g_N : N \longrightarrow \Gamma$ *is not of type* ${}_mI_0, I_b^*, II^*, III^*$ *nor* IV^* (see [Kol] for the notation).

Proof. If L_N is of type ${}_mI_0$, then we have $N = M = S$ and L does not pass through singular points of S. Next, suppose that the support of L_N is a tree of nonsingular rational curves. Then the support of the L_j are also such trees for all j. We claim that an arbitrary irreducible component A of L_j which is not contained in $\text{Supp}(\Theta^{(j)})$ is a (-1)-curve. Otherwise, since $(K_{M_j}.A) \geqq 0$ and $(K_{M_j} + \Theta^{(j)}.A) = 0$, A is a (-2)-curve which is disjoint from $\text{Supp}(\Theta^{(j)})$. Then an irreducible component of L_j intersecting A is either a (-1)-curve or a (-2)-curve of the same type as A. The former case does not occur, because a (-1)-curve of L_j must intersect exactly two irreducible components of L_j and they must belong to $\text{Supp}(\Theta^{(j)})$. Since $\Theta^{(j)} \neq 0$ and $\text{Supp}(L_j)$ is connected, we have a contradiction.

Then we find a branch of the tree Supp(L_M) consisting of nonsingular rational curves $A_1, ..., A_\nu$ and integers $0 < \nu_1 < ... < \nu_p = \nu - 1$ such that

(a) $(A_1.A_2) = ... = (A_{\nu-1}.A_\nu) = 1$,

(b) $A_1, ..., A_{\nu-1}$ do not intersect other irreducible components of L_M ,

(c) A_ν intersects two irreducible components of L_M other than $A_{\nu-1}$,

(d) $A_1, ..., A_{\nu_1-1}, A_{\nu_1+1}, ..., A_{\nu_2-1}, ..., A_{\nu_p-1}$ are irreducible components of Θ,

(e) $A_{\nu_1}, ..., A_{\nu_p}$ are (-1)-curves.

We have $((K_M+\Theta).A) = 0$ for an arbitrary irreducible component A of L_M . Hence by Corollary 3.2, the strings of nonsingular rational curves $\{A_{\nu_{k-1}+1}, ..., A_{\nu_k-1}\}(1 \leqq k \leqq p)$, where we put $\nu_0 = 0$, are contracted to quotient singularites of the same type on S. But then, after successive contractions of (-1)-curves, since A_ν will not be contracted to a point of N, we obtain a nonsingular rational curve of self-intersection number less than -2 on N in the image of the string of rational curves $\{A_1, ..., A_{\nu_1-1}\}$, a contradiction. Q.E.D. for Lemma 4.3.

Case 1. Let us assume that L_N is of type ${}_mI_d$ for $m, d > 0$. Then as in the proof of Lemma 4.3, we see that Supp(L_M) is a cycle of nonsingular rational curves $A_1, ..., A_{d\nu}(= A_0)$ for an integer $\nu \geqq 2$ such that

(a) $(A_k.A_{k+1}) = 1$ and there are no other intersections,

(b) the strings of nonsingular rational curves $\{A_{(k-1)\nu+1}, ..., A_{k\nu-1}\}$ for $k = 1, ..., d$ are contracted to quotient singularities of type $1/r^2(a, r-a)$ for some fixed positive integers r and a with $(r, a) = 1$,

(c) $A_{k\nu}$ for $k = 1, ..., d$ are (-1)-curves.

The multiplicity of $A_{k\nu}$ in L_M is mr by Corollary 3.2. Hence we have ${}_mI_d(r, a)$.

Case 2. We assume that L_N is of type II, i.e., L_N is a rational curve with a cusp. Then τ_1 must be the blow-up of the singular point of $L_N = L_0$. We have $L_1 = E_0' + 2E_1$, where E_0', the strict transform of $E_0 = L_0$, is a (-4)-curve and E_1 is a (-1)-curve. E_0' and E_1 intersect tangentially. If we blow down E_0' , then we obtain II(2).

The center of the next blow-up τ_2 must also be the only singularity of Supp(L_1). We have

$$L_2 = E_0'' + 2E_1' + 3E_2,$$

with $(E_0'')^2 = -5$ and $(E_1')^2 = -2$, where the symbol (′) denotes the strict transform. If we contract E_0'' and E_1' , we obtain II(3).

The center of τ_3 is also the only singularity of Supp(L_2). We have

$$L_3 = E_0^{(3)} + 2E_1'' + 3E_2' + 6E_3,$$

with $(E_0^{(3)})^2 = -6, (E_1'')^2 = -3$ and $(E_2')^2 = -2$. M_3 thus obtained is not a candidate for M, but an arbitrary $M = M_n$ for $n \geqq 4$ is above M_3 . As in the proof of Lemma 4.3, we find a branch of Supp(L_n) consisting of nonsingular rational curves $A_1, ..., A_\nu$ which satisfies the conditions (a) through (e) there and such that A_ν is the strict transform of E_3 . By the same reason as there, this branch is not above E_2' . Hence we may assume that the center of τ_4 is either $E_0^{(3)} \cap E_3$ or $E_1'' \cap E_3$.

By (the proof of) Corollary 2.2, we have $e(L) = e(L_N) = 2$. Hence L is irreducible, and the centers of τ_j for $j > 4$ are all infinitely near over the center of τ_4 and on the strict transforms of E_3 .

If the center of τ_4 is $E_0^{(3)} \cap E_3$, then $n = 7$, and we have II(5). If it is $E_1" \cap E_3$, then $n = 4$, and we have II(4).

Case 3. We assume that L_N is of type III, i.e., $L_N = E_{0a} + E_{0b}$, where E_{0a} and E_{0b} are nonsingular rational curves intersecting tangentially. The first candidate for M appears for $n = 2$; we have

$$L_2 = E_{0a}" + E_{0b}" + 2E_1' + 4E_2$$

and obtain III(2).

Let us consider those M_n with $n > 2$. There is a branch of Supp(L_n) consisting of nonsingular rational curves $A_1, ..., A_\nu$ which satisfies the conditions (a) through (e) in the proof of Lemma 4.3. By the same reason as there, we may assume that this branch is above $E_{0a}"$. As in Case 2, L must have two irreducible components. Suppose that $p = 2$. Then the strict transforms of $E_{0b}", E_2$ and E_1' on M are contracted together to a quotient singularity on S. But the self-intersection numbers of the end components are -4 and -2, and it is impossible by Lemma 3.1.

Hence $p = 1$. Then $r = 4$. We may assume that the centers for τ_3 and τ_4 are over $E_{0a}"$. We obtain on M_4 two strings of nonsingular rational curves $\{E_{0a}^{(3)}, E_3'\}$ and $\{E_{0b}", E_2", E_1'\}$; the sequences of their self-intersection numbers are $(-5, -2)$ and $(-4, -3, -2)$ so that the former is contracted to a point of type $1/9(1, 2)$.

Let us consider what kind of transformations are possible for further blow-ups on the latter sequence of self-intersection numbers. Suppose that we obtain a candidate of M_n for some $n \geqq 5$. Then by Corollary 3.2, the resulting sequence will be of the type

$$(*) \qquad (-c_1, ..., -c_\mu, -1, -c_1, ..., -c_\mu)$$

such that c_1 or $c_\mu = 2$. Hence after the blow-up τ_5 , we must have $(-5, -1, -4, -2)$.

There are two possibilities after τ_6 : $(-5, -2, -1, -5, -2)$ or $(-6, -1, -2, -4, -2)$. The former case gives us III(3). From the latter case, we would have $c_1 \geqq 6, c_{\mu-1} = 4$ and $c_\mu = 2$, a contradiction to Lemma 3.1.

Case 4. We assume that L_N is of type IV, i.e., $L_N = E_{0a} + E_{0b} + E_{0c}$, where the E_{0*} are nonsingular rational curves intersecting at one point. We have three (-3)-curves on M_1 . As in Case 2, the number of irreducible components of L is three. We claim that all three singular points of Supp(L_1) must be blown up. For example, if $E_{0a}' \cap E_1$ is not blown-up, then we obtain a sequence of self-intersection numbers (*) in Case 3 or the following

$$(**) \qquad (-c_1, ..., -c_\mu, -1, -c_1, ..., -c_\mu, -1, -c_1, ..., -c_\mu)$$

such that $c_1 \geqq 3, c_\mu = -(E_{0a})^2 = 3$ and $\mu \geqq 2$, a contradiction to Lemma 3.1. Thus we have IV(2) for $M = M_4$.

Suppose that there is another $M = M_n$ for $n > 4$. Then there is an irreducible component, say E_2, such that all the centers of the further blow-ups over E_2 must be on the strict transforms of $E_1^{(3)}$. In the complement of the branch over E_2 in $\mathrm{Supp}(L_n)$, we find a sequence of self-intersection numbers (**) such that $c_1 \geqq 4$ and $c_\mu \geqq 4$. Hence we must have $\mu = 1$. Q.E.D. for Theorem 4.2.

Theorem 4.4. *Let $g_S : S \to \Gamma$ be as in Theorem 4.2. Then the following canonical bundle formula holds:*

$$K_S = g_S^* \mathbf{d} + \sum_k (m^{(k)} - 1) F^{(k)},$$

where $\mathbf{d}$ *is some divisor on* Γ *and the summation is taken for all the multiple fibers* $L^{(k)} = m^{(k)} F^{(k)}$ *whose multiplicities* $m^{(k)}$ *are defined by the following table* (${}_mI_d$ *is considered as* ${}_mI_d(1,1)$ *):*

type of $L^{(k)}$	${}_mI_d(r, a)$	$II(r)$	$III(r)$	$IV(r)$
$m^{(k)}$	mr	r	r	r

Proof. Let $L_M^{(k)}$ and $L_N^{(k)}$ be the total transforms of $L^{(k)}$ on M and N, respectively. If $\bar{m}^{(k)}$ is the multiplicity of $L_N^{(k)}$, i.e., $L_N^{(k)} = \bar{m}^{(k)} L_{N,\mathrm{red}}^{(k)}$, then by Kodaira's canonical bundle formula ([Ko2]), we can write

$$K_N = g_N^* \mathbf{d} + \sum_k (\bar{m}^{(k)} - 1) L_{N,\mathrm{red}}^{(k)}$$

for some divisor $\mathbf{d}$ on Γ. Since $K_M - \tau^* K_N$ has a support on the exceptional locus of τ, and since $\sigma_* K_M = K_S$, it is enough to check our formula when it is restricted on a small neighborhood $U = U^{(k)}$ of each multiple fiber $L = L^{(k)}$. We set $U_M = \sigma^{-1}(U)$.

For L of type ${}_mI_d(r, a)$, since $\tau^*(K_N + L_{N,\mathrm{red}})\,|_{U_M} = (K_M + L_{M,\mathrm{red}})\,|_{U_M} = \sigma^*(K_S + L_{\mathrm{red}})\,|_{U_M}$, we are done.

For L of other types, by comparing the coefficients of the exceptional curves in the discrepancy $(K_M - \tau^* K_N)\,|_{U_M}$ with those in L_M, we obtain the desired formula. Q.E.D.

We note that the Euler number of a multiple fiber other than the ${}_mI_d(r, a)$ multiplied by its multiplicity is less than 12.

Example 4.5. Let d, m, r and a be positive integers such that $(r, a) = 1$. Let λ_1 and λ_2 be real numbers such that $1 >> \lambda_1 >> \lambda_2 > 0$. Let U_k be an open subset of the affine 3-space $\mathbf{C}^3$ with coordinates (x_k, y_k, z_k) given by $|z_k| < \lambda_1$ and $|t_k| < \lambda_2$, where

$$t_k = \mathbf{e}(-k/dm)(x_k y_k + z_k{}^r),$$

with $\mathbf{e}(\alpha) = \exp(2\pi\sqrt{-1}.\alpha)$. Let $\tilde{X} = \tilde{X}(d, m, r)$ be the union of the U_k for $k \in \mathbf{Z}$ with identifications $U_k - \{y_k = 0\} \to U_{k+1} - \{x_{k+1} = 0\}$ given by

$$(x_{k+1}, y_{k+1}, z_{k+1}) \to (y_k^{-1}, \mathbf{e}(1/dm)x_k {y_k}^2 + (\mathbf{e}(1/dm) - 1)y_k {z_k}^r, z_k),$$

and $U_k \to U_{k+dm}$ by

$$(x_{k+dm}, y_{k+dm}, z_{k+dm}) \to (x_k, y_k, z_k).$$

Let $X = X(d, m, r, a)$ be the quotient of $\tilde{X}$ by the group $G = \mathbf{Z}/mr\mathbf{Z}$ whose generator maps the image of U_k in $\tilde{X}$ to that of U_{k+d} by

$$\begin{aligned} x_{k+d} &\to \mathbf{e}(d(m-1)/2m + a/mr + (-k+1)/m)x_k, \\ y_{k+d} &\to \mathbf{e}(-d(m-1)/2m - a/mr + k/m)y_k, \\ z_{k+d} &\to \mathbf{e}(1/mr)z_k. \end{aligned}$$

We can check that the above action is compatible with the identifications of the U_k, and is free outside the points in the images of the U_k given by $(x_k, y_k, z_k) = (0, 0, 0)$. X has d quotient singular points of type $1/r(a, -a, 1)$ at the images of these points.

Let $\Delta_1 = \{s \in \mathbf{C}; |s| < \lambda_1^{mr}\}$ and $\Delta_2 = \{t \in \mathbf{C}; |t| < \lambda_2\}$. We define a morphism g from X to $\Delta_1 \times \Delta_2$ by

$$(s, t) \to ({z_k}^{mr}, t_k).$$

Composing g with the second projection $h : \Delta_1 \times \Delta_2 \to \Delta_2$, we obtain a morphism $f : X \to \Delta_2$.

The fiber $X_t = f^{-1}(t)$ for $t \neq 0$ is a nonsingular surface with an elliptic surface structure $g_t = g \mid_{X_t}: X_t \to g(X_t)$, whereas $S = f^{-1}(0)$ has d quotient singularities of type $1/r^2(a, r-a)$.

Let us calculate the fibers $g^{-1}(s, t)$. If $s(s^d - t^{dm}) \neq 0$, then the function y_k for any k give a coordinate on $g^{-1}(s, t)$ modulo

$$\begin{aligned} &\prod_{j=0}^{dm-1} (\mathbf{e}(1/dm)x_{k+j}y_{k+j} + (\mathbf{e}(1/dm) - 1){z_{k+j}}^r) \\ &= (-1)^{dm} \prod_{j=0}^{dm-1} ({z_{k+j}}^r - \mathbf{e}((k+j+1)/dm)t_{k+j}) \\ &= (-1)^{dm}(s^d - t^{dm}). \end{aligned}$$

Hence $g^{-1}(s, t)$ is a nonsingular elliptic curve isomorphic to $\mathbf{C}^*/ < (-1)^{dm}(s^d - t^{dm}) >$.

For $t \neq 0, X_t$ has a multiple fiber of type ${}_{mr}I_0$ over $s = 0$ and d singular fibers of type I_1 over $s = \mathbf{e}(j/d)t^m$ for $j = 1, ..., d$. The central fiber S has only one singular fiber of type ${}_mI_d(r, a)$ over $s = 0$.

Proposition 4.6. *Let $f : X \to \Delta$ be a minimal moderate degeneration of surfaces such that $\kappa(f^{-1}(t)) = 2$ for $t \in \Delta^*$. Assume one of the followings:*

(1) $3e(X_t) = K^2_{X_t}$ *for* $t \in \Delta^*$,

(2) $2p_g(X_t) - 4 = K^2_{X_t}$ *for* $t \in \Delta^*$ *and* $\kappa(N) = 2$, *where* N *is a relatively minimal model of* $S = f^{-1}(0)$.

Then f *is smooth.*

Proof. (1): By [Mi2], we have $3e_{\text{orb}}(S) \geqq K_S^2$, where e_{orb} denotes the Euler number as an orbifold. If S is singular, then $e(S) > e_{\text{orb}}(S)$, and we have a contradiction. (b): By Noether's inequality, we have $2p_g(N) - 4 \leqq K_N^2$. If S is singular, then by Lemma 2.4, we have $K^2_{X_t} > K^2_N$, a contradiction. Q.E.D.

§5. Boundedness of indices.

We have the following result on the boundedness of the moduli space of elliptic surfaces in some sense.

Theorem 5.1. *Let* S_0 *be a minimal elliptic surface of general type (i.e.,* $\kappa(S_0) = 1$*) with even first Betti number, and* $\{m_i\}$ *the multiplicities of the multiple fibers of its elliptic fibration. Let* S *be the set of isomorphism classes of all compact complex manifolds of dimension two which are smooth deformations of* S_0 *, and* $f : X \longrightarrow \Delta$ *a minimal permissible degeneration such that* $f^{-1}(t) \in S$ *for* $t \in \Delta^*$. *If* r *is a positive integer which is a common multiple of the* m_i *and 12, then* rK_X *is a Cartier divisor.*

Proof. By [N1, 6.3], the linear system $|nK_X|$ for a sufficiently divisible and large integer n gives a projective morphism $g : X \longrightarrow B$ onto a normal surface B with a projection $h : B \longrightarrow \Delta$ such that $f = h \circ g$ and that the general fibers of g are elliptic curves. By Kodaira's canonical bundle formula, we have $nK_X = g^*(n(K_B + \Theta))$ for a $\mathbf{Q}$-divisor Θ on B defined by

$$\Theta = (1/12)J + \sum a_j D_j + \sum (1 - 1/m_i) F_i,$$

where (1) J is a general member of the linear system on B given by the J-invariant function of the fibers of g, and (2) the D_j (resp. F_i) are divisors on B such that the fibers of g over the general points of the D_j (resp. F_i) are non-multiple singular fibers with Kodaira's coefficients a_j (resp. multiple fibers with multiplicities m_i). Note that the irreducible components of Θ are all horizontal with respect to h. By [N2], the pair (B, Θ) has only log-terminal singularities.

For a positive integer d, let $p_d : \Delta \longrightarrow \Delta$ be the morphism given by $p_d^*(t) = t^d$. By the base change by p_d , we construct the following cartesian daigram:

$$\begin{array}{ccccc} X_d & \xrightarrow{g_d} & B_d & \xrightarrow{h_d} & \Delta \\ \downarrow & & \downarrow{\scriptstyle \pi} & & \downarrow{\scriptstyle p_d} \\ X & \xrightarrow{g} & B & \xrightarrow{h} & \Delta \end{array}$$

We know that X_d has only canonical singularities and the formation of the sheaf $f_*\mathcal{O}_X(mK_X)$ commutes with the base change. Hence the pair $(B_d\ ,\ \Theta_d)$ for $\Theta_d = \pi^*\Theta$ has also only log-terminal singularities.

Lemma 5.2. *B has only rational double points of type A, and* $\Gamma = h^{-1}(0)$ *has only nodes.*

Proof. Let $\mu : B' \longrightarrow B$ be the minimal resolution, and consider the following diagram:

$$\begin{array}{ccc} B'_d & \xrightarrow{\pi'} & B' \\ {\scriptstyle \mu_d}\downarrow & & \downarrow{\scriptstyle \mu} \\ B_d & \xrightarrow{\pi} & B \end{array}$$

where B'_d is the normalization of the fiber product. Let E_k be the exceptional divisors of μ. Then there are positive integers n_k and rational numbers a_k with $a_k > -1$ such that

$$K_{B_d} = \pi^* K_B + (1 - 1/d)\pi^*\Gamma$$
$$K_{B'} = \mu^* K_B + \sum a_k E_k$$
$$\mu^*\Gamma = \Gamma' + \sum n_k E_k$$

where Γ' is the strict transform of Γ. Then

$$\begin{aligned} K_{B'} &= \pi'^* K_{B'} + (1 - 1/d)\pi'^*\Gamma' + \sum (1 - 1/d_k)\pi'^* E_k \\ &= \mu_d^* K_{B_d} + \sum (1 - 1/d_k + a_k - n_k(1 - 1/d))d_k E'_k \end{aligned}$$

for $d_k = d/(d, n_k)$ and $\pi'^* E_k = d_k E'_k$. Since B_d has only quotient singularities, we have

$$1 - 1/d_k + a_k - n_k(1 - 1/d) > -1/d_k$$

for all k. If there exists a k_0 such that $n_{k_0} \geqq 2$, then we obtain a contradiction for $d = n_{k_0}$, since $a_k \leqq 0$. Hence $n_k = 1$ for all k. Then we have also $a_k = 0$ for all k, i.e., B has only rational double points. Since the fundamental cycles are reduced, they are of type A.

We claim that $\Gamma' + \sum E_k$ is a normal crossing divisor on B'. Otherwise, there is a point modificaton $\mu' : B" \longrightarrow B'$ with the exceptional divisor E such that the coefficients of E in $K_{B"}$ and $\mu'^*\mu^*\Gamma$ are 1 and $n \geqq 3$, respectively. If we construct $B"_d$ for $d = n$ by the base change and the normalization as before, then we must have $1 - 1 + 1 - n + 1 > -1$, a contradiction. Hence Γ has only nodes. Q.E.D.

Lemma 5.3. *The support of Θ contains no singular points of Γ.*

Proof. Suppose the contrary. Then by the base change, we may assume that an irreducible component $\Theta^{(1)}$ of Θ is a section of h and passes through a singular point P of Γ. Then P is also a singular point of B. Let e be the positive integer such that the singularity of B at P is of type A_{e-1}. Then B_d has a singularity of type A_{de-1} at the point P_d over P. Let $\Theta_d^{(1)}$ be the irreducible component of Θ_d above $\Theta^{(1)}$, $\nu : B_d^! \longrightarrow B_d$ the minimal resolution, G_k for $k = 1, ..., de-1$ the exceptional divisors over P_d such that $(G_1.G_2) = (G_2.G_3) = ... = (G_{de-2}.G_{de-1}) = 1$, and $\Theta_d^!$ the strict transform of $\Theta_d^{(1)}$ by ν. Then $\Theta_d^!$ intersects G_k for some positive integer k such that $d \leqq k \leqq d(e-1)$. Let c be the coefficient of $\Theta^{(1)}$ in Θ. Then (B_d, Θ_d) is not log-terminal if $cd > 2$, a contradiction. Q.E.D.

By Lemmas 5.2 and 5.3, $m(K_B + \Theta)$ is a Cartier divisor if $m_i | m$ and $12 | m$. By [**K1**, 7.1], we can construct a sequence of birational maps

$$X = X_0 \xrightarrow{\sigma_1} X_1 \xrightarrow{\sigma_2} ... \longrightarrow X_t = X^{\#},$$

consisting of flops (i.e., crepant log-flips) and crepant divisorial contractions, to a normal variety $X^{\#}$ with only canonical singularities, and a morphism $g^{\#} : X^{\#} \longrightarrow B$ such that $g = g^{\#} \circ \sigma_t \circ ... \circ \sigma_1$, and that there does not exist a divisor on $X^{\#}$ which is mapped to a point by $g^{\#}$. Then $mK_{X^{\#}}$ and $g^{\#*}(m(K_B + \Theta))$ coincide in codimension one, hence are equal. Then $mK_{X^{\#}}$ is a Cartier divisor, and so is mK_X (cf. [**K1**, 1.5]). Q.E.D.

Appendix. Extremal rational curves.

Theorem A.1. *Let X be a normal variety with only canonical singularities, $f : X \longrightarrow Y$ a projective birational morphism, and E an irreducible component of the degenerate locus of f. Assume that $-K_X$ is f-nef. Then E is uniruled.*

Proof. We may assume that $f(E)$ is a point and Y is a germ of a complex analytic singularity. Let H be a very ample divisor on X and X_0 the complete intersection of $n-1$ generic members of $|H|$ for $n = \dim E$. Then X_0 is also a normal variety with only canonical singularities and $K_{X_0} = (K_X + (n-1)H)\,|_{X_0}$. If we replace X_0 by a small open neighborhood of $C := E \cap X_0$, then there is a projective bimeromorphic morphism $f_0 : X_0 \longrightarrow Y_0$ whose degenerate locus coincides with C (cf. [**Ka2**, 2.3.3]).

Let $\mu : \bar{E} \longrightarrow E$ be the normalization. Then so is its restriction $\mu_C : \bar{C} \longrightarrow C$. We shall prove that $\deg K_{\bar{C}} < (n-1)(H.C)$. Suppose the contrary. Then, since there is an injective homomorphism $\mu_{C*}\omega_{\bar{C}} \longrightarrow \omega_C$, there exists a Cartier divisor L_0 of degree $(n-1)(H.C)$ on C whose support is contained in the nonsingular locus of C and such that $H^0(\omega_C(-L_0)) \neq 0$. We can extend L_0 to a Cartier divisor L on X_0. Since $L - K_{X_0}$ is f_0-nef, we have $R^1 f_{0*}\mathcal{O}_{X_0}(L) = 0$ by [**N1**, 3.6], hence $H^1(C, L_0) = 0$, a contradiction. Hence $(K_{\bar{E}}.C) = \deg K_{\bar{C}} - (n-1)(H.C) < 0$. By [**MM**], we conclude that E is uniruled. Q.E.D.

By applying the above theorem to the case where X is nonsingular and K_X is relatively numerically trivial, F. Campana and T. Peternell prove the following.

Corollary A.2. *Let X be a nonsingular projective variety of general type. Assume that X is Kobayashi hyperbolic. Then K_X is ample.*

We can extend the theorem to the logarithmic case by the similar proof.

Theorem A.3. *Let X be a normal variety and D a $\mathbf{Q}$-divisor on X such that the pair (X, D) has only weakly log-terminal singularities (cf. [KMM]). Let $f : X \to Y$ be a projective birational morphism such that $-(K_X + D)$ is f-ample. Then an arbitrary irreducible component E of the degenerate locus of f is uniruled. In particular, every log-extremal ray is represented by a rational curve.*

Proof. With the notation of the proof of Theorem A.1, we have $R^1 f_{0*}\mathcal{O}_{X_0}(L) = 0$, since $L - (K_{X_0} + D\,|_{X_0})$ is f_0-ample. Q.E.D.

REFERENCES

[B] E. Brieskorn, *Die Auflösung der rationalen Singularitäten holomorphen Abbildungen*, Math. Ann. **178** (1968), 255–270.

[D] V.I. Danilov, *Birational geometry of toric 3-folds*, Math. USSR. Izv. **21** (1983), 269–279.

[I] S. Iitaka, *Deformations of compact complex surfaces II*, J. Soc. Math. Japan **22** (1970), 247–260.

[Ka1] Y. Kawamata, *Crepant blowing-up of 3-dimensional canonical singularities and its application to degenerations of surfaces*, Ann. of Math. **127** (1988), 93–163.

[Ka2] Y. Kawamata, *Small contractions of four dimensional algebraic manifolds*, Math. Ann. **284** (1989), 595–600.

[KMM] Y. Kawamata, K. Matsuda and K. Matsuki, *Introduction to the minimal model problem*, Adv. St. Pure Math. **10** Algebraic Geometry Sendai 1985 (1987), 283–360.

[KKMS] G. Kempf, F. Knudsen, D. Mumford and B. Saint-Donat, Lect. Notes in Math. **339** (1973), "Toroidal Embeddings I," Springer,.

[Ko1] K. Kodaira, *On compact analytic surfaces II*, Ann. of Math. **77** (1963), 563–626.

[Ko2] K. Kodaira, *On the structure of compact complex analytic surfaces I*, Amer. J. Math. **86** (1964), 751–798.

[MK] J. Morrow and K. Kodaira, "Complex Manifolds," Holt, Rinehart and Winston, 1971.

[Mi1] Y. Miyaoka, *Kähler metrics on elliptic surfaces*, Proc. Japan Acad. **50** (1974), 533–536.

[Mi2] Y. Miyaoka, *The maximal number of quotient singularities on surfaces with given numerical invariants*, Math. Ann. **268** (1984), 159–171.

[MM] Y. Miyaoka and S. Mori, *A numerical criterion for uniruledness*, Ann. of Math. **124** (1986), 65–69.

[Mo] S. Mori, *On 3-dimensional terminal singularities*, Nagoya Math. J. **98** (1985), 43–66.

[MS] D. Morrison and G. Stevens, *Terminal quotient singularities in dimensions three and four*, Proc. AMS. **90** (1984), 15–20.

[N1] N. Nakayama, *The lower semi-continuity of the plurigenera of complex varieties*, Adv. St. Pure Math. **10** Algebraic Geometry Sendai 1985 (1987), 551–590.

[N2] N. Nakayama, *The singularity of the canonical model of compact Kähler manifolds*, Math. Ann. **280** (1988), 509–512.

[P] U. Perrson, Mem. A.M.S. **189** (1977), "On Degeneration of Surfaces,".

[R] M. Reid, *Decomposition of toric morphisms*, Progress in Math. **36** Arithmetic and Geometry II (1983), 395–418, Birkhäuser.
[Si] Y.-T. Siu, *Every K3 surface is Kähler*, Invent. Math. **73** (1983), 139-150.
[St] J. Steenbrink, *Mixed Hodge structure on the vanishing cohomology*.
[T] G. N. Tyurina, *Resolution of singularities of plane deformations of double rational points*, Funct. Anal. Appl. **4** (1970), 68–73.
[W] J. M. Wahl, *Elliptic deformations of minimally elliptic singularities*, Math. Ann. **253** (1980), 241–262.

Department of Mathematics, Faculty of Science, University of Tokyo, Hongo, Tokyo, 113, Japan

Genus Two Fibrations Revisited

A preliminary report

Ulf Persson

Department of Mathematics
Chalmers University of Technology
S-412 96 Gothenburg Sweden

Inspired by the recent reprint [Reid] and the subsequent lecture at the Satellite conference (at Tokyo Metropolitan University) of the ICM, by its somewhat reluctant author, I decided to revive my old, dormant interest in surface geography.

I stress the tentative and preliminary nature of this note, its format and tenor are very much inspired by Reid as well, and can, if so wish be considered as an, admittedly somewhat feeble, pastiche of his mathematical style. (But,needless to add, an imitation not in order to ridicule but rather to flatter)

The work (if it may be so denoted) that went into this effort was done one evening perusing [Reid] and during the descent of a mountain around Komikochi. This may explain its sketchiness.

After a few preliminary versions of this had been written and circulated, I got a letter from Xiao Gang, who kindly pointed out to me that I had not read my own references carefully (A phenomenon prevalent not only among mathematicans). I like to thank him for his intererest and the information concerning examples with K^2 close to the upper limit, which has been incorporated. And finally I like to thank the organisers of the meeting in Bayreuth in April 1990 for letting me participate, and for their kindness in considering this attempt for their proceedings, although it does not faithfully reflect my own oral contribution at the time.

0. Getting Started

The simplest surfaces are the ruled ones (fibered by genus 0 curves), next in line are the elliptic ones. In both cases the Chern invariants (for minimal models) satisfy a simple linear relation. (To be pedantic this should be treated with a grain of salt, the grain being $\mathbf{P}^2$)

The theory of Elliptic surfaces centers around the Kodaira classification of degenerate fibers (A list which one realizes with some horror that one eventually learns by heart including obsolete notation) and the Mordell-Weil group and various subtle interplays. (As a horror example see [P3])

It stands to reason that the next natural object of study are surfaces fibered with genus two curves. (The most comprehensive study of such surfaces is of course to be found in [Xiao] which is a tacit reference for most of what follows)

In this short note we will for notational reasons restrict ourselves to fibrations over $\mathbf{P}^1$: this is of course not a serious restriction, and much of what is said globally can easily be tailored to the more general basecurve. But the assumption will be implicit and the reader should keep it in mind saving himself from confusion

You may without too much sweat write down a list of the possible degenerate fibers (and hope that you never would have occasion to hardwire it into your precious brain). Many authors have done that, and the undertaking is made relatively simple by the observation that every genus two fibration can be realized as a double covering of a ruled surface along a six-section.

But we will not be concerned with the detailed classification but simply note that there are two types of degenerate fibers which are unavoidable, i.e. they are stable. Any elliptic surface which is wiggled may be assumed only to have nodal degenerate fibers (we are leaving aside the multiple fibers for the discussion) the number of which by the Noether formula is divisible by twelve.

In the case of genus two fibrations we have two kinds of singular fibers, the nodal ones and those consisting of two elliptic curves meeting transversally. The existence of the latter type forces the invariants of genus two fibrations to be spread out. The basic problem for genus two fibrations is to get an understanding of in what proportions those singular fibers can occur. (A problem whose solution is more or less immediate in the case of elliptic surfaces)

As we are going to refer to them repeatedly it may be convenient to baptise them (more or less appropriately as is the custom on such occasions) as *nodal* and *elliptic* respectively.

1. Basic Formulas

Any genus two fibration can be thought of as a double covering of a ruled surface branched along a sextic section. (For the formalist: Twist the canonical divisor with as many fibers as you want and consider the associated map). By using Nagata's method of elementary transformations one may assume that the multiplicity of a singular point of the branchlocus (we have sneakily blown down the quotient to a minimal model) is at most three. To get something more serious than a rational double point we must assume that the branches are tangent, getting a so called *infinitely close triple point* . If there are no such serious singularities we are only dealing with simple singularities and the fibrations correspond to the so called Horikawa surfaces lying on the Noether line

$$c_1^2 = 2\chi - 6$$

of surfaces of general type with minimal K^2. Imposing non-simple singularities (in the genus two case this means of course just infinitely close triple points) both c_1^2 and χ

decrease by one, which means that the surface is shifted up one step on the level slopes $c_1^2 - 2\chi$.

The singular point is resolved into an elliptic curve with self-intersection -1 (surely the simplest non-simple surface singularity) and in the fibration we get a companion elliptic indistinguishable from the first in its attributes.

Let us make a simple instructive calculation. A typical Horikawa surface is obtained by taking a ruled surface say $\mathbf{P}^1 \times \mathbf{P}^1$ and a sextic branchcurve of type $6S + 2aF$ where a becomes a discrete parameter which pushes the surface invariants out radially, so to speak.

Standard calculations yield an Eulernumber of $20a - 4$ and a K^2 of $4a - 8$ and hence a χ of $2a - 1$ which in the case of k infinitely close triple points should be modified to $20a - 4 - 11k$, $4a - 8 - k$ and $2a - 1 - k$ respectively.

The Euler characteristic of a nodal fiber as well as an elliptic is -1 while that of a nonsingular is of course -2 thus each degenerate fiber contributes 1 to the Euler characteristic. Thus for rationally based genus two fibrations (for others we have to modify the constant term in an obvious way) we have

$$e = -4 + \#nodal + \#elliptic \tag{1}$$

To talk about the contribution to K^2 of a surface by a degenerate fiber is a bit more far fetched but can be done in some formal way. If there are no elliptic fibers there will be $20a$ nodal fibers, the virgin genus-two fibration (over $\mathbf{P}^1$) has $K^2 = -8$, thus those fibers will contribute a total of $4a$, hence we may think of an individual nodal fibration as $1/5$ (whatever that means) which forces the elliptic contribution to be $7/5$.

Thus we can write (setting N=#nodal, and E=#elliptic)

$$c_1^2 = -8 + {}^1\!/_5 N + {}^7\!/_5 E \tag{1'}$$

and by exploiting Noether

$$\chi = -1 + {}^1\!/_{10} N + {}^1\!/_5 E \tag{1''}$$

We may observe for all it is worth that $N + 2E \equiv 0(5)$; and furthermore as nodal fibers occurs at the branching (down to the base curve) of the branching 6-section that N is even as well. We do stress however that this is only true in the generic situation, the formulas and the divisibility conditions become more involved if we allow more degenerate fibers.

Query: Any reasonable interpretation of those numbers?

A more reasonable thing is to combine the two contributions e and c (to the Eulernumber and the c_1^2 respectively) into the Horikawa invariant $h = {}^5\!/_6 c - {}^1\!/_6 e$ which yields zero for nodal fibers and one for elliptic. (An algebraic interpretation of this invariant is given in [Reid]) and gives

$$c_1^2 - 2\chi = -6 + \sum_F h(F) \tag{2}$$

summed over the (degenerate) fibers. It will also be convenient to refer to fibers with non-vanishing Horikawa invariant as *essential (non-negligible, non-simple)* keeping in mind that they correspond to fibers through essential (or non-negligible, or non-simple singularities) of the sextic branchlocus

2. Specializations

The discussion of degenerate fibers would be sadly incomplete, and leave out some of the flavour of the subject, if no mention were made of further degenerations. In fact basic degenerate fibers can of course coalesce. Typical examples may be

i) two infinitely close triple points of the branch locus occur along the same fiber, in which case the two elliptic curves are separated by a (-2) curve.

ii) The three infinitely close branches tangent to the fiber. In which case the fiber reduces to one elliptic curve (of self intersection -1 of course) and of multiplicity two, with two (-2) curves sticking out (at either end?) like satanic horns.

(*Note:* A local base change of this degenerate fiber will give a so-called double elliptic genus two curve, the fiber will hence be referred to as a *double elliptic*)

iii) The branches are tangent to higher order, say order k (tangency means $k = 2$) in which case there is a chain of (-2) curves a A_{k-2} configuration separating the two elliptic tails. (This is of course a generalization of i)) We will refer to these fibers as *chained elliptic* with an appropriate subscript.

The Horikawa invariants can easily be computed, and the readers willing to delve into the complete story may work it out themselves (or consult [Horikawa] or [Xiao])

Note: The elliptic curves involved need not be smooth, not even irreducible, any Kodaira fiber will do. In fact if the infinitely close triple point consists of three flexed branches, the elliptic curve reduces to an I_0^* fiber with the leg joining the other elliptic component blown up (down to (-3))

For the simple joy of making a table we present the following.

Singularity	*Euler*	K^2	χ	*H-Inv*	*Picard*
Nodal	1	$1/5$	$1/10$	0	0
Elliptic	1	$7/5$	$1/5$	1	1
Double Elliptic	4	2	$1/2$	1	2
Chained$_n$ Elliptic	n+1	$7/5(n+1)$	$1/5(n+1)$	n+1	n+1

Note: The Euler contribution is actually the relative Euler number i.e. the Euler number of the fiber +2 (the last contribution being minus the DEeuler number of the generic fiber). H-Inv is the Horikawa invariant ($5/6 K^2 - 1/6 Euler$) and Picard is of course the contribution to the Picard group.

There is the idea of ***Morsification*** (see [Reid] for more details) in which a special singularity can be resolved into the basic ones (i.e. nodal and elliptic). In this situation, the contributions to the Euler number, K^2 and χ behave additively, as well of course as the Horikawa invariant. (But the Picard number can experience jumps).

With this in mind we see that a double elliptic fiber resolves into one elliptic and three nodal; while the chained$_n$ Elliptic resolves into $(n+1)$ elliptic fibers.

3. Intrinsic characterizations

The canonical divisor of a genus two fibration consists of a transversal part (a bisection) and possibly some vertical stuff. Being a bisection, and the two components of an elliptic fiber being symmetrical, it must pass through their intersection. This leads to the following observation

Lemma *The singular point of an elliptic fiber is always a basepoint of the canonical system*

PROOF : In fact if C is an elliptic curve with $C^2 = -1$ then $KC = 1$ and the restriction of the canonical divisor to C has degree 1 and has to be fixed. Note that if p is the singular point of the elliptic fiber, then the self intersection of either component is $-p$ ◇

Note that the above proof works for any twist of the canonical divisor with fibers, which make sense as it is a relatively local phenomenon.

Thus, contrary to intuition (maybe), a genus two fibration with high K^2 has a canonical divisor with many base points. So those basepoints are something intrinsic to the surface (and one may like to see them say from a curvature point of view) and stable under deformation (as far as the genus two fibrations are stable).

Note: In case of an elliptic fiber with an A_n configuration in the middle, the entire rational configuration becomes fixed under the canonical divisor, the free part of the canonical system is of course a bisection, meeting each tail component of the A_n transversally. In the case of a double elliptic curve, the two intersection points with the two (satanic) horns define an involution, the base point of the canonical system will be one of its fixed points.

One would like to get some *a priori* good estimate on their number. A naive estimate is given by

Observation: *Let $|D|$ be a linear system without fixed components, then the # of fixed points (F) is bounded above by $D^2 - h^0(D) + 1$*

PROOF : One can find a pencil through any $h^0(D) - 1$ (=G) given points, and thus exhibit two divisors meeting in $F + G$ points. ◇

Note that this estimate is useless if applied to $D = K$. To get a useful estimate one needs something of type $F^2 \leq cK^2$ with $c < 1$

4. Canonical and Bicanonical Images

The canonical image (if necessary twisted by fibers) gives a scroll in the appropriate projective space. At elliptic fibers, the base point has to be blown up, and the canonical map will then blow down the two elliptic components. Thus on the level of the ruled surface the branch locus will consist of a fiber on which the residual branch locus has two triple points. The two triple points correspond to the blown down elliptic components (which are in a sense invisible, or at least concealed) and the fiber will of course correspond to the exceptional divisor. Thus in a canonical sense, the canonical form of an elliptic fiber is not the *infinitely close triple point* but the fibral component with two quadruple points. The corresponding forms for, say, double elliptic and chained elliptic can easily be worked out.

The bicanonical image (with the same provisio as above) gives rise to a conic bundle. The bicanonical divisor does not have a fixed point at an elliptic fiber, and the two components are mapped 2:1 to lines. The sextic branch locus will naturally split into two parts. This looks indeed natural, neat and nice, but it does contradict the fact that the branch locus has to be even. Somehow the intersection point, the fixed point of the canonical system cries out to be part of the branching. A moments reflection resolves the potential mystery. The conic bundle is *not* nonsingular, a (-2) curve has been blown down, the hapless image of the former exceptional divisor. In the case of a double elliptic fiber, the image is a double line, and by the same consideration as above, we see that it cannot be non-singular, the singularity corresponding to the fixpoint of the canonical system.

Following the canonical ring approach touted by Reid (see [Reid]) the natural ambiance of the bicanonical image is a $\mathbf{P}(1,1,2)$ bundle (over $\mathbf{P}^1$ under our tacit assumptions). Where $\mathbf{P}(1,1,2)$ simply is the *weighted projective space* with obvious weights. (The weights corresponding to generators of the canonical $(1,1)$ and bicanonical (2) system respectively). The reader slightly uneasy about the notion can simply think of quadric cones. The conic bundle is then simply cut out by a *quadro-linear* relation in each fiber. (type $ay - x_1x_2$ or $ay - x_1^2$ see [Reid]. Interpreting the space as a quadric cone, it is simply a linear relation in the ambient $\mathbf{P}^3$). Wherever it hits the node (the singular point of the space $x_1 = x_2 = 0$) the conic splits up into two components (or one double according to the intersection).

Thus an alternative approach of describing a genus two fibration is to consider a conic bundle (as above) together with a cubic relation to define the sextic branch locus. I do not know of any example where this has been done in practice, the problem is that the objects involved tend to be unwieldy.

In this context it is unavoidable to recall Reids' algeo-geometric interpretation of the Horikawa invariant. Due to the presence of base points for the canonical system and the lack of them for the bicanonical, not every bicanonical divisor can be written as a sum of two canonical divisors. The obstruction (the obvious cokernel) is measured locally (and globally by accumulation) by the Horikawa invariant.

5. Upper Estimates

The basic question concerning genus two fibrations is to determine the kind of Chern-invariants it can have. (The geographical question). Some preliminary results (or remarks) were supplied in [P2] but the most comprehensive results can be found in [Xiao].

A naive way of approaching it is to consider the branch locus and to impose many infinitely close triple points. This leads to the notorious *Orchard problem* which is the essence of *ad hoc* . Typically there will be some spectral phenomenon for high K^2, which will be very difficult to get a hold on.

As for an upper bound one may consider the *cost* of imposing an infinitely close triple point (in terms of deduction from the genus bank, each such point detracts 6 from the genus). This was the simple-minded approach taken in [P2]. The triviality of the argument (which, due to the point of view of specialization vectors, was not so obvious then) can be illustrated by the estimate

$$\#elliptic \leq e(X) + 4$$

given by (1) combining with (2) we obtain

$$K^2 - 2\chi + 6 \leq 12\chi - K^2 + 4$$

translating into the rude and crude

$$c_1^2 < 7\chi.$$

Being slightly more sophisticated, one may use the estimate given not only by topology but by Hodge theory. In fact the Picard number of a surface with k elliptic fibers is at least $2 + k$, thus we have

$$2 + k \leq h^{1,1} \leq 10\chi - K^2.$$

Setting $K^2 = c\chi$ and using (2) we get

$$(c-2)\chi + 8 \leq (10 - c)\chi$$

from which we conclude $c < 6$. Continuing in the same pedestrian manner we apply the same argument again, now setting $K^2 = 6\chi - \alpha$ plugging into (2) yielding

$$c_1^2 \leq 6\chi - 4$$

an estimate to be found in [Xiao] (of course, and much more).

Note: These estimates should only be considered for positive χ for elementary and implicit reasons. (Think of the virgin genus-two fibration!)

Note: I agree with Xiao that it is unlikely that the upper bound is taken (at least asymptotically). In such a case there will be no room for any excess degenerations of fibers. Fibers of type ii) are forbidden as being too demanding of space (in H^2).

The game is now to construct examples with as high K^2 as possible. The first attempt, going beyond mere counting of constants, may have been in [P1] where there is a rather clever construction giving K^2 going up to $4\chi - 4$ using fibers exclusively of type ii) (cf the example below). (Although the construction can probably be traced in earlier literature in different context though.) An even more ingenious construction is to be found in [O-P].

The world record is due (as is to be expected) to Xiao, who in ([Xiao],p 52) manages to get slope 5.8 (a slight modification, adding fibres to the branchlocus, will yield an even simply-connected example). The construction goes beyond the simple-minded Campadelli type constructions that form the basic approach of this paper.

`Exercise` Give an *a priori* upper bound on K^2 if all elliptic fibers are multiple (doubly elliptic) as in the quoted example, using the above methods (*Note:* Either method - naive count by Euler characteristics or by Picard - gives the same answer. It will also prove the conjecture in ([P2] p.27)).

As an illustration it may be worthwhile to give examples of maximal slopes using only double elliptic degenerate fibers.

Example Consider $\mathbf{Z}_2$ operating on $X = C \times B$, where C is doubly elliptic i.e. with an involution σ such that C/σ is elliptic, and B hyperelliptic with the hyperelliptic involution τ, via $(x, b) \mapsto (\sigma x, \tau b)$. If the genus of B is g then there are $2(2g+2)$ fixed points lying pairwise above $2g+2$ degenerate fibers of the quotient $Y = X/\mathbf{Z}_2$. Its invariants are easily computed to be $e(Y) = 8g+4, K^2 = 4g-4$ and $\chi(X) = g$ and thus $K^2 = 4\chi - 4$ the upper bound.

It is easy to see that those quotients have to be irregular, the elliptic curves provide a constant contribution to the Jacobians. One may wonder whether any genus two fibration with high slope is necessarily irregular.(See below)

Given a fibration with high K^2 one can easily construct a whole series. In fact by taking the standard totally ramified covering of the base and lifting one gets a new surface. The chern invariants of the new fibration are independent upon whether the ramified fibers are singular or not, this detail only affects the Picard number.

So if X is a given genus two fibration over $\mathbf{P}^1$ and X^N is gotten by totally ramified basechange of degree N then the formulas (1') and (1") yield

$$\chi(X^N) = N\chi(X) + (N-1)$$

$$c_1^2(X^N) = Nc_1^2(X) + 8(N-1)$$

We can use the above formulas to state the following observation

Observation If the invariants of the genus two fibration X satisfy

$$c_1^2 = (b+8)\chi + b$$

then the same is true for the invariants of X^N.

PROOF : A straightforward calculation using the above identities ◇

Example -1 Consider a pencil of quartics with a given double point. Blowing up the basepoints gives a rational genus two fibration X_{-1} with $\chi = 2$ and $c_1^2 = 9 - 13 = -4$. The surfaces X_{-1}^N will then lie on the line corresponding to $b = -6$ and provide the main series of the Horikawa surfaces.

Example 0 Consider three conics in a pencil, and consider the branchcurve obtained by joining two lines through the four basepoints. The corresponding double cover X_0 will be a genus two fibration (consider the pencil of lines through the intersection point of the two adduced lines) with 2 elliptic fibers (the two lines) and six nodal fibers (the pairwise branching of each conic). The invariants are given by $c_1^2 = 0$ and $\chi = 1$, applying the observation to $b = -4$ we see that the the surfaces X_0^N lie on the line $4\chi - 4$ (borne out by the explicit calculation $\chi(X^N) = 2N - 1$ and $c_1^2(X^N) = 8(N-1)$) (Cf the quotient example above)

Example 0' Consider three bitangent conics in a pencil, the double cover will have a multitude of genus two pencils all with two basepoints. The blow up of basepoints will yield a surface equivalent to the surface X_0 above.

Example 1 Consider the example X_1 by Oort and Peters (see [O-P]) with $\chi = 1$ and $c_1^2 = 1$ (an example famous for being of general type with vanishing geometric genus and small torsion group). The calculation $1 = (b+8) + b$ yields $b = -7/2$ and hence the derivative surfaces X_1^N lie on the line $4.5\chi - 3.5$ This is the best we can do so far. (I guess the surfaces are regular)

Example 2 A genus two fibration X_2 with $\chi(X_2) = 1$ can at most have $c_1^2(X_2^N) = 2$ Such an example would yield a series X_2^N with invariants on the line $5\chi - 3$ To find such an initial example would mean to find 6 elliptic fibres and 8 nodal ones. It is thus a matter of finding a collection of curves on a rational ruled surface, with sectional degrees adding up to six and with twelve triple points lying pairwise on six fibers and eight residual branching. This is a delicate combinatorical problem and it is not *a priori* clear whether it has a solution. (Hence the above reference to the *Orchard problem*). In fact it does have a solution, but not a head-on as indictated above. Xiaos modular constructions ([Xiao] §3) (that have yielded the present record) gives his example of a surface with $p_g = 0$ and $K^2 = 2$ *and* a genus-two fibration. (The torsion group of the surface is incidentally $\mathbf{Z}_3^2$ which was embarrassingly high)

In general the higher χ for an example we can get, with a value of c_1^2 close to the bound given by the Hodge count $(6\chi - 4)$ the higher the slope we can get for our infinite sequence of examples. ($\chi = 2$ and $c_1^2 = 8$ would yield a slope of $16/3$) We may however never hope to get slope 6 this way (this would correspond to $b = -2$ which is impossible)

6. Curves in M_2

Naively genus two fibrations can be considered as curves in M_2, whose compactification consists basically of two divisors of infinity, the nodal curves and the *elliptic* . It would be nice to have a very concrete projective model for M_2.

Now M_2 contains a surface, namely the double elliptic curves. One may (following an old suggestion by Moishezon) consider curves constrained to this surface. The corresponding genus two fibrations are obtained from double coverings of elliptic surfaces along a bisection. It is easy to see that such an elliptic surface does not have any multiple fibers. Let us denote such genus two fibrations, doubly elliptic.

The doubly elliptic fibrations occupy in difficulty and subtleness some middleground between those of a general genus two fibration and an elliptic fibration; and somewhat closer to the last type than the former. Thus they provide a testing ground for various hypothesis and a collection of results that may serve as an inspiration for the general case.

As an example we have the following

Proposition *Given a doubly elliptic fibration then we have*

$$c_1^2 \leq 4\chi - 4$$

It may be convenient to recall how the different degenerate fibers occur in this setting.

Recall the standard analysis of a genus two curve (see e.g. [P4] which works of course over a function field with minor modifications.) It means that either of the elliptic surface which are the quotients come equipped with canonical zero sections. The problem is that they need not be effective in general. (This is a global problem *but* not local) For simplicity (and this does not affect the local picture) we will assume the sections are indeed effective giving us canonical zero sections *and* bisections (defining the branching) invariant under the involution ($z \mapsto -z$). Thus the elliptic quotient surfaces with their branchings are obtained by taking a $\mathbf{P}^1$ bundle with two distinct sections and a trisection, branching along the trisection and one of the sections, and letting the bisection be the inverse image of the other section. (True, the inverse image may not be *even* meaning that we may have to add fibers). Thus the constructions in [P1] are exactly of this type, unbeknownst to its author.

One may now, if one wishes, write down a complete list of the possible degenerate fibers. Such a list is obtained by listing all the possible Kodaira fibers together with the components (if any) that occur in the branching divisor together with the local form of the bisection.

We will not do this systematically, it is routine and not necessary. Thus we will confine ourselves to point out a few typical examples.(see figure which ought to be selfexplanatory)

The main question is how to characterize the essential fibers (i.e. those with non-vanishing Horikawa invariants). Naively one may believe that essential fibers occur exactly when the branching bisection ramifies over the base. But such an impulsive guess leads to puzzling confusion and painful contradictions, the double cover of an elliptic curve over a divisor of multiplicity two does not necessarily split. In fact this happens iff the branching divisor meets the canonical zero section.(compare the two examples in figure with dotted lines)

Even this maybe somewhat of an oversimplification, as we have seen (figure). The example of an I_0^* fiber branched along three simple components and a transversal bisection (which naturally ramifies over the double component) contradicts. This clearly refers to a doubly elliptic fiber (see 2.ii)). In this case the bisection cannot intersect the section at the crucial point. The example also points out that the branching divisior in addition to transversal contribution may contain fibral. The inverse image of each disjoint (-2)component clearly becomes an exceptional divisor upstairs.

As to elliptic fibers with disjoint (-2) curves being part of the branchlocus (and emerging as exceptional divisors after the covering) we see that their number cannot exceed half the eulercharacteristic of the fiber.

With this final remark we are now ready to tackle the proof

PROOF : Letting E be the elliptic surface, its eulercharacteritsic will be $12N$while its canonical class will satisfy $K = (N-2)F$ where F of course is a fiber.(Recall no multiple fibers). We may then compute the characteristic invariants of the double cover branched along an even bisection C

$$c_2 = 24N + KC + C^2$$

$$c_1^2 = 2(K + 1/2C)^2$$

using $K^2 = 0$ and $KC = 2(N-2)$) and the Noether formula we obtain

$$c_1^2 - 4\chi = -6N - 4 + \mu$$

where μ is the number of blowdowns to get the minimal model. From the above analysis of degenerate fibers it is easy to see that $\mu \leq 6N$ (We are exploiting the final remark)◇

Note: One may remark that if all the crucial fibers are I_0^* in the above proof, the surface is indeed of the type exhibited in **Example** (or in [P1] p.498). It might be tempting to conclude that all surfaces attaining the upper bound are of that type. One should not succumb to temptation too readily. The construction in [P1] can be easily modified. Rather than letting the branchlocus pass through all the intersection points of the three sections and producing the vertical infinitely close triple points and the double elliptic fibers in the end; we let it pass through *none* ! Taking advantage of the fact that we then produce two triple points on each fiber concerned, we simply add those fibers to the branchlocus, and a simple calculation (listing the degenerate fibers and exploiting the basic formulas) yields invariants lying on the crucial line.

Conjecture A genus two fibration obtained as a double cover of an elliptic surface cannot satisfy $c_1^2 = 4\chi - 5$

Note: This was a gap left, with some regrets, in [P1] as well as [P2]. If this conjecture would be true, it would illustrate a kind of simple spectral phenomena, which would be much more interesting, (and far more involved in its confirmation), in the general case. The above proof does not rule out the possibility (as we cannot in general assume for fibers F the inequality $1/2e(F) - \mu(F) \leq -2$ whenever $1/2e(F) - \mu(F) < 0$.($\mu(F)$ obviously denoting # exceptional candidates (i.e disjoint (-2) curves in the branchlocus) among the components of the fiber F). But it does restrict the possibilities rather drastically)

7. What Else ?

As we have noted the basic problem on genus two fibration is to exactly determine in what sector they may live. In particular it means finding fibrations with high K^2 (high slope). This calls for some imagination, so far not yet displayed in this paper. What we have tried to do is to suggest a Campadelli approach, i.e. using a high degree of ingenuity to construct a branching curve. (A touchstone is to construct a surface with $\chi = 1$ and $K^2 = 2$ in that way). I fear that we have reached the limits of that approach. Possible alternatives, besides the constructions by Xiao, may be looking at the tricanonical image, by considering the deformations of canonically embedded genus-two curves in $\mathbf{P}^4$, but this seems unwieldy; or to explicitly describe the compactified modulispace M_2 or to use Igusa's explicit description and write down curves in the moduli space. But this does not look too promising either.

The sector problem can be refined. One may ask the same question for simply connected genus two fibrations. An interesting aside is to determine to what extent regular genus two fibrations are simply connected. Counter examples abound, but for low slope ($3\chi - 5$) the two notions are close to being equivalent. (See recent preprint [X] by Xiao)

A more detailed problem is to describe all irreducible families of genus-two fibrations according to the basic chern invariants. (*Botany*) On the Noether line there are two kinds of Horikawa surfaces, those corresponding to connected or disconnected branchloci respectively. This distinction should carry over for higher slopes, as well as other distinctions One may in this context compare the two constructions on the line $c_1^2 = 4\chi - 4$ one irregular the other not.

One should also keep in mind that for low slopes ($^8/_3$) all surfaces are genus two fibrations, so in this region the questions do have a more universal appeal. In this context one may recall the observation.

Observation *(Peters) Most genus two fibrations have odd intersection form*

In fact a surface with positive Horikawa invariant has necessarily an elliptic curve with odd ($= -1$) self intersection. And even among the Horikawa surfaces even intersection form is not automatic. (Consider disconnected branchloci where the infinity section has a self intersection $\not\equiv 0(4)$)

November 1990

Bibliography

[Horikawa] **Horikawa,E.** *On Algebraic Surfaces with Pencils of curves of genus 2 Complex Analysis and Algebraic Geometry C.U.P* 79-90 (1977)

[O-P] **Oort,F.-Peters,C.A.M** *A Campadelli Surface with torsion group* $\mathbf{Z}_2$ *Nederl. Akad. Wetenschap. Indag. Math.* **43** 399-407 (1981)

[P1] **Persson,U.** *A family of Genus two fibrations Algebraic Geometry, Copenhagen 1978 Springer Lecture Notes* **732** 496-503 (1979)

[P2] **Persson,U.** *Chern Invariants of Surfaces of General Type Comp.Math.***43** 3-58 (1981)

[P3] **Persson,U.** *Configurations of Kodaira Fibers on Rational Elliptic Surfaces Math.Zeit* **205** 1-47 (1990)

[P4] **Persson,U.** *Genus two curves with non canonical involutions* preprint (December 1990)

[Reid] **Reid,M.** *Problems on pencils of small genus* preprint (Summer 1990)

[Xiao] **Xiao,G.** *Surfaces fibrées en courbes de genre deux Lecture Notes in Mathematics* **1137** (1985)

[X] **Xiao,G.** π_1 *of Elliptic and Hyperelliptic Surfaces* preprint (August 1990)

Numerically Effective Vector Bundles with Small Chern Classes

Th. Peternell, M. Szurek and J.A. Wiśniewski

Introduction

The aim of this paper is to classify numerically effective holomorphic vector bundles $\mathcal{E}$ on complex projective spaces and smooth quadrics, whose first Chern classes $c_1(\mathcal{E}) = 0$, 1 or 2, the latter case on the projective space $\mathbb{P}^n$ only. The complete lists are given in theorems 1 and 2 below. We call $\mathcal{E}$ to be numerically effective if the relative very ample sheaf $\xi_{\mathcal{E}}$ on $\mathbb{P}(\mathcal{E})$ is numerically effective.

The need for such study arose from our work on classifying ample $(n-1)$-bundles $\mathcal{E}$ on smooth n-dimensional manifolds X such that $c_1(\mathcal{E}) = c_1(X)$, see forthcoming paper [9]. In particular, theorems 1 and 2 from the present paper, together with the comparison lemma from [9] yield the following.

Theorem. Let X be a projective manifold of dimension $n \geq 3$ and $\mathcal{E}$ an ample vector bundle over X with $\operatorname{rank} \mathcal{E} = n-1$. If $c_1(\mathcal{E}) = c_1(X)$ and X is either $\mathbb{P}^n$ or Q_n, then $\mathcal{E} \otimes \mathcal{O}(-1)$ is on $\mathbb{P}_k$ is either decomposable or $k = 3$ and $\mathcal{E}(-1)$ is a null correlation bundle and on Q_k, is either decomposable or a spinor bundle on Q_3 or a spinor bundle $\oplus \mathcal{O}$ on Q_4.

We use Mori theory and examine vector bundles by studying the properties of their contractions $\mathbb{P}(\mathcal{E}) \longrightarrow Y$. Using an estimate of the locus of an extremal ray, [14], we discuss the three types of the contraction (fibre, divisorial and small) separately, calculating in each of the cases the dimensions of appropriate cohomology $H^i(\mathcal{E}(j))$ and then using some more standard vector bundle methods. Ample and spanned vector bundles with small Chern classes were studied by Ballico [3].

0. Mori theory

For an exposition of Mori theory we recommend [6]. Here we recall the facts that are of the most importance for this paper. Let A be the free abelian group generated by 1-cycles on a projective manifold X, modulo numerical equivalence and let $N_1 = A \otimes R$. Then N_1 is a finite dimensional real vector space. Let NE be the smallest cone in N_1 containing all effective 1-cycles and $cl(NE)$ be its closure in (any) metric topology.

Let $NE_{-}\{Z \in cl(NE) | K_X \cdot Z \geq 0\}$ where K_X is the canonical divisor of X. An extremal ray R is a half line in $cl(NE)$ such that

i) $K_X \cdot Z < 0$ if $[Z] \in R \setminus \{0\}$,
ii) for any $Z_1, Z_2 \in cl(NE)$, if $Z_1 + Z_2 \in R$, then $Z_1, Z_2 \in R$.

An extremal rational curve C is a rational curve on X such that $\mathbb{R}^+[C]$ is an extremal ray and $-K_Y \cdot C \leq n+1$. The length $l(R)$ of an extremal ray is defined as follows:

$$l(R) = \min\{-K_X \cdot C | C \text{ is a rational curve whose class is in } R\} \quad ,$$

see [12]. The Mori Cone Theorem says $cl(NE)$ is the smallest closed convex cone containing NE_{-} and all the extremal rays. For any open convex cone U containing $NE_{-} \setminus \{0\}$ there are finitely many extremal rays that do not lie in $U \cup \{0\}$. Every extremal ray is spanned by an extremal rational curve.

The Mori Cone Theorem yields the following bound on $l(R)$:

$$0 < l(R) \leq 1 + \dim X \quad .$$

Whenever we use the term "contraction", we mean it in the sense of the following

Kawamata-Shokurov Contraction Theorem. For any extremal ray R on a manifold X there exists a normal projective variety Y and a surjective morphism $contr_R : X \longrightarrow Y$ with connected fibres that contracts exactly the curves from R, i.e., for any integral curve C on X, $\dim(contr_R C) = 0$ if and only if $[C] \in R$.

A contraction is called to be of fibre type if $\dim Y < \dim X$. Equivalently, the extremal ray R is numerically effective. A divisorial contraction is one where $\dim Y = \dim X$ with the exceptional set $A \subset X$, of dimension $\dim X - 1$. When $\dim A < \dim X - 1$, the contraction is termed small.

An essential result in our study is an estimate of the locus of an extremal ray R. By the locus of R we mean the union of all C, $[C] \in R$. We have

Theorem ([14], 1.1). Assume that the contraction q has a non-trivial fibre of dimension d. Then

$$\dim(\text{locus of } R) + d \geq \dim X + l(R) - 1 \quad .$$

1. Bundles on $\mathbb{P}^n$.

Theorem 1. Let $\mathcal{E}$ be a numerically effective rank-r holomorphic vector bundle on the complex projective space $\mathbb{P}^k$, with $0 \leq c_1(\mathcal{E}) \leq 2$. Then one of the following holds:
(1) $\mathcal{E}$ splits as $\mathcal{E} = \mathcal{O}^r$, $\mathcal{O}(1) \oplus \mathcal{O}^{r-1}$, $\mathcal{O}(1)^2 \oplus \mathcal{O}^{r-2}$ or $\mathcal{O}(2) \oplus \mathcal{O}^{r-1}$;
(2) $\mathcal{E}$ fits in one of the exact sequences

$$\begin{array}{ccccccccc} 0 & \longrightarrow & \mathcal{O}(-1) & \longrightarrow & \mathcal{O}^{r+1} & \longrightarrow & \mathcal{E} & \longrightarrow & 0, \\ 0 & \longrightarrow & \mathcal{O}(-1)^2 & \longrightarrow & \mathcal{O}^{r+2} & \longrightarrow & \mathcal{E} & \longrightarrow & 0, \\ 0 & \longrightarrow & \mathcal{O}(-2) & \longrightarrow & \mathcal{O}^{r+1} & \longrightarrow & \mathcal{E} & \longrightarrow & 0, \quad \text{in particular } r \geq k; \end{array}$$

on $\mathbb{P}^3$ also:

$$0 \longrightarrow \mathcal{O} \longrightarrow \Omega(2) \oplus \mathcal{O}^{r-2} \longrightarrow \mathcal{E} \longrightarrow 0.$$

(3) $\mathcal{E}$ fits in $0 \longrightarrow \mathcal{O} \longrightarrow T(-1) \oplus \mathcal{O}(1) \oplus \mathcal{O}^{r-k} \longrightarrow \mathcal{E} \longrightarrow 0$, where T is the tangent bundle and Ω is its dual.

When $X = \mathbb{P}^k$ or Q_k, $k \geq 3$, we denote the two generators of $Pic(\mathbb{P}(\mathcal{E}))$ by H and ξ. Here H is the pull-back of $p^*(\mathcal{O}(1))$ via the projectivization map $p : \mathbb{P}(\mathcal{E}) \longrightarrow \mathbb{P}^k$, resp. Q_k and $\xi = \xi_{\mathcal{E}}$ is the relative very ample sheaf on $\mathbb{P}(\mathcal{E})$. It is uniquely determined by the conditions $p_*(\xi_{\mathcal{E}}) = \mathcal{E}$ and $\xi_{\mathcal{E}}|F = \mathcal{O}_F(1)$ on fibres of p. For the anticanonical divisor on $\mathbb{P}(\mathcal{E})$ we have

$$-K_{\mathbb{P}(\mathcal{E})} = r\xi + [c_1(\mathbb{P}^k) - c_1(\mathcal{E})]H \quad ,$$

see e.g. [10], (1.3). Hence for a numerically effective bundle with $c_1(\mathcal{E}) = 0$, 1, 2 the projectivization $\mathbb{P}(\mathcal{E})$ is a Fano manifold. All bundles $\mathcal{E}$ on $\mathbb{P}^2$ whose $\mathbb{P}(\mathcal{E})$ is Fano, were classified in [11]. Let us then recall the relevant results.

Theorem [12]. The following is the list of all numerically effective bundles with $0 \le c_1(\mathcal{E}) \le 2$ on $\mathbb{P}^2$:

Bundle	Type of "the other" contraction of $\mathbb{P}(\mathcal{E})$	Dimension of the general non-trivial fibre	$\dim H^0(\mathcal{E})$
$\mathcal{O}^r$	fibre	2	r
$\mathcal{O}(1) \oplus \mathcal{O}^{r-1}$	divisorial	2	$r+2$
$T(-1) \oplus \mathcal{O}^{r-2}$	fibre	1	$r+1$
$\mathcal{O}(2) \oplus \mathcal{O}^{r-1}$	divisorial	2	$r+5$
$\mathcal{O}(1)^2 \oplus \mathcal{O}^{r-2}$	small	2	$r+4^{*}$
$(T(-1) \oplus \mathcal{O}(1) \oplus \mathcal{O}^{r-2})/\mathcal{O}$	divisorial	1	$r+3$
$\mathcal{O}^{r+2}/\mathcal{O}(-1)^2$	divisorial	1	$r+2$
$\mathcal{O}^{r+1}/\mathcal{O}(-2)$	fibre	1	$r+1$

By the "other contraction" we mean here the contraction of the extremal ray on $\mathbb{P}(\mathcal{E})$ that does not contain the lines of the projectivization morphism. Let us point out that since in our situation $Pic(\mathbb{P}(\mathcal{E})) = \mathbb{Z} \oplus \mathbb{Z}$, with the exception of Q_2, apart from the projectivization morphism there always exists one and only one contraction. In the paper [10] the case $k = 3$, $r = 2$ is discussed, hence from now on we may focus on the case $k \ge 4$ or ($k = 3$ and $r \ge 3$). To prove the theorem, we proceed in a number of steps.

Lemma 1. A numerically effective bundle with $c_1(\mathcal{E}) = 0$ on $\mathbb{P}^k$ is trivial.

Proof. Let L be a line in $\mathbb{P}^k$. Because $\xi|L$ is nef with $c_1 = 0$, it is trivial. Hence $\mathcal{E}$ is trivial, too, see e.g. [7], ch. I., 3.2.1.

We shall then assume that $c_2(\mathcal{E}) = 1$ or 2, respectively.

Lemma 2. For any $i > 0$ we have $H^i(\mathcal{E}(-1)) = 0$.

Proof. We have $\xi_{\mathcal{E}(-1)} = \xi_{\mathcal{E}} - H = (r+1)\xi + (k - c_1(\mathcal{E}))H + K_{\mathbb{P}(\mathcal{E})}$. If $k \ge 3$, then $(r+1)\xi + (k - c_1(\mathcal{E}))H$ is ample, hence the conclusion of the lemma follows from $H^i(\mathcal{E}(-1)) = H^i(\xi_{\mathcal{E}(-1)})$ and the Kodaira Vanishing Theorem.

Lemma 3. $\mathcal{E}$ is spanned.

Proof. For $k = 2$ it follows from the given classification list. For $k \ge 3$ it follows by induction from lemma 1 and the exact sequence

$$0 \longrightarrow \mathcal{E}(-1) \longrightarrow \mathcal{E} \longrightarrow \mathcal{E}|H \longrightarrow 0 \quad ,$$

where H is any hyperplane.

Remark. We may assume that $\mathcal{E}$ is not ample. Indeed, if $r \ge 3$, then there are no ample r-bundles on a line with $0 \le c_1(\mathcal{E}) \le 2$. For $r = 2$, the bundle would have to be $\mathcal{O}(1) \oplus \mathcal{O}(1)$ on every line and thus it would split. Hence, being numerically effective and not ample, ξ defines a contraction of an extremal ray in the following way. We take the morphism $\mathbb{P}(\mathcal{E}) \longrightarrow Y'$ defined by the complete linear system $|\xi_{\mathcal{E}}|$. If $X \longrightarrow Y \longrightarrow Y'$ is the Stein factorization, $X \longrightarrow Y$ is the contraction of "the other" extremal ray of $\mathbb{P}(\mathcal{E})$.

The estimate of the dimension of the locus of an extremal ray gives immediately the following

Lemma 4. Let $X = \mathbb{P}(\mathcal{E})$ with the other contraction $q : \mathbb{P}(\mathcal{E}) \longrightarrow Y$. If q is of fibre type, then all its non-trivial fibres are of dimension $d \geq k-2$. If q is of the divisorial type, then all non-trivial fibres are of dimension $d \geq k-1$. If q is small, then all non-trivial fibres are of dimension $d \geq k$.

Proof. We shall discuss the small case only. In this case $\dim(\text{locus of } R) \leq r+k-2$ and $l(R) \geq k+1-c_1(\mathcal{E})$. We have therefore $r+k-3+d = (r+k-2)+d-1 \geq \dim(\text{locus of } R)+d-1 \geq r+k-1+(k+1-c_1(\mathcal{E}))-1 \geq r+2k-3$, i.e. $d \geq k$.

We shall discuss the three types of contraction separately.

Lemma 5. If q is of small type, then $\mathcal{E} = \mathcal{O}(1)^2 \oplus \mathcal{O}^{r-2}$.

Proof. $\mathcal{E}|\mathbb{P}^2$ is numerically effective with the fibres of the other contraction of dimension 2. From our classification list it follows that $\mathcal{E}|\mathbb{P}^2$ is decomposable. Hence $\mathcal{E}$ is decomposable, too, see e.g. [7], ch. I, 2.3.2. It is then easy to see that $\mathcal{E} = \mathcal{O}(1)^2 \oplus \mathcal{O}^{r-2}$.

Lemma 6. Let $\mathcal{E}$ be as above, i.e., nef with $c_1(\mathcal{E}) = 1{,}2$ on $\mathbb{P}^k$. If $r = \operatorname{rank}(\mathcal{E}) < k$, then

1) if $k \geq 4$ or ($k=3$ and $c_1(\mathcal{E}) = 1$), then $\mathcal{E}$ is decomposable,
2) if $k=3$ and $c_1(\mathcal{E}) = 2$, then either $\mathcal{E}$ is $\mathcal{O}(2) \oplus \mathcal{O}$, $\mathcal{E} = \mathcal{O}(1)^2$ or $\mathcal{E} = \mathcal{N}(1)$, where $\mathcal{N}$ is the null-correlation bundle.

For a definition and basic properties of the null-correlation bundle see e.g. [7], ch. I, 4.2.

Proof. Let Z be the zero set of a generic section of $\mathcal{E}$. If Z is empty then $\mathcal{E}$ contains a trivial subbundle and, by induction on the rank of $\mathcal{E}$, this subbundle is a direct factor. If not empty, Z is of codimension $k-r$ and smooth, since $\mathcal{E}$ is generated by global sections. By adjunction, $K_Z = \mathcal{O}_Z(-k-1+c_1(\mathcal{E}))$. Then by the result of Kobayashi and Ochiai, [5], every connected component of Z is either $\mathbb{P}^{k-1}$, Q_{k-1} or $\mathbb{P}^{k-2}$ and consequently in the first two cases $\mathcal{E} = \mathcal{O}(1)$, $\mathcal{O}(2)$ while in the third case $\mathcal{E} = \mathcal{O}(1)^2$ for $k \geq 3$, because the zero set is connected and a complete intersection. The remaining case ($r=2$ and $k=3$) was discussed in [12], see also the classification quoted above.

The case $k=3$ was discussed in [10].

Lemma 7. Let q be of fibre type, $k \geq 4$. Then $\mathcal{E}$ is either as in 2) of the theorem or trivial.

Proof. First we show that $H^0(\mathcal{E}(-1)) = 0$. Let us recall that the contraction of an extremal ray R is of fibre type iff R is numerically effective. If $\mathcal{E}(-1)$ had a section, $\xi_{\mathcal{E}(-1)} = \xi_{\mathcal{E}} - H$ would be effective, so that $(\xi_{\mathcal{E}} - H)R \geq 0$. On the other hand, H is nef, has a positive intersection with R and $\xi_{\mathcal{E}} R = 0$, hence $(\xi_{\mathcal{E}} - H) < 0$. Thus $H^0(\mathcal{E}(-1)) = 0$. As in lemma 3, we then show by induction that $h^0(\mathcal{E}) = h^0(\mathcal{E}|\mathbb{P}^2)$. We may clearly assume that $\mathcal{E}$ is not decomposable. Hence $h^0(\mathcal{E}) = r+1$, $r+2$ or $r+3$, see the classification list on $\mathbb{P}^2$. Let K be the kernel of the evaluation morphism

$$0 \longrightarrow K \longrightarrow H^0(\mathcal{E}) \otimes \mathcal{O} \longrightarrow \mathcal{E} \longrightarrow 0 \quad .$$

K^* is generated by global sections, is of rank 1, 2 or 3 and has $c_1(K^*) = c_1(\mathcal{E})$. From $H^0(K) = 0$ and lemma 6 we get immediately that K^* is $\mathcal{O}(1)$, $\mathcal{O}(2)$ or $\mathcal{O}(1)^2$, hence the lemma follows.

To conclude the discussion of the fibre contraction case, we ought to examine the bundles over $\mathbb{P}^3$.

Lemma 8. Let $\mathcal{E}$ be a numerically effective bundle over $\mathbb{P}^3$ with $1 \le c_1(\mathcal{E}) \le 2$ and with the "other contraction" $q : \mathbb{P}(\mathcal{E}) \longrightarrow Y$ being of fibre type. Then $\mathcal{E}$ is either trivial or of type 2) of theorem 1.

Proof. The case of rank-2 bundles was discussed in [10]. For $r = \operatorname{rank}(\mathcal{E}) \ge 3$ we see that $\mathcal{E}$ is spanned and therefore has a trivial subbundle of rank $r-3$:

$$0 \longrightarrow \mathcal{O}^{r-3} \longrightarrow \mathcal{E} \longrightarrow \mathcal{E}' \longrightarrow 0 \quad .$$

We may then take $\operatorname{rank}(\mathcal{E})$ to be 3, because a posteriori it will occure that the sequence splits. As in lemma 7, we show that $h^0(\mathcal{E}) = 4$, 5 or 6. Since q is of fibre type, $\dim Y < \dim(\mathbb{P}(\mathcal{E})) = 5$. Hence (see remark after lemma 3), in the complete linear system $|\xi_{\mathcal{E}}|$ we may choose a base-point free subsystem of dimension equal to $\dim Y$, i.e. ≤ 4. Consequently, $\mathcal{E}$ may be generated by 5 sections and thus it is either trivial or fits in one of the sequences

$$0 \longrightarrow K_i \longrightarrow \mathcal{O}^{3+i} \longrightarrow \mathcal{E} \longrightarrow 0 \quad ,$$

where K_i, $i = 1,2$ are of rank i. Of course K_1 is a line bundle hence $K_1 = \mathcal{O}(-1)$ or $K_1 = \mathcal{O}(-2)$. As for K_2, its dual K_2^* is spanned, hence the anticanonical divisor of $\mathbb{P}(K_2^*)$ equals $2\xi + 2H$ and then is ample. In the terminology of [10] and [11], K_2^* is a Fano bundle and of course $c_1(K_2^*) = 2$. Because $H^0(K_2) = 0$, theorem 2.1 from [10] gives $K_2^* = \mathcal{N}(1)$, $\mathcal{N}$ being the null-correlation bundle. Then $\mathcal{E} = \Omega(2)$ or $\mathcal{E} = \mathcal{N}(1) \oplus \mathcal{O}$ in virtue of the following

Claim. Any epimorphism $e : \mathcal{O}^5 \longrightarrow \mathcal{N}(1)$ has kernel $\Omega(2)^*$ or $\mathcal{N}(1)^* \oplus \mathcal{O}$.

Proof of claim. If $\ker e$ does not contain a trivial factor, it is the evaluation map and by some easy diagram-chasing (compare e.g. [3]) we get that $\ker e$ is equal to $\Omega(2)$. Clearly there is no epimorphism $\mathcal{O}^3 \longrightarrow \mathcal{N}(1)$. Assume e can be factorized through e_1 : $\mathcal{O}^4 \longrightarrow \mathcal{N}(1)$. Then $\ker e_1$ is a rank-2 vector bundle with $c_1 = -2$, $c_2 = 2$ and its dual is spanned.

From the formula for the anticanonical divisor on $\mathbb{P}(\ker(e^*))$ quoted at the beginning of the paper now follows that $\mathbb{P}(\ker(e^*))$ is a Fano manifold. By [10], 2.1, $\ker e$ must be $\mathcal{N}(1)$.

We then pass to the discussion of the divisorial case.

Lemma 9. Let the "other contraction" $q : \mathbb{P}(\mathcal{E}) \longrightarrow Y$ be of divisorial type. Assume $\mathcal{E}$ is not decomposable and is as above, i.e., nef and $0 \le c_1(\mathcal{E}) \le 2$. Then the restriction of $\mathcal{E}$ to any 2-dimensional plane $P \subset \mathbb{P}^k$ fits into the exact sequence

$$0 \longrightarrow \mathcal{O} \longrightarrow \mathcal{O}^{r-2} \oplus T_{\mathbb{P}^2}(-1) \oplus \mathcal{O}(1) \longrightarrow \mathcal{E}|P \longrightarrow 0 \quad .$$

Proof. From the classification list on $\mathbb{P}^2$ it follows that if $\mathcal{E}|P$ is not as the one in the above sequence, then it occurs in

$$0 \longrightarrow \mathcal{O}(-1)^2 \longrightarrow \mathcal{O}^{r+2} \longrightarrow \mathcal{E}|P \longrightarrow 0 \quad . \tag{1}$$

We exclude this possibility. From (1) we have $h^0(\mathcal{E}(-1)|P) = 0$. We want to prove that the same property holds for any linear projective subspace $V \subset \mathbb{P}^k$ of dimension ≥ 3. For any $m \leq k-1$ the divisor $\xi_{\mathcal{E}} - mH - K_{\mathbb{P}(\mathcal{E})}$ equals $(r+1)\xi_{\mathcal{E}} + (k-m-1)H$ and then is nef and big. Therefore, by the Kawamata-Viehweg vanishing theorem, we get for $i > 0$ and $m \leq k-1$:

$$H^i(\mathcal{E}(-m)) = H^i(\xi_{\mathcal{E}} - mH) = 0 \quad . \tag{2}$$

Similarly, as $\mathcal{E}|V$ keeps the relevant properties when restricted to a general linear projective subspace $V \subset \mathbb{P}^k$, we have

$$H^i(\mathcal{E}(-m)|V) = 0 \quad ,$$

for $m \leq \dim V - 1$. Let us denote by (D_j) the sequence of sheaves that we call divisorial

$$0 \longrightarrow \mathcal{E}(j-1) \longrightarrow \mathcal{E}(j) \longrightarrow \mathcal{E}(j)|H \longrightarrow 0 \quad . \tag{D_j}$$

The vanishing of $H^0(\mathcal{E}(-1)|V)$ now follows by using (D_{-1}) and induction. Using the vanishing properties (2) and similar inductive arguments again we prove that

$$\dim H^0(\mathbb{P}^k, \mathcal{E}) = \dim H^0(\mathbb{P}^2, \mathcal{E}|\mathbb{P}^2) = r+2 \quad .$$

On the other hand, the contrction q is divisorial and given as the Stein factorization of a map determined by the complete linear system $|\xi_{\mathcal{E}}|$, see the remark following lemma 3. Hence the dimension $\dim H^0(\mathbb{P}^k, \mathcal{E})$ must be at least $\dim \mathbb{P}(\mathcal{E}) + 1 = r+k > r+2$.

Lemma 10. As above, let q be of divisorial type and $\mathcal{E}|\mathbb{P}^2$ be as in the conclusion of lemma 9. Then $h^0(\mathcal{E}) = r+k+1$.

Proof. Taking cohomology in the divisorial sequence (D_j), we see inductively on $j = 2, 3, \ldots$ and $\dim V$ that

$$h^0(\mathbb{P}^k, \mathcal{E}(-j)) = h^0(V, \mathcal{E}(-j)|V) = h^0(\mathbb{P}^2, \mathcal{E}(-j)|\mathbb{P}^2) = 0$$

for any linear projective subspace $V \subset \mathbb{P}^k$, $\mathbb{P}^2 \subset V$. The same divisorial sequence (D_{-1}) gives then

$$h^0(\mathcal{E}(-1)) = h^0(\mathcal{E}(-1)|V) = h^0(\mathcal{E}(-1)|\mathbb{P}^2) = 1 \quad ,$$

while (D_0) gives $h^0(\mathcal{E}) = h^0(\mathcal{E}|H) + 1$. The lemma now follows by an obvious induction.

Lemma 11. The dimension $h^i(\mathbb{P}^k, \mathcal{E}(-k)) = 1$ iff $i = k-1$ and is zero otherwise.

Proof. Immediate from lemma 9, vanishing properties (2) and the divisorial sequence (D_{-k+1}), again by induction.

Lemma 12. For $i = 0, 1, 2, \ldots, k$ and $j = 0, -1, -2, \ldots, -k$ the cohomology dimensions $h^i(\mathcal{E}(j))$ are

$$h^0(\mathcal{E}) = r+k+1,$$
$$h^0(\mathcal{E}(-1)) = 1,$$
$$h^{k-1}(\mathcal{E}(-k)) = 1$$

and zero otherwise.

Proof. Immediate from lemmas 10, 11 and the vanishings (2).

Lemma 13. Let $\mathcal{E}$ be a rank-r vector bundle on $\mathbb{P}^k$. Then the following conditions are equivalent:

i) $\mathcal{E}$ has the cohomology as in lemma 12,

ii) there exists an exact sequence

$$0 \longrightarrow \Omega(1) \oplus \mathcal{O}(-1) \longrightarrow \mathcal{O}^{r+k+1} \xrightarrow{\tau} \mathcal{E} \longrightarrow 0 \quad ,$$

iii) there exists an exact sequence

$$0 \longrightarrow \mathcal{O} \longrightarrow T(-1) \oplus \mathcal{O}(1) \oplus \mathcal{O}^{r-k} \longrightarrow \mathcal{E} \longrightarrow 0 \quad .$$

Proof. i) $\Rightarrow$ ii) : Let us recall that the Beilinson spectral sequence E_r^{pq} with E_1 term

$$E_1^{pq} = H^q(\mathbb{P}^k, \mathcal{E}(p)) \otimes \Omega^{-p}(-p)$$

converges to

$$E_\infty^{pq} = 0 \text{ for } p+q \neq 0$$

and $\bigoplus_{p=0}^{n} E_\infty^{-p,p}$ is the associated graded sheaf of a filtration of $\mathcal{E}$, see [7], ch. II, 3.1.3. The data as in lemma 12 then give

$$E_2^{00} = \operatorname{coker}(\Omega(1) \longrightarrow \mathcal{O}^{r+k+1}), \qquad E_2^{-k,k-1} = \mathcal{O}(-1)$$

$$E_{k+1}^{00} = E_\infty^{00} = \operatorname{coker}(\mathcal{O}(-1) \longrightarrow E_2^{00}), \qquad E_\infty^{pq} = 0 \text{ for } (p,q) \neq (0,0) \quad .$$

In other words, if $B := E_2^{00}$, then there is a commutative diagram with exact rows and columns

$$\begin{array}{ccccccccc}
 & & 0 & & 0 & & 0 & & \\
 & & \downarrow & & \downarrow & & \downarrow & & \\
 & & 0 & \longrightarrow & 0 & \longrightarrow & \mathcal{O}(-1) & \xrightarrow{=} & \mathcal{O}(-1) \\
 & & \downarrow & & \downarrow & & \downarrow & & \downarrow \\
0 & \longrightarrow & \Omega(-1) & \longrightarrow & \mathcal{O}^{r+k+1} & \longrightarrow & B & \longrightarrow & 0 \\
 & & \downarrow & & \downarrow & & \downarrow & & \downarrow \\
0 & \longrightarrow & \ker\tau & \longrightarrow & \mathcal{O}^{r+k+1} & \xrightarrow{\tau} & \mathcal{E} & \longrightarrow & 0 \\
 & & & & \downarrow & & \downarrow & & \\
 & & & & 0 & & 0 & &
\end{array}$$

From the snake lemma applied to the two middle vertical columns we obtain an exact sequence

$$0 \longrightarrow \Omega(1) \longrightarrow \ker\tau \longrightarrow \mathcal{O}(1) \longrightarrow 0$$

that splits. This proves that $\mathcal{E}$ is as ii) of the lemma.

ii) $\Rightarrow$ iii) : Dualizing the sequence given in ii) we obtain

$$0 \longrightarrow \mathcal{E}^* \longrightarrow \mathcal{O}^{r+k+1} \longrightarrow T(-1) \oplus \mathcal{O}(1) \longrightarrow 0 \quad .$$

We see then that $h^0(\mathcal{E}^*) \geq r-k-1$. Splitting out trivial factors, we infer that $\mathcal{E}^* = H^0(\mathcal{E}^*) \otimes \mathcal{O} \oplus \mathcal{E}^*$, where $H^0(\mathcal{E}_1^*) = 0$. Hence $\operatorname{rank} \mathcal{E}_1^* \leq k+1$. On the other hand, in view of lemma 6, $\operatorname{rank} \mathcal{E}_1^*$ may be taken $\geq k$.

Then we have the following commutative diagram.

$$
\begin{array}{ccccccccccc}
 & & & & 0 & & 0 & & & & \\
 & & & & \downarrow & & \downarrow & & & & \\
 & & 0 & & 0 & \rightarrow & \mathcal{O}(-1) \oplus \Omega(1) & = & \mathcal{O}(-1) \oplus \Omega(1) & & \\
 & & \downarrow & & \downarrow & & \downarrow & & \downarrow & & \\
0 & = & H^0(\mathcal{E}_1^*) & \rightarrow & \mathcal{O}^{r+k+1} & \rightarrow & \mathcal{O}^{k+1} \oplus \mathcal{O}^{k+1} & \rightarrow & \mathcal{O}^{k+1-r} & \rightarrow & 0 \\
 & & \downarrow & & \downarrow & & \downarrow \sigma & & \downarrow & & \\
0 & \rightarrow & \mathcal{E}_1^* & \rightarrow & \mathcal{O}^{r+k+1} & \rightarrow & T(-1) \oplus \mathcal{O}(1) & \rightarrow & 0 & & \\
 & & & & \downarrow & & \downarrow & & & & \\
 & & & & 0 & & 0 & & & &
\end{array}
$$

with $\sigma = \sigma_1 \oplus \sigma_2$, where σ_1 is the evaluation morphism for $T(-1)$ and σ_2 for $\mathcal{O}(-1)$. The snake morphism $\mathcal{E}_1^* \longrightarrow \mathcal{O}(-1) \oplus \Omega(1)$ then gives the desired exact sequence from iii).

iii) $\Rightarrow$ i) follows from Bott's formula for the cohomology of $\Omega(j)$ and $\mathcal{O}(j)$, see e.g. [7], ch. 1, §1.

This concludes the proof of lemma 13 and hence of theorem 1.

Remark. lemma 13 is stronger then the statement 3) in theorem 1. Let us also notice that iii) gives a "geometric" description of $\mathcal{E}$: it is $T(-1) \oplus \mathcal{O}(1) \oplus \mathcal{O}^{r-k-1}$ or is determined by a choice of a hyperplane (in $\mathbb{P}^k$) and a point not in it.

2. Bundles on Q_k

In this section we study numerically effective bundles $\mathcal{E}$ with $0 \leq c_1(\mathcal{E}) \leq 1$ and arbitrary rank $r \geq 2$. Such bundles are uniform with respect to (straight) lines on quadrics. Indeed, due to the "nefness" the only possible splitting type of $\mathcal{E}$ on a line $L \subset Q_k$ is $\mathcal{O} \oplus \mathcal{O} \oplus \ldots \oplus \mathcal{O}$ or $\mathcal{O}(1) \oplus \mathcal{O} \oplus \ldots \oplus \mathcal{O}$. Hence if $c_1(\mathcal{E}) = 0$, such bundles are trivial, see [13], 3.6.1. Let us mention here that Ballico and Newstead, [1], showed that uniform vector bundles of rank r on a smooth quadric Q with $\dim Q = 2s$ or $2s+1$ are decomposable if $r \leq s+1$ resp. $\leq s+3$.

We want to prove the following

Theorem 2. Let $\mathcal{E}$ be a rank-r vector bundle over a smooth quadric Q_k, $k \geq 3$. Assume that $\mathcal{E}$ is numerically effective and $0 \leq c_1(\mathcal{E}) \leq 1$. Then $\mathcal{E}$ is one of the following:

1) $\mathcal{O}^r$, $\mathcal{O}(1) \oplus \mathcal{O}^{r-1}$,
2) $k \leq r$ and $\mathcal{E}$ fits in the exact sequence

$$0 \longrightarrow \mathcal{O}(-1) \longrightarrow \mathcal{O}^{r+1} \longrightarrow \mathcal{E} \longrightarrow 0 \quad ,$$

3) $k = 3$ or 4 and $\mathcal{E} = E \oplus \mathcal{O}^{r-2}$, where E is a spinor bundle over Q_3 or Q_4.

For the definition and properties of the spinor bundles see e.g. [8], 3. Here we only recall that on Q_{2m-1} the spinor bundle $E = s^*U$ and on Q_{2m} there are two spinor bundles s'^*U,

s''^*U, where U is the universal bundle on $Gr(2^{k-1}-1, 2^{k-1})$ and s, s', s'' are natural embeddings of Q into the Grassmannian defined by the spinor varieties of the quadrics, the varieties parametrizing families of $(m-1)$-planes on Q_{2m-1} or on the two disjoint families of m-planes on Q_{2m}. On $Q_4 = Gr(1,3)$ the two spinor bundles are the universal and the dual of the quotient bundle. On Q_3, the spinor bundle is obtained by restricting the spinor bundles from Q_4.

We begin with discussing the situation on a two-dimensional quadric $Q_2 = \mathbb{P}^1 \times \mathbb{P}^1$.

Lemma 1. If $\mathcal{G}$ is a rank-2 numerically effective bundle on Q_2 with $(0,0) \leq c_1(\mathcal{G}) \leq (1,1)$, then $\mathcal{G}$ is one of the following: $\mathcal{O}^2$, $\mathcal{O} \oplus \mathcal{O}(1,1)$, $\mathcal{O}(1,0) \oplus \mathcal{O}$, $\mathcal{O}(0,1) \oplus \mathcal{O}$, $\mathcal{O}(1,0) \oplus \mathcal{O}(0,1)$ or a spinor bundle, i.e., the restriction of a spinor bundle from Q_3.

For a proof see [11].

Lemma 2. Let $\mathcal{E}$ be a bundle as in the assumption of theorem 2. Then $\mathcal{E}|Q_2$ is spanned.

Proof. Let us denote $\mathcal{E}|Q_2$ by $\mathcal{F}$ and recall the formulas for the Chern classes of a twist of a bundle and the Riemann-Roch formula on Q_2:

$$c_1(\mathcal{E} \otimes \mathcal{O}(p,q)) = (c_1'(\mathcal{E}) + rp, c_1''(\mathcal{E}) + rq) \quad ;$$

$$c_2(\mathcal{E} \otimes \mathcal{O}(p.q)) = 2\binom{r}{2}pq + (r-1)(qc_1''(\mathcal{E}) + pc_1'(\mathcal{E})) + c_2(\mathcal{E}) \quad ,$$

$$\chi(\mathcal{E}) = \frac{1}{2}c_1^2 - c_2 + (c_1' + c_1'') + r = c_1'c_1'' + c_1' + c_1'' - c_2 + r$$

where the pair (c_1', c_1'') denotes the first Chern class.

We show first that $H^i(\mathcal{F}) = 0$ for $i > 0$. Let L be a line from one ruling (say "horizontal") of $Q_2 = \mathbb{P}^1 \times \mathbb{P}^1$. We have the following exact sequence

$$0 \longrightarrow \mathcal{F} \otimes \mathcal{J}_L \longrightarrow \mathcal{F} \longrightarrow \mathcal{F}|L \longrightarrow 0 \quad ,$$

where $\mathcal{J}_L = \mathcal{O}(-1,0)$ is the ideal of the line L. Then $\mathcal{F}|L$ has the decomposition type $(0,\ldots,0,1)$ and from the above formulas we have

$$\chi(\mathcal{F} \otimes \mathcal{J}_L) = 2 - c_2(\mathcal{F}) = 2 - c_2(\mathcal{E}) \quad .$$

We then easily calculate, using nefness of $\xi_{\mathcal{E}}$ and Leray-Hirsch relations that $\xi^{r+1} = c_1(\mathcal{F})^2 - c_2(\mathcal{F}) \geq 0$, i.e., $c_2 \leq 2$. By plugging this into the Euler characteristic formula we obtain immediately $\chi \geq 0$. Let us then consider the push-forward $p_*(\mathcal{F} \otimes \mathcal{J}_L)$ with

$$p_* : \mathbb{P}^1 \times \mathbb{P}^1 \longrightarrow \mathbb{P}^1$$

the projection "down" onto $\mathbb{P}^1$. Because $\mathcal{F} \otimes \mathcal{J}_L$ has the decomposition type $(-1,-1,\ldots,0)$ on the "vertically" ruling lines, its push-forward is a line with positive Euler characteristic, i.e., is $\mathcal{O}(m)$ with $m \geq 1$. Similarily, $R^1p_*(\mathcal{F} \otimes \mathcal{J}_L) = 0$. By Leray's spectral sequence we have then $H^i(\mathcal{F} \otimes \mathcal{J}_L) = 0$ for $i > 0$. The restriction map $\mathcal{F} \longrightarrow \mathcal{F}|L$ then gives an epimorphism on global sections. Since L can be chosen to pass through any arbitrary point and $\mathcal{F}|L$ is spanned, $\mathcal{F}$ is spanned, too.

Returning to a general situation, we have an easy

Lemma 3. $H^i(Q_k, \mathcal{E}(m)) = 0$ for $i > 0$ and $m > -k+1$.

Proof. For such m, $\xi_{\mathcal{E}} + mH - K_{\mathbb{P}(\mathcal{E})}$ is ample, then the lemma follows from the Kodaira Vanishing Theorem and the correspondence between the cohomology

$$H^i(Q_k, \mathcal{E}(m)) = H^i(\mathbb{P}(\mathcal{E}), \xi_{\mathcal{E}} + mH) \quad .$$

Lemma 4. $\mathcal{E}$ is spanned.

Proof. By lemma 2, $\mathcal{E}|Q_2$ is spanned for any two dimensional quadric $Q_2 \subset Q_k$. The conclusion of the lemma then follows from the vanishings of lemma 3 and (the quadric's analogue of) the divisorial sequence (D_0).

Remark. As in section 1, the map associated to the complete linear system $|\xi_{\mathcal{E}}|$ gives, via the Stein factorization, a contraction $q : \mathbb{P}(\mathcal{E}) \longrightarrow Y$ of "the other" ray of the Fano manifold $\mathbb{P}(\mathcal{E})$ for $c_1(\mathcal{E}) = 1$.

Lemma 5. If the contraction q is birational, then $\mathcal{E} = \mathcal{O}(1) \oplus \mathcal{O}^{r-1}$.

Proof. Let $p : \mathbb{P}(\mathcal{E}) \longrightarrow Q_k$ be the projectivization morphism and E be the exceptional set of q. We claim that E is a divisor equivalent to $\xi_{\mathcal{E}} - H$. Indeed, let us take a line $L \subset Q_k$ and restrict the map q to $p^{-1}(L)$. The map q then contracts $\sigma = \mathbb{P}(\mathcal{O}_L^{r-1}) \subset \mathbb{P}(\mathcal{O}(1) \oplus \mathcal{O}^{r-1})$. Thus σ intersects the fibres of p along a linear hyperplane in $\mathbb{P}^{r-1}$, so E is a divisor and $E \equiv \xi_{\mathcal{E}} - mH$ for some $m > 0$. But then E is a section of $\mathcal{O}(\xi_{\mathcal{E}} - mH)$ and thus $\mathcal{E}(-m)$ has a non-zero section which is possible only if $m = 1$. But such a section cannot vanish anywhere. Indeed, if $Z = zero(s)$ were not empty, then on a line L meeting Z at a finite number of points we would have $\mathcal{O}_L(a) \subset \mathcal{O}_L(-1)^{r-1} \oplus \mathcal{O}_L = \xi(-1)$ with $a > 0$, which is not possible. Thus the section does not vanish and $\mathcal{E}$ splits as above.

Lemma 6. If q is of fibre type, then $\dim Y = r-1$, $\dim Y = r$ or $\dim Y = r+1$.

Proof. For any curve C contracted by q we have

$$K_{\mathbb{P}(\mathcal{E})} \cdot C = ((r\xi_{\mathcal{E}} + (k-1)H) \cdot C) = (k-1)H \cdot C \geq k-1 \quad .$$

Therefore, as in section 1, we use the estimate of the extremal locus from [14] to find that the dimension of any fibre does not exceed $k-2$. Obviously, all fibres are of dimension $\leq k$ because no curve can be contracted simultaneously by both q and p. Therefore Y, the "target" of the contraction q, is of dimension $r-1$, r or $r+1$.

Lemma 7. If $\dim Y = r-1$, then $\mathcal{E}$ is trivial.

Proof. Following the proof of lemma 8 in section 1, we may choose from the complete linear system $|\xi_{\mathcal{E}}|$ a base-point free subsystem of dimension $r-1$. Hence $\mathcal{E}$ would be generated by r sections. A rank-r bundle generated by r sections is trivial.

Lemma 8. If $\dim Y = r$, then $\mathcal{E}$ fits in

$$0 \longrightarrow \mathcal{O}(-1) \longrightarrow \mathcal{O}^{r+1} \longrightarrow \mathcal{E} \longrightarrow 0$$

and $r \geq k$.

Proof. Similar to that of lemma 7.

Lemma 9. If $\dim Y = r+1$, then $k = 2, 3$ or 4.

Proof. A general fibre F of q is then of dimension $k-2$ and, by adjunction, the canonical divisor of the fibre is $K_F = -(k-1)H$. By the same arguments as in section 1, i.e., by using Kobayashi-Ochiai's criterion, [5], we see that $F = \mathbb{P}^{k-2}$ and $\mathcal{O}(H)|F = \mathcal{O}(1)$. Thus Q_k contains a linear $(k-2)$-dimensional projective subspace which is possible only when $k \leq 4$.

It remains then to discuss the case of bundles on 3 and 4 dimensional quadrics.

Lemma 10. If $\dim Y = r+1$, then on Q_3 and Q_4 there holds $\mathcal{E} = E \oplus \mathcal{O}^{r-2}$, where E is a spinor bundle.

Proof. As in lemmas 5 and 6, we have an exact sequence

$$0 \longrightarrow K \longrightarrow \mathcal{O}^{r+2} \longrightarrow \mathcal{E} \longrightarrow 0 \quad ,$$

with K a bundle of rank 2 and with $c_1(K) = 1$. We may also assume that $H^0(K) = 0$, since otherwise K splits and we are in the situation of lemma 6. Thus the dual bundle K^* is spanned, not decomposable and uniform with the splitting type $(0,1)$. Then K^* is a spinor bundle, see e.g. [4], 3.1. The lemma now follows by a standard diagram-chasing.

This concludes the proof of theorem 2.

References

[1] Ballico, E., Newstead, P.E.: Uniform Bundles on Quadric Surfaces and some Related Varieties, J. Math. London Soc. II Ser. **31**, 211-223 (1985)

[2] Ballico, E.: Spanned and Ample Vector Bundles with Small Chern Numbers. Pacific J. Math. **140** (2), 209-216 (1989)

[3] Jaczewski, K., Szurek, M., Wiśniewski, J.A.: On the geometry of the Tango bundle. Proc. Conf. Algebraic Geometry Berlin 1985. Teubner-Verlag, Band **92** (1986)

[4] Fritzsche, K.: Linear-Uniforme Vektorraumbündel auf komplexen Quadriken. Ann. Sc. Norm. Super. Pisa (4) **10**, 313-339 (1983)

[5] Kobayashi, Sh., Ochiai, T.: Characterizations of Complex Projective Spaces and Hyperquadrics. J. Math. Kyoto Univ. **13**, 31-47 (1973)

[6] Kollár, J.: Flips, flops and minimal models, To appear.

[7] Okonek, Ch., Schneider, M., Spindler, M.: Vector Bundles on Complex Projective Spaces. Birkhäuser 1981

[8] Ottaviani, G.: Some Extensions of Horrocks Criterion to Vector Bundles on Grassmannians and Quadrics. Annali di Mathematica pura ed applicata (IV), vol. **CLV**, 317-341 (1989)

[9] Peternell, Th., Wiśniewski, J.A.: Ample Vector Bundles with $c_1(X) = c_1(\mathcal{E})$, $\operatorname{rank} \mathcal{E} = \dim X - 1$. To appear.

[10] Szurek, M., Wiśniewski, J.A.: Fano Bundles over $\mathbb{P}^3$ and Q_3, Pacific J. Math. **141**, 197-208 (1990)

[11] Szurek, M., Wiśniewski, J.A.: Fano Bundles on Surfaces. Composition Math. **76**, 295-305 (1990).

[12] Szurek, M., Wiśniewski, J.A.: On Fano Manifolds, which are $\mathbb{P}^k$-bundles over $\mathbb{P}^2$. Nagoya Math. J. **120**, 89-101 (1990).

[13] Wiśniewski, J.A.: Lengths of Extremal Rays and Generalized Adjunction. Math. Z. **200**, 409-427 (1989)

[14] Wiśniewski, J.A.: On Contractions of Extremal Rays of Fano Manifolds. To appear.

Thomas Peternell
Mathematisches Institut
Universität Bayreuth
Postfach 10 12 51
8580 Bayreuth

Michal Szurek
Instytut Matematyki
Uniwersytet Warszawski
Palac Kultury 9 p
00-901 Warszawa

Jaroslaw A. Wiśniewski
Instytut Matematyki
Uniwersytet Warszawski
Palac Kultury 9 p
00-901 Warszawa

On the rank of non-rigid period maps in the weight one and two case

C.A.M. PETERS *,
University of Leiden

1. Introduction

A variation of polarized Hodge structures over a quasi-projective smooth complex manifold S can be thought of as a holomorphic horizontal locally liftable map

$$f : S \to \Gamma\backslash D,$$

where Γ is the monodromy-group of the variation and D is a period domain (see section 2). In this note we find an upperbound on the rank of period maps which admit a non-trivial deformation in the case of weight one and two (our techniques only apply to these weights). See section 3 for precise results. Suffices to say that this upperbound is sharp and the bound can be attained using families of projective varieties.

2. Preparations

In order to make this note as self contained as possible, we shall recall a few relevant facts from [P]. First, we repeat the definition of a polarized Hodge structure of weight w. We start with

- A free $\mathbf{Z}$-module $H := H_{\mathbf{Z}}$ of rank N,
- A Hodge vector $\mathbf{h} = (h^0, h^1, \ldots, h^w) \in \mathbf{N}^{w+1}$ with $h^j = h^{w-j}$, $j = 0, \ldots w$ and $\sum_{j=1}^{w} h^j = N$,
- An integral $\mathbf{Z}$-bilinear form Q, which is $(-1)^w$-symmetric and which has signature $\sum_{j=1}^{w}(-1)^j h^j$.

A Hodge structure of weight w on H with Hodge vector $\mathbf{h}$ is a direct sum decomposition $H_{\mathbf{C}} := H \otimes \mathbf{C} = \oplus_{j=0}^{w} H^{j,w-j}$ with $H^{j,w-j} = \overline{H}^{w-j,j}$ and dim $H^{j,w-j} = h^j$. The form Q polarizes this Hodge structure if

i. $Q(H^{j,w-j}, H^{w-k,k}) = 0$ for $j \neq k$.

ii. $(-1)^{j(j+1)/2}(\sqrt{-1})^{-w} Q(h, \bar{h}) > 0$ if $h \neq 0$.

The Hodge structures on H with Hodge vector $\mathbf{h}$ polarized by Q are parametrized by points of a period domain $D = D(\mathbf{h}, Q)$. It is a homogeneous domain for the action of the Lie group $G_{\mathbf{R}}$ of isometries (with respect to Q) of the vector space $H_{\mathbf{Z}} \otimes \mathbf{R}$. The domain D is open in its compact dual $\check{D}$, a projective variety homogeneous under the

---

* Research partially supported by the Max Planck Institut für Mathematik, Bonn and the University of Utah, Salt Lake City

group G of isometries of Q acting on $H_{\mathbb{C}}$. So, if we fix a reference Hodge structure $F \in D$ and we let B be the isotropy group with respect to G we have the principal fibration

$$B \to G \to \check{D}.$$

The tangent bundle of $\check{D}$ is the associated bundle under the adjoint representation of B on Lie$G$/Lie$B$. To define the horizontal tangent bundle, observe that the choice of the reference Hodge structure F corresponding to the decomposition $\oplus_{i=0}^{w} H^{i,w-i}$ induces a weight zero Hodge structure on

$$\mathfrak{g} = \mathrm{Lie}G$$

by setting

$$\mathfrak{g}^{j,-j} = \{X \in \mathfrak{g} \mid XH^{i,w-i} \subset H^{i+j,w-i-j}\}.$$

The horizontal tangent space $T_F^{\mathrm{hor}}(\check{D})$ is given by Lie$B + \mathfrak{g}^{-1,1}/$Lie$B$. There is an almost canonical identification

$$\iota : \mathfrak{g}^{-1,1} \xrightarrow{\cong} T_F^{\mathrm{hor}}(\check{D})$$

which one gets by sending $X \in \mathfrak{g}^{-1,1}$ to $\exp(tX) \cdot F|_{t=0}$. In the sequel we will always use this identification, e.g when we write down Lie-brackets of tangent vectors.

Suppose that we are given a quasi-projective smooth complex variety S and a representation $\sigma : \pi_1(S) \to G_{\mathbb{Z}}$ whose image, the monodromy group, is denoted by Γ. From it we can form a locally constant system H_S on S with fibres isomorphic to (H, Q). A polarized variation of Hodge structures of type $\mathbf{h}$ over S polarized by Q is given by a so-called period map, i.e. a holomorphic map $f : S \to \Gamma\backslash D$ which comes from a σ-equivariant holomorphic map $\tilde{f}$ from the universal cover $\tilde{S}$ to D which is horizontal i.e whose derivative sends tangents along $\tilde{S}$ to horizontal tangents along D, i.e for any $s \in S$, setting $F = \tilde{f}(s)$ we have:

> Using the identification ι, the subset $(d\tilde{f})T_s(\tilde{S})$ of $T_F^{\mathrm{hor}}(D)$ defines a subspace $\mathfrak{a}$ of $\mathfrak{g}^{-1,1}$.

In fact, it follows [C-T, Proposition 5.2] that $\mathfrak{a}$ is an abelian subspace.

Our next topic is the curvature of the natural $G_{\mathbb{R}}$-invariant metric $\langle\,,\rangle$ on D which on horizontal tangents is given by $\langle X, Y\rangle = -\mathrm{Trace}(XY^*)$. The asterisk means that one takes the transpose conjugate with respect to Q and the natural complex structure on $\mathfrak{g}$. For a proof of the next Lemma, see [P, section 1].

(2.1) **Lemma** *The holomorphic bisectional curvature tensor H at F evaluated on commuting non zero horizontal vectors X and Y of length one is equal to*

$$-\langle [X^*, Y], [X^*, Y]\rangle$$

and hence is non-positive.

Finally, we need to recall some facts related to (small) deformations of period maps.

(2.2) **Definition** *A deformation of a period map $f : S \to \Gamma\backslash D$ consists of a locally liftable horizontal map $\mathbf{f} : S \times T \to \Gamma\backslash D$ extending f in the obvious way.*

Every deformation of a period map f has its associated infinitesimal deformation $\delta \in H^0(f^*T(\Gamma\backslash D))$. Now, since $\mathbf{f}$ is itself horizontal, using [C-T, Proposition 5.2] again, it follows that any two vectors tangent to $S \times T$ at (s,t) map to two commuting (horizontal) tangents in the tangent space to $\Gamma\backslash D$ at $\mathbf{f}(s,t)$. So we can apply the curvature estimates not only to tangents which are images of tangents to S under the period map, but also to those tangents in $\Gamma\backslash D$ which correspond to values of the sections in $f^*T(\Gamma\backslash D)$ which are infinitesimal deformations of f. Indeed these values give certain tangent vectors to $\Gamma\backslash D$. We conclude that for all $s \in S$, $X \in T_s(S)$ and any infinitesimal deformation $\delta \in H^0(S, f^*T(\Gamma\backslash D))$ the holomorphic bisectional curvature $H(X, \delta(u))$ for the induced metric connection on $f^*T(\Gamma\backslash D)$ is non-positive. We now invoke

(2.3) **Lemma** *Suppose U, M are manifolds, $f : U \to M$ a holomorphic map and $\delta \in H^0(U, f^*(M)$. Fix a Riemannian metric g on M, inducing one on $f^*T(M)$ denoted by the same letter. Assume that*
(i) The function $G(u) := g(\delta(u), \delta(u)$ is bounded.
(ii) U does not admit bounded plurisubharmonic functions.
(iii) For all u, $X \in T_u(U)$ the holomorphic bisectional curvature $H(X,Y)$ of the metric connection ∇ for g in the directions X and $Y := \delta(u)$ is non-positive,
then
δ is a flat section and $H(X,Y) = 0$.

This follows immediately from the formula (we normalize so that X and Y have length one):

$$\partial_X \bar{\partial}_{\bar{X}} G|_u = g(\nabla_X(\delta), \nabla_X(\delta))|_u - H(X,Y).$$

From this lemma we can infer that the infinitesimal deformations are flat sections of the bundle $f^*\mathrm{End}T^{\mathrm{hor}}(\Gamma\backslash D)$ (see [P, Theorem 3.2]) and so, upon taking values at F, we get a subspace of $\mathfrak{g}^{-1,1}$. Recalling that $\mathfrak{a}$ corresponds to the full tangent space to S at a point of S, by lemma 2.1 this formula also shows :

(2.4) **Corollary** *The tangent space to deformations of a period map $f : S \to \Gamma\backslash D$ is contained in*

$$\mathfrak{b} := \{Y \in \mathfrak{g}^{-1,1} \mid [\, Y^*, \mathfrak{a}] = 0\}.$$

3. The results

Let us now introduce for any $Y \in \mathfrak{g}^{-1,1}$ the following notation

$\mathfrak{a}(Y) :=$ maximal abelian subspace $\mathfrak{a}'$ of $\mathfrak{g}^{-1,1}$ with $[Y^*, X] = 0 \ \forall X \in \mathfrak{a}'$.
$a(Y) := \dim \mathfrak{a}(Y)$
$a = a(\mathfrak{g}^{-1,1}) := \max a(Y)$ (maximum over $Y \in \mathfrak{g}^{-1,1}$, $Y \neq 0$).

Clearly $a(Y)$ is an upper bound for the rank of a period map which admits non trivial deformations in the direction of Y and so a bounds the rank of period maps deformable in any direction. Consequently, any period map of rank $\geq a+1$ has to be rigid.

In this note we determine the number a as a function of the Hodge numbers, but only for weight one and two. The result can be summarized as follows

(3.1) **Theorem** *(i) In weight one with Hodge vector* (g, g) *one has*

$$a = \frac{1}{2}g(g-1).$$

There exists a quasi-projective variation of rank a which has exactly 1 deformation parameter.
(ii) In weight two with Hodge vector (p, q, p) *we have*

$$a = \begin{cases} 1 & \text{if } p = 1 \\ q-1 & \text{if } p = 2 \\ (p-1)[\frac{1}{2}(q-1)] + \epsilon & \text{if } p \geq 3 \end{cases}$$

where $\epsilon = 1$ *if* q *is even and* $\epsilon = 0$ *if* q *is odd.*
There exists a quasi-projective variation of rank a which has exactly 1 deformation parameter.
(iii) Any period map having rank $\geq a+1$ *is rigid.*

Remark. The variations of rank a can all be constructed from 2-cohomology of projective families of smooth complex algebraic varieties. See the remark at the end of section 6.
Remark. Malcev's technique in principle only gives non-trivial bounds in weights one and two, because with this technique one cannot exploit the fact that the deformation tangent vectors commute with the tangent vectors to the base of the parameter space - after suitable identifications with endomorphisms of H. The method however works also for certain very degenerate sequences of Hodge numbers, e.g. in the even weight case if all Hodge numbers $h^{w-2j-1,2j+1}$ vanish. The result in this case is almost identical; one has to view the number a in the preceding theorem as a function of one, resp. two variables for weight one, resp two and substitute $g = h^{m,m-1}$, resp $p = h^{m-1,m+1}$, $q = h^{m,m}$ if the weight is $2m-1$, resp. $2m$.

Examples

- Any family of g-dimensional polarized abelian varieties having $\frac{1}{2}g(g-1)+1$ or more moduli is rigid.

- Any family of K3-surfaces or Enriques surfaces, whose period map has rank two or more is rigid.

Remark. Sunada in [S] considers holomorphic maps from a compact complex variety to a smooth compact quotient of a bounded symmetric domain by a discrete group. His results are formulated somewhat differently, but it covers the two cases of the preceding theorem, where D is a bounded symmetric domain (but Sunada's techniques need that S be projective and smooth). More recently Noguchi in [N] used techniques from hyperbolic geometry to arrive at the bounds of the two previously given Examples.

4. A variation of Malcev's theorem

In this section we derive our main technical tool, which is a variation of [C-K-T, Theorem 3.1].

(4.1) **Theorem** *Let $\mathfrak{g}$ be a the complexification of a real semi-simple Lie algebra $\mathfrak{g}_{\mathbb{R}}$. Assume that there exists an ordering of the roots relative to some Cartan subalgebra such that complex conjugation maps the root space for a positive root α to the root space of $-\alpha$. Let $\mathfrak{s}$ be a subalgebra of $\mathfrak{g}$ which is a direct sum of positive root spaces and let $\mathfrak{a}$, $\mathfrak{b}$ two abelian subspaces of $\mathfrak{s}$ such that $[\mathfrak{a}, \overline{\mathfrak{b}}] = 0$, where the bar denotes complex conjugation. Then $\mathfrak{s}$ contains two abelian subspaces $\lambda(\mathfrak{a}), \lambda(\mathfrak{b})$ which are direct sums of positive root spaces with $\dim \lambda(\mathfrak{a}) = \dim \mathfrak{a}$, $\dim \lambda(\mathfrak{b}) = \dim \mathfrak{b}$ and such that $[\lambda(\mathfrak{a}), \overline{\lambda(\mathfrak{b})}] = 0$.*

Proof: One has to modify the proof of [C-K-T, Theorem 3.1] slightly. Let $\{\alpha_1, \ldots, \alpha_n\}$ be an ordering of the positive roots. We let X_j be a root vector for the root α_j. We can find a basis $\{A_1, \ldots, A_a\}$, resp. $\{B_1, \ldots, B_b\}$ of $\mathfrak{a}$, resp. $\mathfrak{b}$ such that

$$\begin{aligned} A_j &= X_{k_j} + \text{linear combin. of root vectors for roots} > \alpha_{k_j} \\ & \qquad 1 \le k_1 < \ldots < k_a \le n, \\ B_j &= X_{l_j} + \text{linear combin. of root vectors for roots} < \alpha_{l_j} \\ & \qquad n \ge l_1 > \ldots > l_b \ge 1, \end{aligned}$$

Since $0 = [A_i, A_j] = [X_{k_i}, X_{k_j}] +$ a linear combination of root vectors for roots $> \alpha_{k_i} + \alpha_{k_j}$, it follows that $[X_{k_i}, X_{k_j}] = 0$ and similarly we find that $[X_{l_i}, X_{l_j}] = 0$. Finally, since complex conjugation is assumed to reverse the sign of the roots, we find $[A_i, \overline{B_j}] = [X_{k_i}, \overline{X_{l_j}}] +$ root spaces belonging to roots $> \alpha_{k_i} - \alpha_{l_j}$ we can also conclude that $[X_{k_i}, \overline{X_{l_j}}] = 0$. In this last argument $k_i - l_j$ can become 0 and then the corresponding vector $[X_{k_i}, \overline{X_{l_j}}] = 0$ need not be a root vector, but possibly lies in the Cartan subalgebra. For given A_i and B_j this happens at most once and does not affect the argument. We take now for $\lambda(\mathfrak{a})$, resp. $\lambda(\mathfrak{b})$ the space spanned by the X_{k_i}, resp. X_{l_i}, i.e the space of the *leading root vectors*, resp. the *terminal root vectors*. ∎

We apply Malcev's theorem (Theorem 4.1) to the real Lie algebra $\mathfrak{g}_{\mathbb{R}}$ introduced in section 2. It is shown in [C-K-T, Section 5] that a Cartan subalgebra exists which is of Hodge type $(0,0)$ and that there exists an ordering of the roots such that for each $p>0$, resp $p<0$ the Hodge component $\mathfrak{g}^{p,-p}$ is a direct sum of root vectors of positive roots , resp. negative roots and the complex conjugate of a root vector in $\mathfrak{g}^{p,-p}$ belongs to $\mathfrak{g}^{-p,p}$ so that we can indeed apply Malcev's theorem with $\mathfrak{s}=\mathfrak{g}^{-1,1}$. Now $a(Y)$ is the dimension of the largest abelian subspace $\mathfrak{a}'$ consisting of vectors commuting with $Y^*=-\overline{Y}$. The previous theorem allows us to assume that Y is a root vector and so we obtain:

(4.2) **Corollary** *We have* $a:=\max a(Y)$ *(maximum over root vectors* $Y\in\mathfrak{g}^{-1,1}, Y\neq 0$*).*

5. Bounds for the rank of non-rigid period maps

We recall some conventions from [C-K-T]. If we choose any basis for $H\otimes\mathbb{C}$ we decompose it in blocks according to the Hodge decomposition, where blocks range from $(0,0)$ (left upper corner) to $(w,w), w=1$ or 2 (the lower right corner). A matrix A placed in block (p,q) is denoted by $A[p,q]$. E_{ij} denotes a matrix with 1 in position (i,j) and no other non zero entries. In the course of deriving an upper bound for $a(\mathfrak{g}^{-1,1})$ we repeat the computations from [C-K-T] for a good Hodge frame, a corresponding Cartan subalgebra and root vectors for $\mathfrak{g}^{-1,1}$.
We first compute a in the weight one case.

(5.1) **Lemma** $a=\frac{1}{2}g(g-1)$

Proof: There is a *Hodgeframe* for $H\otimes\mathbb{C}$, i.e. a basis of $H\otimes\mathbb{C}$ consisting of a basis $\{e_1,\dots,e_g\}$ for $H^{0,1}$ and its complex conjugate for $H^{1,0}$ such that the matrix for $\sqrt{-1}Q$ is equal to $I_g[0,1]-I_g[1,0]$. Introduce for $k=1,\dots g$ the diagonal matrices $Y_k:=E_{kk}[0,0]-E_{kk}[1,1]$. These form a basis of the Cartan subalgebra of $\mathfrak{g}$. The root vectors spanning $\mathfrak{g}^{-1,1}$ are the $\frac{1}{2}g(g+1)$ symmetric matrices $Y_{ij}=E_{ij}[1,0]+E_{ji}[1,0]$ since

$$[Y_{ij},Y_k]=(\delta_{ik}+\delta_{jk})Y_{ij}.$$

Now for every symmetric $g\times g$-matrix X the condition $[\,X[1,0],Y_{ij}]=0$ is equivalent to X having zero i-th row (and column) and zero j-th row (and column). If $i=j$ we find $\frac{1}{2}g(g-1)$ for the maximal dimension of spaces of such X. ∎

Now we treat the case of weight two. A Hodge frame, in this case consists of a basis for $H^{2,0}$, its conjugate for $H^{0,2}$ and a real basis for $H^{1,1}$ such that the matrix for Q has the form $-I_p[0,2]+I_q[1,1]-I_p[2,0]$. For our purposes however it is better to use a different frame. Starting from such a Hodge frame we modify the middle part, say $\{f_1,\dots,f_q\}$ as follows. If $q=2t$, we take $\{f_1+\sqrt{-1}f_{t+1},\dots,f_t+\sqrt{-1}f_{2t},\ f_1-\sqrt{-1}f_{t+1},\dots,f_t-\sqrt{-1}f_{2t}\}$. In this case $Q=-I_p[0,2]+M[1,1]-I_p[2,0]$, where $M=\begin{pmatrix}0_t & I_t\\ I_t & 0_t\end{pmatrix}$.
If $q=2t+1$ we do essentially the same except that we retain one real basis vector for

$H^{1,1}$ and it take as our last basis vector for $H^{1,1}$. This modifies M in the preceding formula slightly ; it becomes $M = \begin{pmatrix} 0_t & I_t & 0 \\ I_t & 0 & 0 \\ 0 & 0 & 1 \end{pmatrix}$.

We have

(5.2) **Lemma** *If $q = 2t$ we have*

$$a = \begin{cases} 1 & \text{if } p = 1 \\ q-1 & \text{if } p = 2 \\ (p-1)(t-1)+1 & \text{if } p \geq 3 \text{ and } q \geq 2 \end{cases}$$

Proof: First observe that in case $p = 2$ the bound from [C-K-T] gives the result by subtracting off one from their bound, allowing for the extra deformation parameter. The other less trivial bounds are obtained as follows.

The diagonal matrices

$$Y_k(0) = E_{kk}[0,0] + E_{kk}[2,2], \ k = 1, \dots, p$$

and

$$Y_k(1) = (E_k - E_{t+k})[1,1], \ k = 1, \dots, t$$

give a basis for the Cartan subalgebra and the matrices

$$Y_{ij} = E_{ij}[1,0] + E_{ji}[2,1]M, \ i = 1, \dots, p, \ j = 1, \dots, q$$

give a basis for the root vectors in $\mathfrak{g}^{-1,1}$ since

$$[Y_k(j), Y_{ij}] = (\delta_{kj} - \delta_{t+j\,k})Y_{ij}.$$

The complex conjugate of Y_{ij} is equal to $\overline{Y}_{ij} = E_{ji}M[0,1] + E_{ij}[1,2]$ and if we have $X' = X[1,0] + X^T M[2,1] \in \mathfrak{g}^{-1,1}$, the condition that $[X', \overline{Y}_{ij}] = 0$ means that X has zeros in rows $i, t+i$ and column j except in the entry (i,j). In other words, the problem reduces to the abelian subspace problem for Hodge numbers $p-1$, $2t-1$, $p-1$, and the main theorem of [C-K-T] tells us this maximum is $(p-1)(t-1)+1$ if $p \geq 3$ and it is of course zero if $p = 1$. Taking into account the possibly non-zero entry (i,j) yields the desired upper bound. ∎

(5.3) **Lemma** *If $q = 2t+1$ we have*

$$a = \begin{cases} 0 & \text{if } q = 1 \\ 1 & \text{if } p = 1 \\ q-1 & \text{if } p = 2 \\ (p-1)t & \text{if } p \geq 3 \text{ and } q \geq 3 \end{cases}$$

Proof: The only change with the previous case is that there is an extra element $Y_q = E_{q,q}[1,1]$ in the Cartan subalgebra which leads to extra root vectors

$$Y_{qi} = E_{qi}[1,0] + E_{iq}M[2,1],$$

as one can easily check. The new root vectors however do not change any of the computations we did in the case where q is even. ∎

6. Construction of non-rigid period maps of maximal rank

We introduce some basic variations.

1) A weight one variation.
We have the tautological variation $\mathbb{A}_g$ of weight one over $\mathfrak{h}_g$. If we take a torsion free subgroup Γ of finite index in $Sp_g\mathbb{Z}$ not containing -Id, this variation descends to a variation on $\Gamma\backslash\mathfrak{h}_g$, which quasi-projective by [B-B]. This variation we denote by $\overline{\mathbb{A}}_g$.

2) A variation of weight 2 with $p = 1$.
Let H be a lattice with form Q of signature $(2,q)$ and consider the tautological variation $\mathbb{B}_q$ of weight two over

$$B_q := \{[F] \in \mathbb{P}(H \otimes \mathbb{C}) \mid Q(F,F) = 0, Q(F,\overline{F}) > 0\}.$$

As in 1) this variation descends to a variation $\overline{\mathbb{B}}_q$ over a suitable quasi-projective smooth quotient of B_q.

3) A variation of weight 2 with Hodge numbers $\{p, 2q, p\}$.
Over

$$B_{p,q} = \{Z \in \mathbb{C}^{p,q} \mid \overline{Z}^T Z < I_{q,q}\}$$

there exists a variation of weight 2 and this also descends to a variation $\overline{\mathbb{B}}_{p,q}$ over a suitable quasi-projective quotient (see [C-K-T, Section 7]).

The construction of a variation realizing the bound in Lemma 5.1 is easy. One takes the variation $\overline{\mathbb{A}}_{g-1}$ and takes the direct sum with a constant Hodge structure with Hodge numbers $h^{0,1} = h^{1,0} = 1$. This actually has 1 deformation parameter (compare with the variation $\mathbb{A}_1$).

In case of weight two we use the following remark repeatedly. The tensorproduct of $\overline{\mathbb{A}}_1$ with a fixed weight one Hodge structure with Hodge vector $(1,1)$ gives a weight two variation with Hodge numbers $(1,2,1)$ over a smooth quasi-projective curve and his has 1 deformation parameter. Let us denote this variation with $\overline{\mathbb{B}}'$. The construction for the bound in Lemma 5.2 proceeds as follows. For $p = 1$ and $q \geq 2$ we take the variation $\overline{\mathbb{B}}'$ and take the direct sum with $q-2$ copies of the trivial Hodge structure of pure type $(1,1)$. For $p = 2$ and $q \geq 2$ we take $\overline{\mathbb{B}}_{q-2}$ which has a parameter space of dimension $q-2$

and Hodge numbers $\{1, q-2, 1\}$. Now take the direct sum with $\overline{\mathbb{B}}'$. In total we have a base of dimension $q-1$ and 1 deformation parameter. If $p \geq 3$ a similar construction applies: instead of $\overline{\mathbb{B}}_{q-2}$ one takes $\overline{\mathbb{B}}_{p-1,t-1}$ and then proceeds as before.
In case of odd q (Lemma 5.3) the constructions are similar. The last construction needs a modification: one starts with $\overline{\mathbb{B}}_{p-1,t}$ and takes the direct sum with a constant Hodge structure with Hodge numbers $\{1,1,1\}$. If we view it as a fibre of the 1-parameter variation $\mathbb{B}_1$ it is clear that also here we have an extra deformation parameter.

Remark All of these variations occur as variations on primitive 2-cohomology of projective families of smooth algebraic varieties. For the weight one variation this is trivial, and for $\mathbb{B}_q$ one can take families of K3-surfaces with Picard number q for $q \leq 19$ and products of these for higher values of q. For $\overline{\mathbb{B}}_{p,q}$ one can realize them using a generalized Prym construction [C-S].

Bibliography

[B-B] W.L. Baily and A. Borel: Compactifications of arithmetic quotients of bounded symmetric domains, Ann. of Math. (2) **84** (1966), 442-528.
[C-T] J. A. Carlson, D. Toledo: Integral manifolds, harmonic mappings, and the abelian subspace problem, in: Springer Lect. Notes in Math. **1352**, 1989.
[C-K-T] J. A. Carlson, A. Kasparian, D. Toledo: Variations of Hodge structure of maximal dimension, Duke Math. J. **58** (1989) 669–694.
[C-S] J.A. Carlson, C. Simpson: Shimura varieties of weight two Hodge structures in *Hodge theory*, Springer Lecture Notes in Mathematics **1246** (1987) 1-15, Springer Verlag, Berlin etc.
[N] J. Noguchi: Moduli spaces of holomorphic mappings into hyperbolically embedded complex spaces and locally symmetric spaces, Invent. Math. **93** (1988) 15-34.
[P] C. A. M. Peters: Rigidity for variations of Hodge structure and Arakelov-type finiteness theorems, Comp. Math. **75**,(1990) 113–126.
[S] T. Sunada: Holomorphic mappings into a compact quotient of symmetric bounded domain. Nagoya Math. J. **64** (1976) 159-175.

Dept. of Mathematics
Univ. of Leiden, Postbus 9512, 2300 AL Leiden Netherlands

revised April 7, 1991

The geometry of the special components of moduli space of vector bundles over algebraic surfaces of general type.

A. N. Tyurin
Steklov Institute of Mathematics
ul. Vavilova 42, Moscow, GSP–1, 117966, USSR.

The essence of this article is a complementary to my talk at the Conference "Complex Algebraic Varieties" 2-6 april 1990 in Bayreuth. I would like to express my thanks to K. Hulek, T. Peternell, M. Schneider and F.-O. Schreyer for the invitation and personally Frank-Olaf Schreyer for help and fruitful discussions.

My aim is to describe the birational type and geometry of several important components of moduli space of vector bundles over regular algebraic surfaces of general type. The investigation of their geometry proved to be useful in Donaldson theory of smooth structures (see [D]). The algebraic geometric approach to these problems very much reminds the Brill-Noether theory of special linear series on a curve. At any case it is the next field of the applications of The Strong Connectedness Theorem of Fulton and Lazarsfeld.

Now, S denotes a smooth regular surface over $\mathbf{C}$ and H a polarization of S. First of all we discuss the discrete invariants associated with each vector bundle E on S or, more generally, with each torsion free sheaf F.

Consider the $\mathbb{Z}$-module

$$\begin{array}{ccccc} \mathbb{Z} & \oplus & \operatorname{Pic} S & \oplus & \mathbb{Z} \\ \cup\!\!\!\!\!\smile & & \cup\!\!\!\!\!\smile & & \cup\!\!\!\!\!\smile \\ (\;r & , & c & , & \chi\;) \end{array} \tag{1}$$

with a symmetric bilinear form $\langle\;,\;\rangle$:

$$\langle (r,c,\chi),(r',c',\chi')\rangle = rr'\left[\frac{c}{r}\frac{c'}{r'} - \frac{\chi}{r} - \frac{\chi'}{r'} - \left(\frac{c}{r}+\frac{c'}{r'}\right)\frac{K_S}{2} + \chi(\mathcal{O}_S)\right] \tag{2}$$

and with a skew-symmetrical form $\langle\!\langle\;,\;\rangle\!\rangle$:

$$\langle\!\langle (r,c,\chi),(r',c',\chi')\rangle\!\rangle = \frac{K_S}{2} rr'\left(\frac{c}{r} - \frac{c'}{r'}\right) \tag{3}$$

where $K_S \in \operatorname{Pic} S$ is the canonical class of S and $c \cdot K_S$ the intersection of K_S with $c \in \operatorname{Pic} S$.

For a sheaf F the vector

$$v(F) = (\operatorname{rk} F, c_1(F), \chi(F)) \tag{4}$$

is called the *Mukai vector* of F. It is a discrete invariant of F.

By the Riemann-Roch theorem it is easy to see that for two sheaves F, F'

$$\langle v(F), v(F')\rangle + \langle\langle v(F), v(F')\rangle\rangle = \sum_{i=0}^{2}(-1)^{i+1}\operatorname{rk}\operatorname{Ext}^i(F, F') \tag{5}$$

The first problem of the classification of vector bundles is the following **Realization Problem:** Describe the set

$$\operatorname{St}_H(S) = \{v(E) \mid E \ E \text{ is } H\text{-stable}\} \tag{6}$$

Why is this problem so important?

Examples—motivation.
1. Miyaoka-trick. Consider the vector

$$M(S) = (2, K_S, 0) \tag{7}$$

in the plane

$$\mathbb{Z} \oplus \mathbb{Z} \cdot K_S \oplus \{0\} \subset \mathbb{Z} \oplus \operatorname{Pic} S \oplus \mathbb{Z} \tag{8}$$

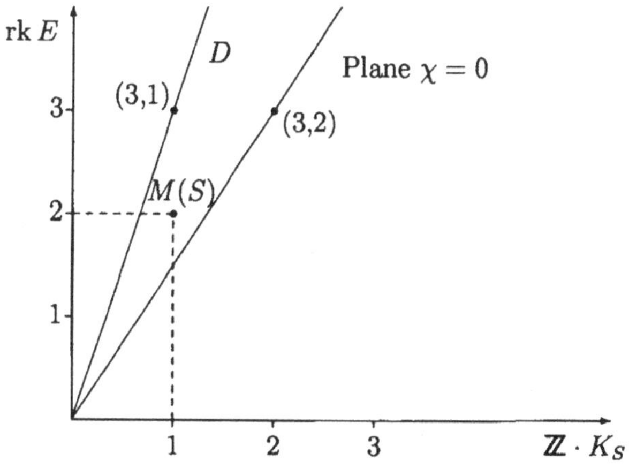

Fig.1

Then it is easy to see that

$$M(S) \in \operatorname{St}_H(S) \Rightarrow K_S^2 \leq 2c_2(S) \tag{9}$$

Indeed the Bogomolov inequality and the Riemann-Roch theorem

$$c_1(E)^2 \leq 4c_2(E)$$

$$c_2(E) = \frac{1}{2}(c_1(E)^2 - c_1(E)\cdot K_S) + \frac{1}{6}(K_S^2 + c_2(S)) - \chi(E)$$

give the right side of (9).

2. Similarly, if the intersection

$$D \cap \mathrm{St}_H(S) \neq \emptyset$$

where $D = \{(r, d \cdot K_S, 0) \mid r > 0, \frac{1}{3} \leq \frac{d}{s} \leq \frac{2}{3}\}$ (see Fig. 1) then $K_S^2 \leq 3c_2(S)$ for the same reasons.

3. The assertion

$$N \gg 0 \Rightarrow (2, 0, -N) \in \mathrm{St}_H(S) \tag{10}$$

is the crucial step in proving the non-vanishing of the Donaldson polynomials of algebraic surfaces.

Remark. Although the Realization Problem is difficult, we have

Schwarzenberger's theorem ([S]). For every $v = (r, c, \chi)$ with $r \geq 2$ there exists a vector bundle E on S such that $v(E) = v$.

(But this vector bundle E is not simple, and thus it is not stable).

Now, the Realization Problem is solved for $S = \mathbb{P}^2$ by Drezet and Le Potier (see [D-L]), for $S = \mathbb{P}^1 \times \mathbb{P}^1$ by Rudakov (see [R]) and in important particular cases for K3 surfaces by Mukai (see [Mu]). (Here I mean the complete description of $\mathrm{St}_H(S)$ but one has a lot of the important cases of the realization).

Definition 1. A sheaf F on S is called

1) *simple* if $h^\circ(\mathrm{End}\, F) = 1$

2) *regular* if $h^\circ(\mathrm{End}\, F \otimes K_S) = p_g$

The set $\mathrm{St}_H(S)$ (6) has ths subset

$$\mathrm{St}_H^r(S) = \{v(E) \mid E \text{ is stable and regular}\} \tag{11}$$

Definition 2. A torsion-free H-stable sheaf F has a *fine moduli* if there exists a complete irreducible scheme M, a smooth point $[F] \in M$ and a flat family F on $S \times M$ such that

a) $F = \mathcal{F}|_{S \times [F]}$

b) the Kodaira-Sprencer homomorphism

$$TM_{[F]} \longrightarrow \mathrm{Ext}(F, F)$$

is an isomorphism.

c) for any two points $m_1 \neq m_2 \in M$

$$\mathcal{F}|_{S \times m_1} \neq \mathcal{F}|_{S \times m_2}$$

The irreducible algebraic variety $M = M(F)$ is called the *moduli space* of F and the flat family $\mathcal{F}$ is called the *universal family* of F. The universal family $\mathcal{F}$ is defined uniquely up to $\otimes p_M^*(L)$, where $L \in \mathrm{Pic}\, M$, p_M is the projection from $S \times M$ to M.

Now one has only one criterion for F to have a fine moduli in terms of the cohomology of the sheaf itself and its Mukai vector.

The Mukai-Maruyama criterion. A torsion free stable sheaf F has a fine moduli if

1) F is regular

2) g.c.d.$(\operatorname{rk} F, \{c_1(F) \cdot \operatorname{Pic} S\}, \chi(F)) = 1$ (see [M], [Mu]).

Remark. We can replace the usual number $c_1(F) \cdot H$ to g.c.d.$\{c_1(F) \cdot \operatorname{Pic} S\}$ by the fact that for every $D \in \operatorname{Pic} S$ the exists the number n_0 such that

1) $n \geq n_0 \Rightarrow nH + D$ is a polarization

2) F is H-stable $\Rightarrow F$ is $(nH + D)$-stable.

It is easy to see that if a *regular* stable F has a fine moduli and $M = M(F)$ is a moduli space of F then

$$\dim M(F) = \|v(F)\|^2 + p_g + 1 \tag{12}$$

where $\|v(F)\| = \langle v(F), v(F) \rangle$ (see (2)).

Now we will describe the process of the enlargement of the moduli spaces of stable vector bundles.

The construction of the 1-enlargement. Let E be a stable vector bundle on S with Mukai-vector $v(F) = (r, c, \chi)$, $r \geq 2$.

Let $p \in S$ be a point and $\mathcal{O}_p$ is the structural sheaf of p.

Consider some epimorphism $\varphi \colon E \to \mathcal{O}_p$ and the exact sequence of sheaves:

$$0 \longrightarrow F_1 \longrightarrow E \xrightarrow{\varphi} \mathcal{O}_p \longrightarrow 0 \tag{13}$$

Then F_1 is a torsion free and stable sheaf with

$$v(F) = (r, c, \chi - 1) \tag{14}$$

It is easy to see that we can take p and φ such that for every $L \in \operatorname{Pic} S$

$$\operatorname{rk} \operatorname{Hom}(E, E \otimes L) = k > 0 \Rightarrow \operatorname{rk} \operatorname{Hom}(F_1, F_1 \otimes L) < k \tag{15}$$

(compare with (2.58) of [T_2]).

Hence $\{E$ is regular$\} \Rightarrow \{F_1$ is regular$\}$.

Consider the moduli space $M = M(E)$ of E and the moduli space $M_1 = M_1(F)$ of F_1 (may be non-fine).

Proposition 1. If E is regular then the generic point $[E_1] \in M(F_1)$ is a stable vector bundle.

This follows from Lemma 6.2 of [A].

Definition 3. 1) Any stable vector bundle E_1 from $M(F_1)$ is called an 1-enlargement of E.

2) The moduli space $M_1 = M(F_1) = M(E_1)$ is called 1-enlargement of the moduli space $M = M(E)$.

The iteration of 1-enlargement gives to us N-enlargement M_N. From this we have

$$v = (r, c, \chi) \in \operatorname{St}^r_H(S) \Rightarrow \forall N \in \mathbb{Z}^+ \quad v' = (r, c, \chi - N) \in \operatorname{St}^r_H(S) \tag{16}$$

Similarly, if

$$\mathrm{St}_H^k(S) = \{v(E) \mid E \text{ is stable, } h^\circ(\operatorname{ad} E \otimes K_S) \geq k\}$$

then

$$v = (r, c, \chi) \in \mathrm{St}_H^k(S) \Rightarrow \forall N - k \in \mathbb{Z}^+ \quad v' = (r, c, \chi - N) \in \mathrm{St}_H^r(S)$$

Hence describing $\mathrm{St}_H(S)$ up to a finite set is describing the slopes $\frac{c}{r}$ of (r, c, χ) (by the Bogomolov inequality).

Now we can confirm the assertion: if a surface S admits some geometrical structure then a fine moduli space M admits this structure, too. To make this assertion rigorous one needs to refine what is a structure.

Let us begin with the Hodge-structure of S and M that is with the $(2,0)$-forms.

Definition 4. 1) Any skew-symmetric homomorphism

$$\omega: TB \to T^*B, \quad \omega^* = -\omega \tag{17}$$

on the tangent bundle of a smooth algebraic variety B into the cotangent bundle is called an algebraic symplectic structure on B.
2) If B' is singular then by an algebraic symplectic structure we understand the rule by which each morphism $B \xrightarrow{f} B'$ of any smooth B induced (17) on B and for different morphisms these structures correlate well. (Mumford trick, see [Mum].)

If B is complete then ω is any holomorphic $(2,0)$-form.

Now let M be a fine moduli space and $\mathcal{F}$ on $S \times M$ be a universal family. Consider the $(2,2)$-Künneth-component of $c_2(\mathcal{F})$

$$c_2^{(2,2)}(\mathcal{F}) \in H^2(S, \mathbb{Z}) \otimes H^2(M, \mathbb{Z}) \tag{18}$$

as the homomorphism

$$c_2^{(2,2)}(\mathcal{F}): H^2(S, \mathbb{Z}) \to H^2(M, \mathbb{Z}) \tag{19}$$

where $H^2(S, \mathbb{Z}) = H^2(S, \mathbb{Z})^* = H_2(S, \mathbb{Z})$ by Poincaré-duality. The cycle $c_2(\mathcal{F})$ is algebraic hence the Künneth component (18) of it has the Hodge-type $(2,2)$.

Thus the homomorphism (19) preserves the Hodge-structure on S and defines the homomorphism

$$\mu = \left[c_{(2,2)}(\mathcal{F}) \otimes \mathbb{C}\right]^{(2,0)} : H^{2,0}(S) \to H^{2,0}(M) \tag{20}$$

Hence each symplectic structure on S

$$\omega: TS \to T^*S, \quad \omega \in H^{2,0}(S)$$

gives the structure

$$\mu(\omega): TB \to T^*B \tag{21}$$

on M.

It is easy to see that

$$\left[c_{(2,2)}(\mathcal{F}) \otimes \mathbb{C}\right]^{(2,0)} = \left[c_{(2,2)}(\mathcal{F} \otimes p_M^* L)\right]^{(2,0)}$$

Hence μ (20) does not depend on the universal family $\mathcal{F}$.

Now we give the description of the homomorphism (17) in any point $[E] \in M$. Let

$$\operatorname{End} E = \mathcal{O}_S \oplus \operatorname{ad} E \tag{22}$$

Then the fibre of TM over $[E]$

$$TM_{[E]} = H^1(\operatorname{ad} E), \quad T^*M_{[E]} = H^1(\operatorname{ad} E \otimes K_S) \tag{23}$$

Let interpret ω as a section s_ω of the canonical line bundle K_S. Let C be 1-dimensional zero-set of s_ω. Then one has the exact sequence

$$0 \longrightarrow \mathcal{O}_S \xrightarrow{s_\omega} K_S \longrightarrow K_S|_C \longrightarrow 0 \tag{24}$$

By the adjunction formula $K_S|_C = \theta$ is a theta-characteristic on C.

$$\theta^2 = K_S^2|_C = \omega_C \tag{25}$$

From (24)one has

$$h^\circ(\theta) = p_g - 1 \tag{26}$$

If E is regular then the part of the long cohomological sequence of the sequence

$$0 \longrightarrow \operatorname{ad} E \longrightarrow \operatorname{ad} E \otimes K_S \longrightarrow \operatorname{ad} E \otimes \theta \longrightarrow 0 \tag{27}$$

obtained by tensoring sequence (24) with $\operatorname{ad} E$ gives

$$\begin{array}{ccccccccc} & & TM_{[E]} & & T^*M_{[E]} & & & & \\ & & \| & & \| & & & & \\ 0 \longrightarrow H^\circ(\operatorname{ad} E|_C \otimes \theta) & \longrightarrow & H^1(\operatorname{ad} E) & \longrightarrow & H^1(\operatorname{ad} E \otimes K_S) & \longrightarrow & H^1(\operatorname{ad} E|_C \otimes \theta) & \longrightarrow & 0 \end{array} \tag{28}$$

In the following assertions the symbol "XX-enlargement" means "N-enlargement with $N \gg 0$".

Proposition 2. For each component M of the moduli space of stable vector bundles on S there exists XX-enlargement M_N such that M_N is regular and fine.

This follows from (15) with $L = K_S$ and the condition 2) of the Mukai-Maruyama criterion.

Proposition 3. For each component M of the moduli space and $\omega \in H^\circ(K_S)$ there exists XX-enlargement M_N such that

$$\operatorname{corank} \mu(\omega) = \begin{cases} 0 \text{ if } p_g \text{ is odd,} \\ 1 \text{ if } p_g \text{ is even.} \end{cases}$$

This follows from geometrical properties of the moduli space of vector bundles on one-dimensional subscheme $C \subset S$ (2.14). Let θ be a theta-characteristic on C. Then for each family of vector bundles $\mathcal{E}$ on $C \times B$ such that for every $b \in B$

$$h^\circ(\operatorname{ad} \mathcal{E}|_{C \times b} \otimes \theta) > \begin{cases} 0 \text{ if } \theta \text{ is even} \\ 1 \text{ if } \theta \text{ is odd} \end{cases} \tag{29}$$

the Kodaira-Spencer homomorphism

$$TB_b \xrightarrow{k} H^1(\operatorname{ad}\mathcal{E}|_{C\times b}) \tag{30}$$

is not surjective.

Now, by (15) with $L = K_S^2$ there exists XX-enlargement M_N such that for generic $[E_N] \in M_N$

$$H^\circ(\operatorname{ad} E_N \otimes K_S^2)^* = H^2(\operatorname{ad} E_N \otimes K_S^*) = 0 \tag{31}$$

Then from the exact sequence

$$0 \longrightarrow \operatorname{ad} E_N \otimes K_S^* \longrightarrow \operatorname{ad} E_N \longrightarrow \operatorname{ad} E_N|_C \longrightarrow 0 \tag{32}$$

one has

$$H^1(\operatorname{ad} E_N) \longrightarrow H^1(\operatorname{ad} E_N|_C) \longrightarrow H^2(\operatorname{ad} E \otimes K_S^*) = 0 \tag{33}$$

If in the family $\{E_N|_C\}$, $[E_N] \in M_N$ of vector bundles on C we have (29) for every $E_N|_C$ then the epimorphism (33) contradicts (30) and we are done.

Definition 5. A component M of the moduli space of the stable bundle on S is called 1-special if for the generic vector bundle E of M one has

1) $h^1(E) = h^1(E \otimes K_S) = h^2(E) = h^2(E \otimes K_S) = 0$

2) $\chi(E) = h^\circ(E) = \operatorname{rk} E - 1$

3) The canonical homomorphism

$$H^\circ(E) \otimes \mathcal{O}_S \xrightarrow{\text{can}} E \tag{34}$$

can be extended to the exact secuence

$$0 \longrightarrow H^\circ(E) \otimes \mathcal{O}_S \longrightarrow E \longrightarrow J_\xi \otimes \det E \longrightarrow 0 \tag{35}$$

or equivalent, $E \otimes K_S$ has unique K-block

$$0 \longrightarrow H^\circ(E) \otimes K_S \longrightarrow E \otimes K_S \longrightarrow J_\xi \otimes \det E \otimes K_S \longrightarrow 0 \tag{36}$$

(see [T$_3$]).

Let M° be the Zariski-open set of M containing F with all of these conditions. Then one has the embedding

$$M^\circ \overset{\varphi}{\hookrightarrow} \widetilde{S}^{(c_2(E))}, \quad \varphi(E) = \xi \text{ (see (35))} \tag{37}$$

where $\widetilde{S}^{(d)}$ is the Hilbertscheme of 0-dimensional subscheme of S length d.

Indeed, the extension (35) is given by the cocycle

$$\begin{aligned} e \in H^\circ(E) \otimes \operatorname{Ext}^1(J_\xi \otimes \det E, \mathcal{O}_S) &= H^1(J_\xi \otimes \det E \otimes K_S)^* \otimes H^\circ(E) \\ &= \operatorname{Hom}(H^1(J_\xi \otimes \det E \otimes K_S), H^\circ(E)) \end{aligned}$$

which is the coboundary homomorphism δ for the sequence (36):

$$\begin{array}{ccccccc} H^1(E\otimes K_S) & \longrightarrow & H^1(J_\xi\otimes E\otimes K_S) & \xrightarrow{\delta} & H^\circ(E)\otimes H^2(K_S) & \longrightarrow & H^2(E\otimes K_S) \\ \| & & & & \| & & \| \\ 0 & & & & H^\circ(E) & & 0 \end{array}$$

(see [T₃] for the details).

Let $\overline{M} \subset \widetilde{S}^{(c_2)}(S)$ is the closure of $\varphi(M^\circ)$ in the smooth complete variety $\widetilde{S}^{(c_2)}$.

Definition 6. The variety $\overline{M}$ is called the *birational approximation* of M.

Remark. As an algebraic cycle of $\widetilde{S}^{(c_2)}$, $\overline{M}$ can be interpreted as Segre-class of some Grassmannization of the standard vector bundle $\mathcal{E}$ on $\widetilde{S}^{(c_2)}$:

$$\xi \in \widetilde{S}^{(d)}, \quad \mathcal{E}|_\xi = H^\circ(\mathcal{O}_\xi \otimes \det E \otimes K_S) \tag{38}$$

Example. ([D]) Let $\operatorname{rk} E = 2$ then from (36)

$$\begin{aligned} h^1(J_\xi \otimes \det E \otimes K_S) &= 1 \\ h^\circ(J_\xi \otimes \det E \otimes K_S) &= \chi(E\otimes K_S) - p_g \end{aligned} \tag{39}$$

Consider the homomorphism:

$$\begin{aligned} &H^\circ(\det E \otimes K_S)\otimes \mathcal{O}_{\widetilde{S}^{(d)}} \xrightarrow{\text{res}} \mathcal{E} \\ &\text{res}|_\xi = H^\circ(\det E\otimes K_S) \xrightarrow{\text{res}} H^\circ(\mathcal{O}_\xi \otimes \det E\otimes K_S) \end{aligned} \tag{40}$$

Then,

$$\overline{M} = \text{support coker res} \tag{41}$$

Now, by Riemann-Roch theorem ($c = c_1(\det E)$)

$$h^\circ(\det E\otimes K_S) = \frac{c_1(c_1+K)}{2} + p_g + 1$$

$$d = c_2(E) = \frac{c_1(c_1-K)}{2} + 2p_g + 1$$

From this we have

$$\operatorname{codim} \overline{M} = c_1 \cdot K - p_g + 1 \tag{42}$$

in $\widetilde{S}^{(d)}$.

Corollary. Let S be a K3-surface. Then the 1-special M is birational equivalent to $\widetilde{S}^{(d)}$.

It is not hard to verify using the Strong Connectedness Theorem for degeneracy loci of Fulton and Lazarsfeld that the union of all 1-special components is connected.

Now we can prove

Proposition 4. For each component M of the moduli space of stable bundles on a regular surface S there exists an XX-enlargement M_N which is 1-special.

Torsing by H we can suppose that the condition 1) is valid and E is generated by the sections. Consider the process of 1-enlargement (13). It is easy to see that we can take a point $p \in S$ and φ such that

$$h^{\circ}(F_1) = h^{\circ}(E) - 1$$
$$h^1(F_1) = h^2(F_1) = h^1(F_1 \otimes K_S) = h^2(F_1 \otimes K_S) = 0 \tag{43}$$

Hence after $N = h^{\circ}(E) - \operatorname{rk} E + 1$ such steps we will get the conditions 2) and 3) of Definition 5.

Now consider fine 1-special M and the birational approximation $\overline{M}$ of M. For each symplectic structure ω on S one has the symplectic structure $\mu(\omega)$ on M (21) and $\omega^{(d)}$ on $S^{(d)}(S)$. It is easy to see that

$$\mu(\omega) = \varphi^* \omega^{(d)}|_{\overline{M}} \tag{44}$$

that is the birational isomorphism

$$M \leftrightarrow \overline{M} \tag{45}$$

is $\mu(\omega)$-symplectic for each $\omega \in H^{2,0}(S)$.

Now, let, as usual

$$CH^2(S) = \frac{\text{the free abelian group of points on } S}{\text{cycles rationally equivalent to zero}}$$

be the Chow group of cycles and $CH^2_d(S)$ be the cycles of degree d. For each component M of the moduli space of stable bundles we have the map

$$\begin{array}{c} M \xrightarrow{c} CH^2_{c_2}(S) \\ c(E) = c_2(E) \end{array} \tag{46}$$

and for 1-special M this map is the composition:

$$\begin{array}{ccc} M & \longrightarrow & \overline{M} \subset \tilde{S}^{(d)} \\ & {}^{c}\searrow & \downarrow p \\ & & CH^2_d(S) \end{array} \tag{47}$$

Conjecture A (An analogue Abel theorem). If for a regular surface S the canonical map $\varphi \colon S \to \mathbb{P}^{p_g - 1}$ is an embedding then for each component M of the moduli space of stable bundles there exists an XX-enlargement for which the map c (46) is of degree one.

To make this assertion rigorous we have to use the analogue of Mumford's trick from [Mum]: "$\deg c = 1$" is equivalent to "if we have a decomposition of c

$$\begin{array}{ccc} M & \xrightarrow{c} & CH^2(S) \\ {}_{f}\searrow & & \swarrow {}_{c'} \\ & V & \end{array} \tag{48}$$

where V is a finite dimensional algebraic variety and f is an algebraic morphism, then f is finite and $\deg f = 1$".

At the moment we can only prove

Theorem. If the canonical map $\varphi: S \to \mathbb{P}^{p_g - 1}$ of a regular surface S is an embedding then for every component M of the moduli space of stable bundles there exists an XX-enlargement for which the map c from (46) is finite. (See §3, ch. III, of [T₁]).

Now, we can prepare the set of the candidates of the birational approximations of the 1-special moduli spaces beforehand: for each $L \in \operatorname{Pic} S$ and $d > 0$ consider the homomorphism

$$H^\circ(L \otimes K_S) \otimes \mathcal{O}_{\widetilde{S}^{(d)}} \xrightarrow{\text{res}} \mathcal{E} \tag{49}$$

where $\mathcal{E}$ is the vector bundle (38) with $\det E = L$ and the subscheme

$$M_0(r, L, d) = \{\xi \in \widetilde{S}^{(d)} \mid \operatorname{rk} \operatorname{coker} \operatorname{res}_\xi = r - 1\} \tag{50}$$

By [T₃] one has the family of the extensions (36) with $M_0(r, L, d)$ as a base. It is almost a fine moduli space. The question is: is a generic extension (36) H-stable?

References

[A] V. Artamkin, Izvestiya AN SSSR v. 54, N3, 1990, 435–468.

[D] S. K. Donaldson "Instantons in Yang-Mills theory", "Interface of Mathematics and Particle Physics", Oxford U. P. 1990.

[D-L] J.-M. Drezet and J. Le Potier "Fibrés stables et fibrés exceptionnels sur $\mathbb{P}^2$", Ann. Ec. Norm. Sup. 1985, t. 18, p. 193–244.

[M] M. Maruyama "Moduli of stable sheaves, II", J. Math. Kyoto Univ. 18–3 (1978), 557–614.

[Mu] S. Mukai "On the moduli space of bundles on K3 surface *I*", Tata Inst. Fund. Res. Studies v. 11 in Math. 1987.

[Mum] D. Mumford "Rational equivalence of 0-cycles on surfaces". J. Math. Kyoto Univ. 9–2 (1969), 195–204.

[R] A. N. Rudakov "The exceptional vector bundles on $\mathbb{P}^1 \times \mathbb{P}^1$", Proceedings of USSR-USA-convention, Chicago, 1989.

[S] R. L. E. Schwarzenberger "Vector bundles on algebraic surfaces", Proc. London Math. Soc. (3) 11 (1961), 601–622.

[T_1] A. N. Tyurin "Symplectic structures...", Math. USSR Izvestiya v. 33 (1989), N1.

[T_2] A. N. Tyurin "Algebraic geometric aspects...", Russian Math. Survey (44/3), 1989.

[T_3] A. N. Tyurin "Cycles, curves..." Duke Math. J. 54 (1987), 1–26.

Programme

A. Beauville:	Cohomology of topologically trivial line bundles
T. Fujita:	On the ampleness of adjoint bundles of ample vector bundles
C. Peskine:	The artinian Gorenstein ring associated to a line bundle of a surface in P^3
P.M.H. Wilson:	Some questions on Calabi-Yau manifolds
M. Beltrametti:	On the adjunction theoretic classification of projective varieties, I
V.V. Shokurov:	Log-terminal 3-fold flips
F. Catanese:	Symmetric products of elliptic curves and classification of surfaces with $q=1$
D. Barlet:	Convexity of the cycle space and applications
O. Debarre:	Towards a stratification of the moduli space of abelian varieties
Ch. Peters:	Biholomorphic sectional curvature and deformations of period maps
Y. Miyaoka:	Bounding the number of rational points on curves over function fields
U. Persson:	Configurations of elliptic fibers on rational and K3 surfaces
Y. Kawamata:	Moderate degenerations of algebraic surfaces
G. Ellingsrud:	On the cohomology of some moduli spaces of stable sheaves on P^2
S. Kosarew:	Hodge theory on convex algebraic manifolds
C. Okonek:	Donaldson-Floer invariants and singularities
V.A. Iskovskih:	On rationality of conic bundles
J.P. Demailly:	A numerical criterion for very ample line bundles
P. Ionescu:	Polarised and non-minimal projective manifolds

A. Sommese: On the adjunction theoretic classification of projective varieties, II

A.N. Tyurin: The geometry of moduli spaces of vector bundles on surfaces and curves

E. Viehweg: Moduli of polarised manifolds

MATHEMATICHES INSTITUT
DER UNIVERSITÄT BAYREUTH

Participants

Alcati A., Milano
Banica C., Bukarest
Barlet D., Nancy
Barlow R., Warwick
Beauville A., Paris
Beltrametti M., Genua
Birkenhake C., Erlangen
Bohnhorst G., Osnabrück
Braun R., Bayreuth
Campana F., Nancy
Catanese F., Pisa
Coanda I., Bukarest
Colombo E., Pavia
Debarre O., Paris
Decker W., Saarbrücken
Demailly J.P., Grenoble
Ellingsrud G., Bergen
Fania M.L., L'Aquila
Fischer P., Bayreuth
Flenner H., Göttingen
Fujita T., Tokyo
Goldmann H., Bayreuth
Grauert H., Göttingen
Hilgert J., Erlangen
Horst C., München
Hulek K., Hannover
Hunt B., Kaiserslautern
Imbens H.J., Bonn,
Ionescu P., Bukarest
Iskovskih V.A., Moskau
Kawamata A., Tokyo
Kahn C., Hamburg,
Kerner H., Bayreuth
Kosarew I., Bonn
Kosarew S., Grenoble
Kurke H., Berlin
Lange H., Erlangen
Martens G., Erlangen
Miele M., Erlangen
Miyaoka Y., Tokyo
Ogata S., Bonn
Okonek C., Bonn
Olivia C., Bonn
Ottaviani G., Rom
Paranjape K., Bombay
Persson U., Göteborg
Peskine C., Paris
Peternell Th., Bayreuth
Peters C., Leiden
Pirola G., Pavia
Pöhlmann T., Bayreuth
Popp H., Mannheim
Remmert R., Münster
Ruppert W., Erlangen
Sankaran G., Cambridge
Schanz W. Erlangen
Schindler B., Erlangen
Schneider M., Bayreuth
Schottenloher M., München
Schreyer F.-O., Bayreuth
Seiler W.K., Mannheim
Shokurov V.V., Moskau
Sommese A., Notre Dame
Spandaw J., Leiden
Spindler H., Osnabrück
Terasoma T., Mannheim
Timmerscheidt K., Essen
Trautmann G., Kaisersl.
Tyurin A.N., Moskau
Viehweg E., Essen
Weinfurtner R., Bayreuth
Wilson P.-M.H., Cambridge
Wisniewski J., Warschau
Zuo, Bonn

Lecture Notes in Mathematics

For information about Vols. 1–1312
please contact your bookseller or Springer-Verlag

Vol. 1313: F. Colonius, Optimal Periodic Control. VI, 177 pages. 1988.

Vol. 1314: A. Futaki, Kähler-Einstein Metrics and Integral Invariants. IV, 140 pages. 1988.

Vol. 1315: R.A. McCoy, I. Ntantu, Topological Properties of Spaces of Continuous, Functions. IV, 124 pages. 1988.

Vol. 1316: H. Korezlioglu, A.S. Ustunel (Eds.), Stochastic Analysis and Related Topics. Proceedings, 1986. V, 371 pages. 1988.

Vol. 1317: J. Lindenstrauss, V.D. Milman (Eds.), Geometric Aspects of Functional Analysis. Seminar, 1986–87. VII, 289 pages. 1988.

Vol. 1318: Y. Felix (Ed.), Algebraic Topology – Rational Homotopy. Proceedings, 1986. VIII, 245 pages. 1988.

Vol. 1319: M. Vuorinen, Conformal Geometry and Quasiregular Mappings. XIX, 209 pages. 1988.

Vol. 1320: H. Jürgensen, G. Lallement, H.J. Weinert (Eds.), Semigroups, Theory and Applications. Proceedings, 1986. X, 416 pages. 1988.

Vol. 1321: J. Azéma, P.A. Meyer, M. Yor (Eds.), Séminaire de Probabilités XXII. Proceedings. IV, 600 pages. 1988.

Vol. 1322: M. Métivier, S. Watanabe (Eds.), Stochastic Analysis. Proceedings, 1987. VII, 197 pages. 1988.

Vol. 1323: D.R. Anderson, H.J. Munkholm, Boundedly Controlled Topology. XII, 309 pages. 1988.

Vol. 1324: F. Cardoso, D.G. de Figueiredo, R. Iório, O. Lopes (Eds.), Partial Differential Equations. Proceedings, 1986. VIII, 433 pages. 1988.

Vol. 1325: A. Truman, I.M. Davies (Eds.), Stochastic Mechanics and Stochastic Processes. Proceedings, 1986. V, 220 pages. 1988.

Vol. 1326: P.S. Landweber (Ed.), Elliptic Curves and Modular Forms in Algebraic Topology. Proceedings, 1986. V, 224 pages. 1988.

Vol. 1327: W. Bruns, U. Vetter, Determinantal Rings. VII, 236 pages. 1988.

Vol. 1328: J.L. Bueso, P. Jara, B. Torrecillas (Eds.), Ring Theory. Proceedings, 1986. IX, 331 pages. 1988.

Vol. 1329: M. Alfaro, J.S. Dehesa, F.J. Marcellan, J.L. Rubio de Francia, J. Vinuesa (Eds.): Orthogonal Polynomials and their Applications. Proceedings, 1986. XV, 334 pages. 1988.

Vol. 1330: A. Ambrosetti, F. Gori, R. Lucchetti (Eds.), Mathematical Economics. Montecatini Terme 1986. Seminar. VII, 137 pages. 1988.

Vol. 1331: R. Bamón, R. Labarca, J. Palis Jr. (Eds.), Dynamical Systems, Valparaiso 1986. Proceedings. VI, 250 pages. 1988.

Vol. 1332: E. Odell, H. Rosenthal (Eds.), Functional Analysis. Proceedings, 1986–87. V, 202 pages. 1988.

Vol. 1333: A.S. Kechris, D.A. Martin, J.R. Steel (Eds.), Cabal Seminar 81–85. Proceedings, 1981–85. V, 224 pages. 1988.

Vol. 1334: Yu.G. Borisovich, Yu.E. Gliklikh (Eds.), Global Analysis – Studies and Applications III. V, 331 pages. 1988.

Vol. 1335: F. Guillén, V. Navarro Aznar, P. Pascual-Gainza, F. Puerta, Hyperrésolutions cubiques et descente cohomologique. XII, 192 pages. 1988.

Vol. 1336: B. Helffer, Semi-Classical Analysis for the Schrödinger Operator and Applications. V, 107 pages. 1988.

Vol. 1337: E. Sernesi (Ed.), Theory of Moduli. Seminar, 1985. VIII, 232 pages. 1988.

Vol. 1338: A.B. Mingarelli, S.G. Halvorsen, Non-Oscillation Domains of Differential Equations with Two Parameters. XI, 109 pages. 1988.

Vol. 1339: T. Sunada (Ed.), Geometry and Analysis of Manifolds. Proceedings, 1987. IX, 277 pages. 1988.

Vol. 1340: S. Hildebrandt, D.S. Kinderlehrer, M. Miranda (Eds.), Calculus of Variations and Partial Differential Equations. Proceedings, 1986. IX, 301 pages. 1988.

Vol. 1341: M. Dauge, Elliptic Boundary Value Problems on Corner Domains. VIII, 259 pages. 1988.

Vol. 1342: J.C. Alexander (Ed.), Dynamical Systems. Proceedings, 1986–87. VIII, 726 pages. 1988.

Vol. 1343: H. Ulrich, Fixed Point Theory of Parametrized Equivariant Maps. VII, 147 pages. 1988.

Vol. 1344: J. Král, J. Lukes, J. Netuka, J. Vesely´ (Eds.), Potential Theory – Surveys and Problems. Proceedings, 1987. VIII, 271 pages. 1988.

Vol. 1345: X. Gomez-Mont, J. Seade, A. Verjovski (Eds.), Holomorphic Dynamics. Proceedings, 1986. VII, 321 pages. 1988.

Vol. 1346: O.Ya. Viro (Ed.), Topology and Geometry – Rohlin Seminar. XI, 581 pages. 1988.

Vol. 1347: C. Preston, Iterates of Piecewise Monotone Mappings on an Interval. V, 166 pages. 1988.

Vol. 1348: F. Borceux (Ed.), Categorical Algebra and its Applications. Proceedings, 1987. VIII, 375 pages. 1988.

Vol. 1349: E. Novak, Deterministic and Stochastic Error Bounds in Numerical Analysis. V, 113 pages. 1988.

Vol. 1350: U. Koschorke (Ed.), Differential Topology Proceedings, 1987, VI, 269 pages. 1988.

Vol. 1351: I. Laine, S. Rickman, T. Sorvali (Eds.), Complex Analysis, Joensuu 1987. Proceedings. XV, 378 pages. 1988.

Vol. 1352: L.L. Avramov, K.B. Tchakerian (Eds.), Algebra – Some Current Trends. Proceedings. 1986. IX, 240 Seiten. 1988.

Vol. 1353: R.S. Palais, Ch.-l. Teng, Critical Point Theory and Submanifold Geometry. X, 272 pages. 1988.

Vol. 1354: A. Gómez, F. Guerra, M.A. Jiménez, G. López (Eds.), Approximation and Optimization. Proceedings, 1987. VI, 280 pages. 1988.

Vol. 1355: J. Bokowski, B. Sturmfels, Computational Synthetic Geometry. V, 168 pages. 1989.

Vol. 1356: H. Volkmer, Multiparameter Eigenvalue Problems and Expansion Theorems. VI, 157 pages. 1988.

Vol. 1357: S. Hildebrandt, R. Leis (Eds.), Partial Differential Equations and Calculus of Variations. VI, 423 pages. 1988.

Vol. 1358: D. Mumford, The Red Book of Varieties and Schemes. V, 309 pages. 1988.

Vol. 1359: P. Eymard, J.-P. Pier (Eds.) Harmonic Analysis. Proceedings, 1987. VIII, 287 pages. 1988.

Vol. 1360: G. Anderson, C. Greengard (Eds.), Vortex Methods. Proceedings, 1987. V, 141 pages. 1988.

Vol. 1361: T. tom Dieck (Ed.), Algebraic Topology and Transformation Groups. Proceedings. 1987. VI, 298 pages. 1988.

Vol. 1362: P. Diaconis, D. Elworthy, H. Föllmer, E. Nelson, G.C. Papanicolaou, S.R.S. Varadhan. École d'Été de Probabilités de Saint-Flour XV–XVII. 1985–87 Editor: P.L. Hennequin. V, 459 pages. 1988.

Vol. 1363: P.G. Casazza, T.J. Shura, Tsirelson's Space. VIII, 204 pages. 1988.

Vol. 1364: R.R. Phelps, Convex Functions, Monotone Operators and Differentiability. IX, 115 pages. 1989.

Vol. 1365: M. Giaquinta (Ed.), Topics in Calculus of Variations. Seminar, 1987. X, 196 pages. 1989.

Vol. 1366: N. Levitt, Grassmannians and Gauss Maps in PL-Topology. V, 203 pages. 1989.

Vol. 1367: M. Knebusch, Weakly Semialgebraic Spaces. XX, 376 pages. 1989.

Vol. 1368: R. Hübl, Traces of Differential Forms and Hochschild Homology. III, 111 pages. 1989.

Vol. 1369: B. Jiang, Ch.-K. Peng, Z. Hou (Eds.), Differential Geometry and Topology. Proceedings, 1986–87. VI, 366 pages. 1989.

Vol. 1370: G. Carlsson, R.L. Cohen, H.R. Miller, D.C. Ravenel (Eds.), Algebraic Topology. Proceedings, 1986. IX, 456 pages. 1989.

Vol. 1371: S. Glaz, Commutative Coherent Rings. XI, 347 pages. 1989.

Vol. 1372: J. Azéma, P.A. Meyer, M. Yor (Eds.), Séminaire de Probabilités XXIII. Proceedings. IV, 583 pages. 1989.

Vol. 1373: G. Benkart, J.M. Osborn (Eds.), Lie Algebras. Madison 1987. Proceedings. V, 145 pages. 1989.

Vol. 1374: R.C. Kirby, The Topology of 4-Manifolds. VI, 108 pages. 1989.

Vol. 1375: K. Kawakubo (Ed.), Transformation Groups. Proceedings, 1987. VIII, 394 pages, 1989.

Vol. 1376: J. Lindenstrauss, V.D. Milman (Eds.), Geometric Aspects of Functional Analysis. Seminar (GAFA) 1987–88. VII, 288 pages. 1989.

Vol. 1377: J.F. Pierce, Singularity Theory, Rod Theory, and Symmetry-Breaking Loads. IV, 177 pages. 1989.

Vol. 1378: R.S. Rumely, Capacity Theory on Algebraic Curves. III, 437 pages. 1989.

Vol. 1379: H. Heyer (Ed.), Probability Measures on Groups IX. Proceedings, 1988. VIII, 437 pages. 1989.

Vol. 1380: H.P. Schlickewei, E. Wirsing (Eds.), Number Theory, Ulm 1987. Proceedings. V, 266 pages. 1989.

Vol. 1381: J.-O. Strömberg, A. Torchinsky, Weighted Hardy Spaces. V, 193 pages. 1989.

Vol. 1382: H. Reiter, Metaplectic Groups and Segal Algebras. XI, 128 pages. 1989.

Vol. 1383: D.V. Chudnovsky, G.V. Chudnovsky, H. Cohn, M.B. Nathanson (Eds.), Number Theory, New York 1985–88. Seminar. V, 256 pages. 1989.

Vol. 1384: J. Garcia-Cuerva (Ed.), Harmonic Analysis and Partial Differential Equations. Proceedings, 1987. VII, 213 pages. 1989.

Vol. 1385: A.M. Anile, Y. Choquet-Bruhat (Eds.), Relativistic Fluid Dynamics. Seminar, 1987. V, 308 pages. 1989.

Vol. 1386: A. Bellen, C.W. Gear, E. Russo (Eds.), Numerical Methods for Ordinary Differential Equations. Proceedings, 1987. VII, 136 pages. 1989.

Vol. 1387: M. Petković, Iterative Methods for Simultaneous Inclusion of Polynomial Zeros. X, 263 pages. 1989.

Vol. 1388: J. Shinoda, T.A. Slaman, T. Tugué (Eds.), Mathematical Logic and Applications. Proceedings, 1987. V, 223 pages. 1989.

Vol. 1000: Second Edition. H. Hopf, Differential Geometry in the Large. VII, 184 pages. 1989.

Vol. 1389: E. Ballico, C. Ciliberto (Eds.), Algebraic Curves and Projective Geometry. Proceedings, 1988. V, 288 pages. 1989.

Vol. 1390: G. Da Prato, L. Tubaro (Eds.), Stochastic Partial Differential Equations and Applications II. Proceedings, 1988. VI, 258 pages. 1989.

Vol. 1391: S. Cambanis, A. Weron (Eds.), Probability Theory on Vector Spaces IV. Proceedings, 1987. VIII, 424 pages. 1989.

Vol. 1392: R. Silhol, Real Algebraic Surfaces. X, 215 pages. 1989.

Vol. 1393: N. Bouleau, D. Feyel, F. Hirsch, G. Mokobodzki (Eds.), Séminaire de Théorie du Potentiel Paris, No. 9. Proceedings. VI, 265 pages. 1989.

Vol. 1394: T.L. Gill, W.W. Zachary (Eds.), Nonlinear Semigroups, Partial Differential Equations and Attractors. Proceedings, 1987. IX, 233 pages. 1989.

Vol. 1395: K. Alladi (Ed.), Number Theory, Madras 1987. Proceedings. VII, 234 pages. 1989.

Vol. 1396: L. Accardi, W. von Waldenfels (Eds.), Quantum Probability and Applications IV. Proceedings, 1987. VI, 355 pages. 1989.

Vol. 1397: P.R. Turner (Ed.), Numerical Analysis and Parallel Processing. Seminar, 1987. VI, 264 pages. 1989.

Vol. 1398: A.C. Kim, B.H. Neumann (Eds.), Groups – Korea 1988. Proceedings. V, 189 pages. 1989.

Vol. 1399: W.-P. Barth, H. Lange (Eds.), Arithmetic of Complex Manifolds. Proceedings, 1988. V, 171 pages. 1989.

Vol. 1400: U. Jannsen. Mixed Motives and Algebraic K-Theory. XIII, 246 pages. 1990.

Vol. 1401: J. Steprans, S. Watson (Eds.), Set Theory and its Applications. Proceedings, 1987. V, 227 pages. 1989.

Vol. 1402: C. Carasso, P. Charrier, B. Hanouzet, J.-L. Joly (Eds.), Nonlinear Hyperbolic Problems. Proceedings, 1988. V, 249 pages. 1989.

Vol. 1403: B. Simeone (Ed.), Combinatorial Optimization. Seminar, 1986. V, 314 pages. 1989.

Vol. 1404: M.-P. Malliavin (Ed.), Séminaire d'Algèbre Paul Dubreil et Marie-Paul Malliavin. Proceedings, 1987–1988. IV, 410 pages. 1989.

Vol. 1405: S. Dolecki (Ed.), Optimization. Proceedings, 1988. V, 223 pages. 1989. Vol. 1406: L. Jacobsen (Ed.), Analytic Theory of Continued Fractions III. Proceedings, 1988. VI, 142 pages. 1989.

Vol. 1407: W. Pohlers, Proof Theory. VI, 213 pages. 1989.

Vol. 1408: W. Lück, Transformation Groups and Algebraic K-Theory. XII, 443 pages. 1989.

Vol. 1409: E. Hairer, Ch. Lubich, M. Roche. The Numerical Solution of Differential-Algebraic Systems by Runge-Kutta Methods. VII, 139 pages. 1989.

Vol. 1410: F.J. Carreras, O. Gil-Medrano, A.M. Naveira (Eds.), Differential Geometry. Proceedings, 1988. V, 308 pages. 1989.

Vol. 1411: B. Jiang (Ed.), Topological Fixed Point Theory and Applications. Proceedings. 1988. VI, 203 pages. 1989.

Vol. 1412: V.V. Kalashnikov, V.M. Zolotarev (Eds.), Stability Problems for Stochastic Models. Proceedings, 1987. X, 380 pages. 1989.

Vol. 1413: S. Wright, Uniqueness of the Injective III_1 Factor. III, 108 pages. 1989.

Vol. 1414: E. Ramirez de Arellano (Ed.), Algebraic Geometry and Complex Analysis. Proceedings, 1987. VI, 180 pages. 1989.

Vol. 1415: M. Langevin, M. Waldschmidt (Eds.), Cinquante Ans de Polynômes. Fifty Years of Polynomials. Proceedings, 1988. IX, 235 pages.1990.

Vol. 1416: C. Albert (Ed.), Géométrie Symplectique et Mécanique. Proceedings, 1988. V, 289 pages. 1990.

Vol. 1417: A.J. Sommese, A. Biancofiore, E.L. Livorni (Eds.), Algebraic Geometry. Proceedings, 1988. V, 320 pages. 1990.

Vol. 1418: M. Mimura (Ed.), Homotopy Theory and Related Topics. Proceedings, 1988. V, 241 pages. 1990.

Vol. 1419: P.S. Bullen, P.Y. Lee, J.L. Mawhin, P. Muldowney, W.F. Pfeffer (Eds.), New Integrals. Proceedings, 1988. V, 202 pages. 1990.

Vol. 1420: M. Galbiati, A. Tognoli (Eds.), Real Analytic Geometry. Proceedings, 1988. IV, 366 pages. 1990.

Vol. 1421: H.A. Biagioni, A Nonlinear Theory of Generalized Functions, XII, 214 pages. 1990.

Vol. 1422: V. Villani (Ed.), Complex Geometry and Analysis. Proceedings, 1988. V, 109 pages. 1990.

Vol. 1423: S.O. Kochman, Stable Homotopy Groups of Spheres: A Computer-Assisted Approach. VIII, 330 pages. 1990.

Vol. 1424: F.E. Burstall, J.H. Rawnsley, Twistor Theory for Riemannian Symmetric Spaces. III, 112 pages. 1990.

Vol. 1425: R.A. Piccinini (Ed.), Groups of Self-Equivalences and Related Topics. Proceedings, 1988. V, 214 pages. 1990.

Vol. 1426: J. Azéma, P.A. Meyer, M. Yor (Eds.), Séminaire de Probabilités XXIV, 1988/89. V, 490 pages. 1990.

Vol. 1427: A. Ancona, D. Geman, N. Ikeda, École d'Eté de Probabilités de Saint Flour XVIII, 1988. Ed.: P.L. Hennequin. VII, 330 pages. 1990.

Vol. 1428: K. Erdmann, Blocks of Tame Representation Type and Related Algebras. XV. 312 pages. 1990.

Vol. 1429: S. Homer, A. Nerode, R.A. Platek, G.E. Sacks, A. Scedrov, Logic and Computer Science. Seminar, 1988. Editor: P. Odifreddi. V, 162 pages. 1990.

Vol. 1430: W. Bruns, A. Simis (Eds.), Commutative Algebra. Proceedings. 1988. V, 160 pages. 1990.

Vol. 1431: J.G. Heywood, K. Masuda, R. Rautmann, V.A. Solonnikov (Eds.), The Navier-Stokes Equations – Theory and Numerical Methods. Proceedings, 1988. VII, 238 pages. 1990.

Vol. 1432: K. Ambos-Spies, G.H. Müller, G.E. Sacks (Eds.), Recursion Theory Week. Proceedings, 1989. VI, 393 pages. 1990.

Vol. 1433: S. Lang, W. Cherry, Topics in Nevanlinna Theory. II, 174 pages.1990.

Vol. 1434: K. Nagasaka, E. Fouvry (Eds.), Analytic Number Theory. Proceedings, 1988. VI, 218 pages. 1990.

Vol. 1435: St. Ruscheweyh, E.B. Saff, L.C. Salinas, R.S. Varga (Eds.), Computational Methods and Function Theory. Proceedings, 1989. VI, 211 pages. 1990.

Vol. 1436: S. Xambó-Descamps (Ed.), Enumerative Geometry. Proceedings, 1987. V, 303 pages. 1990.

Vol. 1437: H. Inassaridze (Ed.), K-theory and Homological Algebra. Seminar, 1987–88. V, 313 pages. 1990.

Vol. 1438: P.G. Lemarié (Ed.) Les Ondelettes en 1989. Seminar. IV, 212 pages. 1990.

Vol. 1439: E. Bujalance, J.J. Etayo, J.M. Gamboa, G. Gromadzki. Automorphism Groups of Compact Bordered Klein Surfaces: A Combinatorial Approach. XIII, 201 pages. 1990.

Vol. 1440: P. Latiolais (Ed.), Topology and Combinatorial Groups Theory. Seminar, 1985–1988. VI, 207 pages. 1990.

Vol. 1441: M. Coornaert, T. Delzant, A. Papadopoulos. Géométrie et théorie des groupes. X, 165 pages. 1990.

Vol. 1442: L. Accardi, M. von Waldenfels (Eds.), Quantum Probability and Applications V. Proceedings, 1988. VI, 413 pages. 1990.

Vol. 1443: K.H. Dovermann, R. Schultz, Equivariant Surgery Theories and Their Periodicity Properties. VI, 227 pages. 1990.

Vol. 1444: H. Korezlioglu, A.S. Ustunel (Eds.), Stochastic Analysis and Related Topics VI. Proceedings, 1988. V, 268 pages. 1990.

Vol. 1445: F. Schulz, Regularity Theory for Quasilinear Elliptic Systems and – Monge Ampère Equations in Two Dimensions. XV, 123 pages. 1990.

Vol. 1446: Methods of Nonconvex Analysis. Seminar, 1989. Editor: A. Cellina. V, 206 pages. 1990.

Vol. 1447: J.-G. Labesse, J. Schwermer (Eds), Cohomology of Arithmetic Groups and Automorphic Forms. Proceedings, 1989. V, 358 pages. 1990.

Vol. 1448: S.K. Jain, S.R. López-Permouth (Eds.), Non-Commutative Ring Theory. Proceedings, 1989. V, 166 pages. 1990.

Vol. 1449: W. Odyniec, G. Lewicki, Minimal Projections in Banach Spaces. VIII, 168 pages. 1990.

Vol. 1450: H. Fujita, T. Ikebe, S.T. Kuroda (Eds.), Functional-Analytic Methods for Partial Differential Equations. Proceedings, 1989. VII, 252 pages. 1990.

Vol. 1451: L. Alvarez-Gaumé, E. Arbarello, C. De Concini, N.J. Hitchin, Global Geometry and Mathematical Physics. Montecatini Terme 1988. Seminar. Editors: M. Francaviglia, F. Gherardelli. IX, 197 pages. 1990.

Vol. 1452: E. Hlawka, R.F. Tichy (Eds.), Number-Theoretic Analysis. Seminar, 1988–89. V, 220 pages. 1990.

Vol. 1453: Yu.G. Borisovich, Yu.E. Gliklikh (Eds.), Global Analysis – Studies and Applications IV. V, 320 pages. 1990.

Vol. 1454: F. Baldassari, S. Bosch, B. Dwork (Eds.), p-adic Analysis. Proceedings, 1989. V, 382 pages. 1990.

Vol. 1455: J.-P. Françoise, R. Roussarie (Eds.), Bifurcations of Planar Vector Fields. Proceedings, 1989. VI, 396 pages. 1990.

Vol. 1456: L.G. Kovács (Ed.), Groups – Canberra 1989. Proceedings. XII, 198 pages. 1990.

Vol. 1457: O. Axelsson, L.Yu. Kolotilina (Eds.), Preconditioned Conjugate Gradient Methods. Proceedings, 1989. V, 196 pages. 1990.

Vol. 1458: R. Schaaf, Global Solution Branches of Two Point Boundary Value Problems. XIX, 141 pages. 1990.

Vol. 1459: D. Tiba, Optimal Control of Nonsmooth Distributed Parameter Systems. VII, 159 pages. 1990.

Vol. 1460: G. Toscani, V. Boffi, S. Rionero (Eds.), Mathematical Aspects of Fluid Plasma Dynamics. Proceedings, 1988. V, 221 pages. 1991.

Vol. 1461: R. Gorenflo, S. Vessella, Abel Integral Equations. VII, 215 pages. 1991.

Vol. 1462: D. Mond, J. Montaldi (Eds.), Singularity Theory and its Applications. Warwick 1989, Part I. VIII, 405 pages. 1991.

Vol. 1463: R. Roberts, I. Stewart (Eds.), Singularity Theory and its Applications. Warwick 1989, Part II. VIII, 322 pages. 1991.

Vol. 1464: D. L. Burkholder, E. Pardoux, A. Sznitman, Ecole d'Eté de Probabilités de Saint- Flour XIX-1989. Editor: P. L. Hennequin. VI, 256 pages. 1991.

Vol. 1465: G. David, Wavelets and Singular Integrals on Curves and Surfaces. X, 107 pages. 1991.

Vol. 1466: W. Banaszczyk, Additive Subgroups of Topological Vector Spaces. VII, 178 pages. 1991.

Vol. 1467: W. M. Schmidt, Diophantine Approximations and Diophantine Equations. VIII, 217 pages. 1991.

Vol. 1468: J. Noguchi, T. Ohsawa (Eds.), Prospects in Complex Geometry. Proceedings, 1989. VII, 421 pages. 1991.

Vol. 1469: J. Lindenstrauss, V. D. Milman (Eds.), Geometric Aspects of Functional Analysis. Seminar 1989-90. XI, 191 pages. 1991.

Vol. 1470: E. Odell, H. Rosenthal (Eds.), Functional Analysis. Proceedings, 1987-89. VII, 199 pages. 1991.

Vol. 1471: A. A. Panchishkin, Non-Archimedean L-Functions of Siegel and Hilbert Modular Forms. VII, 157 pages. 1991.

Vol. 1472: T. T. Nielsen, Bose Algebras: The Complex and Real Wave Representations. V, 132 pages. 1991.

Vol. 1473: Y. Hino, S. Murakami, T. Naito, Functional Differential Equations with Infinite Delay. X, 317 pages. 1991.

Vol. 1474: S. Jackowski, B. Oliver, K. Pawałowski (Eds.), Algebraic Topology, Poznań 1989. Proceedings. VIII, 397 pages. 1991.

Vol. 1475: S. Busenberg, M. Martelli (Eds.), Delay Differential Equations and Dynamical Systems. Proceedings, 1990. VIII, 249 pages. 1991.

Vol. 1476: M. Bekkali, Topics in Set Theory. VII, 120 pages. 1991.

Vol. 1477: R. Jajte, Strong Limit Theorems in Noncommutative L_2-Spaces. X, 113 pages. 1991.

Vol. 1478: M.-P. Malliavin (Ed.), Topics in Invariant Theory. Seminar 1989-1990. VI, 272 pages. 1991.

Vol. 1479: S. Bloch, I. Dolgachev, W. Fulton (Eds.), Algebraic Geometry. Proceedings, 1989. VII, 300 pages. 1991.

Vol. 1480: F. Dumortier, R. Roussarie, J. Sotomayor, H. Żoładek, Bifurcations of Planar Vector Fields: Nilpotent Singularities and Abelian Integrals. VIII, 226 pages. 1991.

Vol. 1481: D. Ferus, U. Pinkall, U. Simon, B. Wegner (Eds.), Global Differential Geometry and Global Analysis. Proceedings, 1991. VIII, 283 pages. 1991.

Vol. 1482: J. Chabrowski, The Dirichlet Problem with L^2-Boundary Data for Elliptic Linear Equations. VI, 173 pages. 1991.

Vol. 1483: E. Reithmeier, Periodic Solutions of Nonlinear Dynamical Systems. VI, 171 pages. 1991.

Vol. 1484: H. Delfs, Homology of Locally Semialgebraic Spaces. IX, 136 pages. 1991.

Vol. 1485: J. Azéma, P. A. Meyer, M. Yor (Eds.), Séminaire de Probabilités XXV. VIII, 440 pages. 1991.

Vol. 1486: L. Arnold, H. Crauel, J.-P. Eckmann (Eds.), Lyapunov Exponents. Proceedings, 1990. VIII, 365 pages. 1991.

Vol. 1487: E. Freitag, Singular Modular Forms and Theta Relations. VI, 172 pages. 1991.

Vol. 1488: A. Carboni, M. C. Pedicchio, G. Rosolini (Eds.), Category Theory. Proceedings, 1990. VII, 494 pages. 1991.

Vol. 1489: A. Mielke, Hamiltonian and Lagrangian Flows on Center Manifolds. X, 140 pages. 1991.

Vol. 1490: K. Metsch, Linear Spaces with Few Lines. XIII, 196 pages. 1991.

Vol. 1491: E. Lluis-Puebla, J.-L. Loday, H. Gillet, C. Soulé, V. Snaith, Higher Algebraic K-Theory: an overview. IX, 164 pages. 1992.

Vol. 1492: K. R. Wicks, Fractals and Hyperspaces. VIII, 168 pages. 1991.

Vol. 1493: E. Benôt (Ed.), Dynamic Bifurcations. Proceedings, Luminy 1990. VII, 219 pages. 1991.

Vol. 1494: M.-T. Cheng, X.-W. Zhou, D.-G. Deng (Eds.), Harmonic Analysis. Proceedings, 1988. IX, 226 pages. 1991.

Vol. 1495: J. M. Bony, G. Grubb, L. Hörmander, H. Komatsu, J. Sjöstrand, Microlocal Analysis and Applications. Montecatini Terme, 1989. Editors: L. Cattabriga, L. Rodino. VII, 349 pages. 1991.

Vol. 1496: C. Foias, B. Francis, J. W. Helton, H. Kwakernaak, J. B. Pearson, H_∞-Control Theory. Como, 1990. Editors: E. Mosca, L. Pandolfi. VII, 336 pages. 1991.

Vol. 1497: G. T. Herman, A. K. Louis, F. Natterer (Eds.), Mathematical Methods in Tomography. Proceedings 1990. X, 268 pages. 1991.

Vol. 1498: R. Lang, Spectral Theory of Random Schrödinger Operators. X, 125 pages. 1991.

Vol. 1499: K. Taira, Boundary Value Problems and Markov Processes. IX, 132 pages. 1991.

Vol. 1500: J. P. Serre, Lie Algebras and Lie Groups. VII, 164 pages. 1992.

Vol. 1501: A. De Masi, E. Presutti, Mathematical Methods for Hydrodynamic Limits. IX, 196 pages. 1991.

Vol. 1502: C. Simpson, Asymptotic Behavior of Monodromy. V, 139 pages. 1991.

Vol. 1503: S. Shokranian, The Selberg-Arthur Trace Formula (Lectures by J. Arthur). VII, 97 pages. 1991.

Vol. 1504: J. Cheeger, M. Gromov, C. Okonek, P. Pansu, Geometric Topology: Recent Developments. Editors: P. de Bartolomeis, F. Tricerri. VII, 197 pages. 1991.

Vol. 1505: K. Kajitani, T. Nishitani, The Hyperbolic Cauchy Problem. VII, 168 pages. 1991.

Vol. 1506: A. Buium, Differential Algebraic Groups of Finite Dimension. XV, 145 pages. 1992.

Vol. 1507: K. Hulek, T. Peternell, M. Schneider, F.-O. Schreyer (Eds.), Complex Algebraic Varieties. Proceedings, 1990. VII, 179 pages. 1992.